T0328636

GOVERNANCE OF THE WORLD'S MINERAL RESOURCES

GOVERNANCE OF THE WORLD'S MINERAL RESOURCES

Beyond the Foreseeable Future

THEO HENCKENS

ELSEVIER

Elsevier
Radarweg 29, PO Box 211, 1000 AE Amsterdam, Netherlands
The Boulevard, Langford Lane, Kidlington, Oxford OX5 1GB, United Kingdom
50 Hampshire Street, 5th Floor, Cambridge, MA 02139, United States

Notices
Knowledge and best practice in this field are constantly changing. As new research
and experience broaden our understanding, changes in research methods, professional
practices, or medical treatment may become necessary.

Practitioners and researchers must always rely on their own experience and knowledge
in evaluating and using any information, methods, compounds, or experiments
described herein. In using such information or methods they should be mindful of
their own safety and the safety of others, including parties for whom they have a
professional responsibility.

To the fullest extent of the law, neither the Publisher nor the authors, contributors, or
editors, assume any liability for any injury and/or damage to persons or property as a
matter of products liability, negligence or otherwise, or from any use or operation of
any methods, products, instructions, or ideas contained in the material herein.

Library of Congress Cataloging-in-Publication Data
A catalog record for this book is available from the Library of Congress

British Library Cataloguing-in-Publication Data
A catalogue record for this book is available from the British Library

ISBN: 978-0-12-823886-8

For information on all Elsevier publications visit our website at
https://www.elsevier.com/books-and-journals

Publisher: Candice Janco
Acquisitions Editor: Peter Llewellyn
Editorial Project Manager: Hannah Makonnen
Production Project Manager: Joy Christel Neumarin Honest Thangiah
Cover Designer: Mark Rogers

Typeset by TNQ Technologies

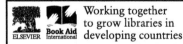

Working together
to grow libraries in
developing countries

www.elsevier.com • www.bookaid.org

For my wife Nelleke, who inspired me to start this discovery, but could not witness its completion.
For my partner Anni, who gave me the spirit to continue.
For my children Josée and Guy and my grandchildren.

Contents

Preface

Resources of molybdenum are adequate to supply world needs for the foreseeable future.

This quote from the Mineral Commodity Summaries, Molybdenum, January 2017, of the unsurpassed United States Geological Survey, is repeated for a number of other mineral resources. The US Geological Survey, however, does not define how many years the foreseeable future is.

The time horizon of private companies is mostly limited to a maximum of between 5 and 10 years. After all, circumstances can change so quickly that planning beyond a horizon of 5 years can seem senseless. Mining companies are an exception to this rule as, for the sake of their continuity, they need to have reserves for 15 to 25 years ahead. For states, the time horizon is longer, as infrastructure (roads, railways, waterways, airports, electricity supply) is planned with a horizon of 50 years or even longer.

For humanity, the time horizon is—or needs to be—much longer, if we take the discussions and measures in the context of climate change into consideration; this concerns a horizon of at least 50 to 100 years. For nonrenewable resources such as mineral resources, the time horizon needs to be longer than a century, similar to the time horizon of climate change, if we want our grandchildren and their grandchildren to have affordable access to essential mineral resources.

The purpose of this book is to provide an understanding of the availability and scarcity of mineral resources in the future. The main questions that we will try to answer are whether it is possible to keep providing sufficient mineral resources to a world population of 10 billion people; whether this is possible for a period of at least 200 years, but preferably 1000 years; whether it is possible to do this at a service level that is equal to the service level that mineral resources delivered to developed countries in 2020; whether it is possible to combine this with affordable mineral resource prices; and whether it is possible to achieve all of this even for the geologically scarcest mineral resources.

The book identifies scarce mineral resources and provides tools and options for global, national, and industrial decision-makers to prevent their uncontrolled exhaustion. A sound mineral resources policy at a global scale is the key factor for safeguarding raw materials for future generations. The book is also aimed at students and teachers of technical, economic, environmental, and law faculties, which educate and deliver our future leaders on whom the development and implementation of sound mineral resources governance will depend.

Finally, this book is based on research within the framework of my PhD, which crystallized in a thesis in 2016[1] and in a number of other scientific papers[2-10] that were published both before and after my PhD. Of course, my insights and knowledge on the subject of mineral resources have developed over time. Often, this was a result of comments and suggestions of reviewers of the various papers and of my coauthors, in particular Peter Driessen, Ernst Worrell, Cedric Ryngaert, Ekko van Ierland, and Frank Biermann, who I thank deeply for their contributions. It is for this reason that I use the prefix "we" instead of "I" in this book.

[1] Henckens, M.L.C.M., 2016, Managing raw materials scarcity: safeguarding the availability of geologically scarce mineral resources for future generations. http://dspace.library.uu.nl/handle/1874/339827.

[2] Henckens, M.L.C.M., Driessen, P.P.J., Worrell, E., 2014, Metal scarcity and sustainability, analyzing the necessity to reduce the extraction of scarce metals. Resources, Conservation and Recycling 93, 1—8.

[3] Henckens, M.L.C.M., Driessen, P.P.J., Worrell, E., 2015, Towards a sustainable use of primary boron. Resources, Conservation and Recycling 103, 9—18.

[4] Henckens, M.L.C.M., Driessen, P.P.J., Worrell, E, 2016, How can we adapt to geological scarcity of antimony? Resources, Conservation and Recycling 108, 54—62.

[5] Henckens, M.L.C.M., Driessen, P.P.J., Worrel, E., 2018, Molybdenum resources: Their depletion and safeguarding for future generations. Resources, Conservation and Recycling 134, 61—69.

[6] Henckens, M.L.C.M., Van Ierland, E.C., Driessen, P.P.J., Worrell, E., 2016, Mineral resources: Geological scarcity, market price trends, and future generations. Resources Policy 49, 102—111.

[7] Henckens, M.L.C.M., Ryngaert, C.M.J., Driessen, P.P.J., Worrell, E., 2018, Normative principles and the sustainable use of geologically scarce mineral resources. Resources Policy 59, 351—359.

[8] Henckens, M.L.C.M., Driessen, P.P.J., Ryngaert, C.M.J., Worrell, E., 2016, The set-up of an international agreement on the conservation and sustainable use of geologically scarce mineral resources. Resources Policy 49, 92—101.

[9] Henckens, M.L.C.M., Biermann, F.H.B., Driessen, P.P.J., 2019, Mineral resources governance: A call for the establishment of an International Competence Center on Mineral Resources Management. Resources, Conservation and Recycling 141, 255—263.

[10] Henckens, M.L.C.M., Worrell, E., 2020, Reviewing the availability of copper and nickel for future generations. The balance between production growth, sustainability and recycling rates. Journal of Cleaner Production 264, 121460.

CHAPTER 1

Introduction

Mineral resources (the physical foundation of society) have given their name to prehistoric periods: Stone Age, Bronze Age, and Iron Age. Without mineral resources there would be no wealth or progress. But since the beginning of the 20th century, their use has been increasing faster than ever (see Fig. 1.1).

For a long time in history, stone, wood, iron, copper, and tin were the main raw materials for buildings, transport, tools, and weaponry. Wood is a renewable resource. In practice, stone can be considered as a non-exhaustible (quasi-renewable) resource as well, because of its abundance.

From the Industrial Revolution of the 18th century, the use of nonrenewable resources increased quickly. This was the result of a continuous search for improving the properties of materials: higher strength, lower weight, better corrosion resistance, higher (or lower) conductivity of electricity and heat, lower costs, etc. The Industrial Revolution coincided with a technological revolution. Technology development worked as a self-reinforcing flywheel, resulting in an accelerating sequence of new discoveries. New elements were identified (e.g., cobalt in 1737, zinc in 1746, nickel in 1751, tungsten in 1783, magnesium and aluminum in 1808). New applications were developed (crucible steel in 1765, stainless steel in 1912, aluminum production in 1890, and super alloys in 1947) (Ashby, 2009).

Gradually, the dependence of humanity on nonrenewable mineral resources has increased: from a total dependence on nonexhaustible materials to a near-total dependence on exhaustible resources (Ashby, 2009).

The GDP of the average world citizen in 2018 was about USD 12,000. The GDP of the average Northern American in that year was about USD 65,000 (International Monetary Fund, 2019). Let us assume that the GDP of the average world citizen in 2100 will have increased to the same level as that of the average American in 2018. This means an annual growth of about 2.15%, which does not seem unrealistic considering the global GDP growth in the (recent) past. Given the expected increase in the world

Governance of the world's mineral resources
ISBN 978-0-12-823886-8
https://doi.org/10.1016/B978-0-12-823886-8.00015-4

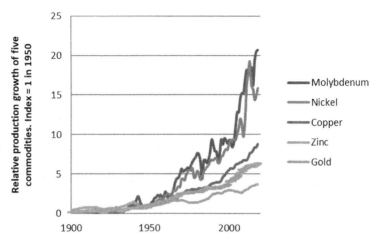

Figure 1.1 Global production of commodities (US Geological Survey, 2015, 2018, 2019).

population from 7.8 billion people in 2020 to about 9.8 billion people in 2050 and 11.2 billion people in 2100 (United Nations, 2017), the global GDP in 2100 could be about eight times higher than in 2018.

Because the use of mineral resources—at least to a certain degree—is related to GDP growth, it can be expected that the demand for mineral resources will grow strongly until the end of the 21st century. The production growth rates of seven mineral resources in the period between 1985 and 2015 are presented in Table 1.1. The actual growth rates are more capricious. Years with a production increase alternate with years with a production decrease.

Table 1.1 Annual production growth rates of seven elements between 1985 and 2015 extrapolated to the year 2100. The data are derived from statistical data of the United States Geological Survey.

Element	Annual production growth rate between 1980 and 2015 (%)	Estimated annual production in 2100, expressed as a magnitude of production in 2018 (times)
Bismuth	3.8	21
Nickel	3.1	12
Copper	2.8	11
Gold	2.7	9
Silver	2.5	8
Molybdenum	2.2	6
Antimony	2.2	6

Table 1.1 shows the outcome of extrapolating the production growth rates in the period between 1980 and 2015 to the year 2100. For instance, the annual copper production in 2100 will be 11 times higher than in 2018.

Of course, in reality, production rates will develop differently. For some minerals the rate will be slower, for other minerals it may be even faster. The transition to a low-carbon economy will have a certain, though limited, impact on the growth of overall metal use (De Koning et al., 2018).

Observing the potential production growth of the minerals in Table 1.1 the obvious question is whether or not this is sustainable. The answer to this question is inconclusive and differs per resource. Moreover, the use of a mineral resource may strongly change over time. The use may accelerate or decline due to new applications (e.g., in batteries) or to disappearing types of use, for example, the use of silver in the photographic sector. It is important to monitor the development closely and to anticipate timely on a situation of resources depletion, especially in cases of a high usage of a resource in comparison to the extractable amount in the Earth's crust. Failure to do so may result in serious economic consequences in the future. Depletion may lead to a situation in which the scarcest resources will become so expensive that the services they provide will hardly be attainable any more in the future for poor nations and poor people. That situation might contribute to further increasing in the economic inequality between nations and between people and could increase the risk that nations might consider using force to obtain resources. In his book *The Looting Machine*, Burgis (2015) gives a frightening picture of the hunt for minerals in Africa, already existing today. He reports on "the systematic theft of Africa's wealth by warlords, tycoons and smugglers." Diamond (2004) warns of a collapse of our civilization, because it is almost inconceivable that the Earth's resources could support raising all the world's citizens to the level of wealth of the citizens of the industrialized world.

Historical concerns about availability of resources

Thinking and warning about the scarcity of resources and its consequences is not new. In 1798, Malthus predicted that population growth could outpace the production capacity of fertile land, which would then lead to mass starvation, epidemic diseases, and wars for resources. The only solution he saw was "moral restraint" by poor people who are aware that they cannot support a family.

Later, Malthus' pessimistic view on the limited resources for an increasing global population was followed by influential essays from Ricardo (1817) and Mill (1848). In his essay, Ricardo included mineral scarcity as factor limiting

Continued

Historical concerns about availability of resources—cont'd

population growth. Mill, however, recognized the possibilities of new technology for increasing the productivity of exploited land and other resources.

At the end of the 19th century there was broad concern about resource availability, especially in the United States. This was reflected by the so-called Conservation Movement that was active between 1890 and 1920 (Tilton and Coulter, 2001). The origin of the concerns about scarce resources was especially connected to rapid industrialization and the development of vast wild lands. President Theodore Roosevelt was one of the prominent members/supporters of the Conservation Movement. The Movement promoted the wise use of resources, which entailed using renewable resources in place of nonrenewable resources, more abundant nonrenewable resources in place of less abundant nonrenewable resources, and recycled products in place of primary resources (Tilton and Coulter, 2001).

After World War II, new concerns about the long-term availability of resources were born in connection with the substantial resource use related to reconstruction. In the United States, this led in 1952 to the creation of the President's Material Policy Commission (or Paley Commission, after its chairman). One of the spinoffs of the work of this Commission was the sponsoring of organizations and studies on growth and scarcity. A very influential book in this framework was published by Barnett and Morse (1963). Their findings and views on the relationship between economic growth and depletion of nonrenewable resources stood in sharp contrast with the view that had prevailed thus far, which was that technological development had completely compensated for increasingly scarce nonrenewable resources and could be expected to do so in the future as well. The strong potential of technology development to solve scarcity problems was also emphasized in books by Maurice and Smithson (1984) and Diamandis and Kotler (2012). In 1979, a number of scientists reconsidered and nuanced Barnett and Morse's vision (Smith, 1982).

In 1972, the book *Limits to Growth* (Meadows et al., 1972) was published for the so-called Club of Rome. In contradiction to Barnett and Morse, they came to the conclusion that per capita food and industrial output would collapse as a result of exhaustion of mineral resources by 2050. In 1992, Meadows and his coauthors updated their advice to the Club of Rome (Meadows et al., 1992), basically confirming their original point of view.

The above-described differences of view on scarcity reflect the discussion between the so-called resource optimists and resource pessimists. The resource pessimists support the so-called fixed stock paradigm: the Earth is finite and so the amount of mineral resources is finite as well. However, as demand will not stop growing, it is only a matter of time before supply is no longer able to meet

---

Historical concerns about availability of resources—cont'd

demand. Resource pessimists include, among others, Meadows et al. (1972, 1992), Kesler (1994), Diederen (2009), and Bardi (2013).

The resource optimists support the so-called opportunity cost paradigm. The optimists do not deny that mineral resources will deplete gradually, but they have a strong belief that humanity will be able to cope with the effects of depletion. When demand outpaces supply, the price of the material will rise and—simultaneously—the pressure to find substitutes or alternatives for the depleted mineral will increase. When the real price for a mineral commodity is rising, society has to consider what to give up in order to obtain an additional ton of that scarce commodity. According to the resource optimists, the market will automatically solve the problem. Moreover, in their view, differing from what is the case for oil, natural gas, and coal, for example, most mineral resources are not destroyed by being used. Recycling and reuse are possible. Finally, the total geological stocks are enormous. It will always be possible to extract minerals, although the costs will be considerable. Many resource optimists can be found in the mining industry and linked institutions, including Hodges (2019), Gunn (2011), Simon (1980, 1981), Adelman (1990), and Beckerman (1995), Maurice and Smithson (1984) and Diamandis and Kotler (2012). Resource pessimists and resource optimists, acknowledge that mineral resources are exhaustible. The difference between them is that the optimists trust that humankind will timely find a clever solution for the replacement of depleting resources by substitutes, as has always been the case thus far. The pessimists are of the opinion that humanity should not deliberately deprive future generations of scarce resources, and that future generations must have the same levels of opportunity and freedom as the current generation.

First, in Chapter 2, mineral resources are investigated with respect to their availability for extraction over the long term. An overview of data and different approaches to quantify the ultimately available resources is provided.

The time period within which a resource is exhausted is determined not only by the extractable amount of that resource as such but also by the size and trend (increase or decrease) of the extraction rate and the increase thereof. In Chapter 3 we provide an indication of the exhaustion periods of different mineral resources, using different scenarios.

Apart from physical depletion, there are also other circumstances which can create can create temporary or permanent scarcity of a raw material on the world market. Many mineral resources are produced in only a limited number of countries. Those countries may create scarcity deliberately in order to increase prices or to promote their own industry, or for other geopolitical purposes. It may also happen that relatively large proportions of certain raw materials are produced in countries with weak governments, which makes continuous production for the world market uncertain. If—at the same time—these resources are essential for global society and industry and they are not easy to replace with other raw materials, they are defined as critical raw materials. The subject of critical materials is treated in Chapter 4.

What, precisely, is the (moral) responsibility of a generation with regard to the conservation of mineral resources for the well-being of future generations(assuming there is a responsibility)? Chapter 5 provides an answer to the question of how we can define the concept of a "sustainable extraction rate."

In Chapter 6 we investigate the role of the price mechanism with regard to the availability of mineral resources. We examine whether the price mechanism of the free market system will automatically, timely, and permanently lead to such a reduction of the extraction of geologically scarce mineral resources that sufficient resources will always be guaranteed, even for future generations.

In Chapter 7 we explore, for 13 scarce mineral resources (antimony, bismuth, boron, chromium, copper, gold, indium, molybdenum, nickel, silver, tin, tungsten, and zinc), whether and, if so, how a combination of substitution, improvement of recovery efficiency, material efficiency, and recycling, and a reduction of in-use-dissipation can realistically (because technically and economically feasible) lead to sustainable use of these materials, while simultaneously increasing and maintaining the global service level of these resources to a level similar to that prevailing in developed countries in 2020.

The availability of technical solutions for making the use of a resource sustainable does not automatically mean that these solutions will be implemented by humanity. For the necessary technical steps to be taken it is essential to adopt and implement suitable policy measures. Mineral resources governance is the subject of Chapter 8.

Finally, in Chapter 9 we propose and discuss setting up an international agreement on the conservation and sustainable use of geologically scarce mineral resources.

References

Adelman, M.A., 1990. Mineral depletion, with special reference to petroleum. Rev. Econ. Stat. 72 (1), 1—10.

Ashby, M.F., 2009. Materials and the Environment, Eco-Informed Material Choice, 2009. Elsevier.

Bardi, U., 2013. Plundering the Planet, 2013, 33rd Report for the Club of Rome.

Barnett, H., Morse, C., 1963. Scarcity and Growth, the Economics of Natural Resource Availability. Johns Hopkins University Press, for Resources for the Future, Baltimore, MD.

Beckerman, W., 1995. Small Is Stupid. Duckworth, London.

Burgis, T., 2015. The Looting Machine, Warlords, Tycoons, Smugglers and the Systematic Theft of Africa's Wealth. Harper Collins Publishers.

De Koning, A., Kleijn, R., Huppes, G., Sprecher, B., Van Engelen, G., Tukker, A., 2018. Metal supply constraints for a low carbon economy? Resour. Conserv. Recycl. 129, 202—208.

Diamandis, P.H., Kotler, S., 2012. Abundance: The Future Is Better than You Think. Free Press, New York. ISBN number: 978-1-4516-9576-2.

Diamond, J., 2004. Collapse. North Point Press.

Diederen, A., 2009. Metal Minerals Scarcity: A Call for Managed Austerity and the Elements of Hope. TNO.

Gunn, G., 2011. Mineral Scarcity — a Non-issue? British Geological Survey.

Hodges, C.A., 2019. Mineral resources, Environmental issues and Land use, 1995. Science 268 (5215), 1305—1312. New Series.

International monetary Fund, March 30, 2019. World Economic Outlook, October 2018.

Kesler, S.E., 1994. Mineral Resources, Economics and the Environment. Macmillan, New York.

Malthus, T., 1798. An Essay on the Principles of Population.

Maurice, C., Smithson, C.W., 1984. The Doomsday Myth, 10,000 Years of Economic Crises. Hoover Institution Press, Stanford University.

Meadows, D.H., et al., 1972. The Limits to Growth. Universe Books, New York.

Meadows, D.H., et al., 1992. Beyond the Limits. Chelsea Green Publishing, Post Mills, VT.

Mill, J.S., 1848. Principles of Political Economy.

Ricardo, D., 1817. Principles if Political Economy and Taxation.

Simon, J.L., 1980. Resources, population, environment: an oversupply of false bad news. Science 208, 1431—1438.

Simon, J.L., 1981. The Ultimate Resource. Princeton University Press, Princeton, NJ.

Smith, K., 1982. Scarcity and Growth Reconsidered, second ed. Resources for the Future Inc. John Hopkins Press, Ltd London.

Tilton, J.E., Coulter, W.J., 2001. Manuscript for the workshop on the long-run-availability of mineral commodities. In: sponsored by Mining, Minerals and Sustainable Development Project and Resources for the Future, Washington DC, April21-23, 2001.

United Nations, 2017. World Population Prospects. UN Department of Economic and Social Affairs.

US Geological Survey, 2015. Historical Statistics for Mineral and Material Commodities in the United States.

US Geological Survey, January 2018. Mineral Commodity Summaries.

US Geological Survey, February 2019. Mineral Commodity Summaries.

CHAPTER 2

The availability of mineral resources[1]

2.1 Introduction

The Earth consists of the following layers: crust, mantle, outer core, and inner core. We will focus on mineral resources in the crust. The crust is 30—50 km thick under the continents, but much shallower under the oceans: 5—10 km. The Earth's crust accounts for less than 1% of the Earth's volume. A total of 98.5% of the crust consists of only eight elements (see Fig. 2.1).

The remaining 110 elements of the periodic system, which include all the elements in which we are interested in this book, contribute only 1.5% to the weight of the Earth's crust. The most important of these (in weight) are titanium (0.44%), hydrogen (0.14%), manganese (0.10%), and phosphorus (0.12%). Most of the minerals which are the subject of this book are so-called minor elements and have an average concentration in the Earth's crust of less than 1000 ppm (0.1%).

Despite their low concentrations, the amounts of the minor elements in the Earth's crust are still enormous. The mass of the upper kilometer of the continental Earth's crust is about 40×10^{10} Mt. For copper, with an average occurrence in the Earth's crust of 50 ppm, this results in an amount of 20,000 billion tons of copper in the upper kilometer of the continental Earth's crust.

Elements are not homogeneously distributed in the Earth's crust. The distribution of grade (weight percentage) and tonnage of the major elements (with a crustal occurrence of >0.1%) is assumed to be normal bell-shaped, whereas most of the minor elements (with a crustal occurrence

---

[1] This chapter is partly based on the publication of Henckens MLCM, Driessen PPJ, Worrell E, Metal scarcity and sustainability, analyzing the necessity to reduce the extraction of scarce metals, Resources, Conservation and Recycling, 93, 2014, 1—8.

Governance of the world's mineral resources
ISBN 978-0-12-823886-8
https://doi.org/10.1016/B978-0-12-823886-8.00010-5

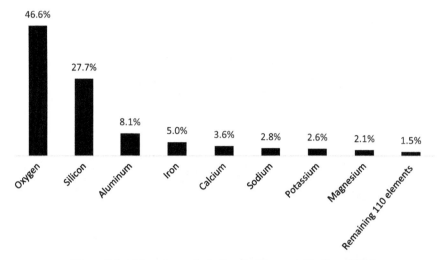

Figure 2.1 Major elements in the Earth's crust (Darling, 2020).

of <0.1%) are assumed to be bimodally distributed (Tilton, 2001; Skinner, 2001; Phillips, 1977; Van Vuuren et al., 1999) (see Fig. 2.2). The right-hand side of the bimodal curve represents the *minor* element ores. The supposed bimodal distribution of the *minor* elements in their ores explains the relatively high enrichment factors of these elements compared to the

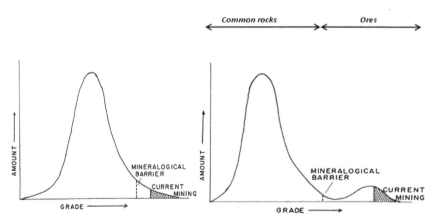

Figure 2.2 Grade and tonnage distribution of major elements (>0.1 wt.%) in the Earth's crust in the left-hand figure, and minor elements (<0.1 wt.%) in the right-hand figure. The distribution is log normal. *(Derived from Skinner, B.J., 2001. Long run availability of minerals. In: Keynote Talk at a Workshop "Exploring the Resource Base", Resources for the Future, Washington DC, April 22–23, 2001.)*

Table 2.1 Typical ore grades and enrichment factors for some mineral resources (Rankin, 2011, p. 88; UNEP, 2011). The upper crustal occurrences are derived from Table 2.2.

	Typical minimum ore grades (ppm)	Upper crustal occurrence[a] (ppm)	Enrichment factor (rounded)
Major elements in the Earth's crust			
Aluminum	320,000	81,000	4
Iron	250,000	48,000	7
Titanium	100,000	4900	20
Minor elements in the Earth's crust			
Copper	5000	50	100
Nickel	10,000	68	200
Zinc	40,000	72	600
Gold	4	0.0025	2000
Lead	40,000	14	3000
Tin	50,000	2.9	20,000

[a]Occurrence is average concentration.

relatively low enrichment factors of the *major* elements with a (supposed) modal distribution. The enrichment factor of an element in an ore is the ratio between its ore grade and the upper crustal abundance of the element. Obviously, the highest grades of a mineral are exploited first. This concerns only a minor part of the total amount of an element in the Earth's crust.

Table 2.1 presents enrichment factors of some major and some minor elements in the Earth's crust. We see that for minor elements, enrichment factors between 100 and 20,000 are common, whereas for major elements, enrichment factors are much lower, typically between 4 and 20. This means that the ores of minor elements can be considered as real treasures for humanity compared to the enriched deposits of the major elements. Once the ores of a minor element have been exhausted, these treasures are lost forever. Although a very large quantity of the minor element will still be left in the Earth's crust, this is at a much lower concentration. Although, in theory, it is technically possible to extract these dilute occurrences, it will be costly, not only from a financial perspective, but also from an environmental impact angle.

The following flows arise when mining a raw material:
- Overburden: Waste rock or materials overlying an ore or mineral body that are displaced during mining without being processed
- Gangue: Material surrounding or closely mixed with a desired mineral in an ore deposit (see Fig. 2.3)

Figure 2.3 Crystals of cassiterite, the main tin mineral, in a matrix of quartz, the gangue (Wikipedia, 2020). *(Photo retrieved via Wikipedia Commons on 2-7-2020, http://www.mindat.org/photo-229273.html, Ralph Bottril.)*

- Tailings: Ore stripped of valuable minerals
- Concentrate: Product produced by a metal ore mine
- Slag: Glass-like by-product left after separation of desired mineral from its ore (i.e., by smelting)

The limit grade between extractability and nonextractability of a resource is not fixed. In technical terms, everything is extractable. But whether extraction is economically feasible depends on a combination of factors, the most important being ore grade, depth, location, energy price, and willingness to pay for the extracted material (Tilton, 2001, 2003; Allwood et al., 2011). It is therefore important to differentiate between reserves and potentially extractable resources. The US Geological Survey (2017) provides the following definitions:
- Resources: A concentration of naturally occurring solid, liquid, or gaseous material in or on the Earth's crust in such form and amount that economic extraction of a commodity from the concentration is currently or potentially feasible.

- Identified resources are resources whose location, grade, quality, and quantity are known or estimated from specific geologic evidence. Identified resources include economic, marginally economic, and subeconomic components.
- The reserve base is the part of an identified resource that meets specified minimum physical and chemical criteria related to current mining and production practices, including those for grade, quality, thickness, and depth.
- Reserves are the part of the reserve base that could be economically extracted or produced at the time of determination.
  See Fig. 2.4.

Part of the world has not yet been extensively explored for reasons of political instability or topographical inaccessibility. Resources may lie beneath a cover of rocks and may be so deep that the prospecting technology presently in use cannot yet detect them (Skinner, 2001). In principle, deep resources are available, albeit at considerably higher exploitation costs. However, due to the increasing temperature and rock pressure, mining becomes more complex with increasing depth. Additionally, exploration is expensive and therefore has a relatively short time horizon. Mining companies focus exploration on deposits with the highest concentrations and not on the lower subeconomic grades.

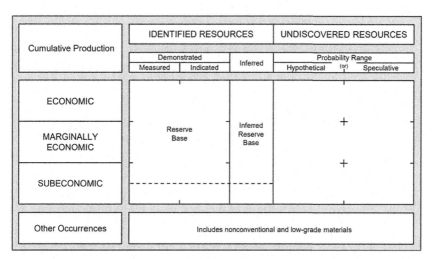

Figure 2.4 Mineral resources classification. *(Derived from US Geological Survey, 2017. Mineral Commodity Summaries 2017, Appendix C.)*

Gold is extracted until much lower grades (in the order of 10,000 times lower!) than zinc, because the market is willing to pay much more for gold than for zinc. In August 2018, the market price of gold was about $39,000/kg and the price of zinc was $2.50/kg. The limit-grade for extractability is called the *"mineralogical barrier"*. It is not a physical barrier, but primarily an economic barrier. Exploitation becomes much more costly once the enriched deposits are depleted, especially for minor elements with a bimodal distribution in the Earth's crust. It can be expected that the extraction costs of elements with a bimodal distribution will increase quite abruptly after exhaustion of the ore bodies. From the grade distribution of Fig. 2.2 it can be concluded that the exploitable resources of *major* elements will not be abruptly exhausted. The extraction costs of *major* elements will increase more gradually and will be moderated by technological progress.

Several authors have tried to determine the order of magnitude of the limit grade of mineral resources extraction. In the next sections, four different approaches will be discussed. They concern the availability of resources in the upper kilometers of the Earth's continental crust. In Sections 2.5—2.8 we will present rough data on the availability of resources in deeper parts of the Earth's crust, in the sea, and on extraterrestrial bodies, such as the moon.

In the scientific literature, four different approaches can be identified for estimating the extractable resources: three are generic approaches and the fourth is specific per element.

(1) The approach assuming that the ultimately extractable amount of a resource is proportional to the average crustal abundance of that resource (Rankin, 2011; UNEP, 2011; Skinner, 1976; Graedel and Nassar, 2015; Mookherjee and Panigrahi, 1994; Nishiyama and Adachi, 1995; Erickson 1973);

(2) The approach based on the global extrapolation of the relationship between the results of a thorough and relatively recent assessment of the amount of extractable resources of a number of minerals in the United States on the one hand, and the amount of the same resources as identified by the US Geological Survey in the United States on the other hand;

(3) The approach based on the so-called tectonic diffusion model as developed by Kesler and Wilkinson (2008);

(4) The approach based on research of the exploration data of specific resources, for example, regarding
  - gold by Jowitt and Mudd (2014) and by Mudd (2007)

- lithium by Mohr et al. (2012)
- nickel by Mudd and Jowitt (2014)
- copper by Mudd and Jowitt (2018), Singer (2017), Sverdrup et al. (2014b), US Geological Survey (2014)
- lead by Mudd et al. (2017)
- zinc by Mudd et al. (2017)
- platinum group metals (PGM) by Mudd et al. (2018) and by Sverdrup and Ragnarsdottir (2016)
- cobalt by Mudd et al. (2013)
- silver by Sverdrup et al. (2014a)
- lithium by Sverdrup (2016)
- aluminum by Sverdrup et al. (2015)
- rare earth elements (REE) by Weng et al. (2015).

The results of the three generic approaches are presented in the next three sections.

2.2 Availability of a resource in relation to the crustal abundance of an element

On the basis of data on known deposits of minerals, there are strong arguments that the combined size of extractable ore deposits of a particular mineral is *"directly proportional to the crustal occurrence of the mineral"* (Rankin, 2011). This is based on the following observations:

- The reserve base of the elements is roughly proportional to their crustal abundance. See also Graedel and Nassar (2015), Mookherjee and Panigrahi (1994), Nishiyama and Adachi (1995);
- The size of the largest known deposit of each element is proportional to the average crustal abundance of the element (Skinner, 1976);
- The number of known deposits of over one million tons of a given element is proportional to the average crustal abundance of the element (Skinner, 1976).

The relationship between the global amount of extractable resources of an element and the crustal abundance of that element was presented by a study by Graedel and Nassar (2015). They plotted US Geological Survey Reserve Base estimates as a function of upper crustal abundance in a log-log graph. The logarithmic correlation is linear and strong: $\log(RB) = 0.7432 \log(UCA) + 0.2510$ with $r^2 = 0.83$ (*RB*, reserve base; *UCA*, upper crustal abundance) (Mookherjee and Panigrahi, 1994, derived from Graedel and Nassar, 2015). See Fig. 2.5, line 1.

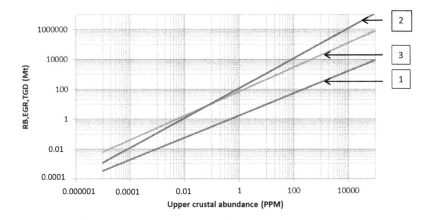

Figure 2.5 Line 1: the relation between reserve base (RB) and upper crustal abundance (UCA) is provided by Mookherjee and Panigrahi (1994), cited by Graedel and Nassar (2015). Log RB = 0.7432 log(UCA) + 0.2510. Line 2: the relation between extractable global resources (in the upper 3 km of the Earth's crust) and UCA is log(EGR) = log(UCA) + 2.08. See Section 2.2. Line 3: the relation between total global deposits (TGD) and UCA: log(TGD) = 0.81 log(UCA) + 1.86. See Section 2.3.

Observing the apparent correlation between extractable resources and crustal abundance, a working group of the International Resource Panel of UNEP (2011) concluded that 0.01% of the total amount of a material in the top 1 km of the Earth's continental crust is a *"reasonable estimate for the upper limit for the Extractable Global Resources"* of that material. The figure of 0.01% is based on earlier work of Skinner (1976) and Erickson (1973). From Erickson (1973) and Rankin (2011) it can be inferred that the estimate of the EGR of an element ranges between 0.01% and 0.001% of the total amount of that element in upper part of the continental crust. Deviating from the above approach of UNEP, we will suppose that, in the long term, technological developments will make it economically feasible to extract scarce mineral resources to a depth of 3 km in the Earth's crust instead of to 1 km, although very few mines are currently deeper than 1 km (see Section 2.5).

Assuming that the amount of extractable global resources is 0.01% of the total amount of a material in the top 3 km of the continental Earth's crust, the extractable global resources can be calculated as follows. Given that the radius of the Earth is 6371 km, the volume of the top 3 km of the crust is $4/3 \ \pi \ 6371^3 - 4/3 \ \pi \ 6368^3 = 1529 \times 10^6 \ km^3$. The continental crust occupies about 30% of the Earth's surface. Therefore, the volume of the upper 3 km of the continental crust is $0.30 \times 1529 \times 10^6 \, km^3 = 459 \times 10^6 \, km^3$. The average density of the Earth's continental crust is $2.7 \, t/m^3$ (Rankin, 2011, p. 300) or $2.7 \times 10^9 \, t/km^3$.

The International Resource Panel was launched by the United Nations Environment Program (UNEP) and the European Union in 2007. Its purpose is to build and share knowledge to improve the use of resources worldwide. Its remit is (1) "to provide independent, coherent, and authoritative scientific assessments of policy relevance on the sustainable use of natural resources and, in particular, their environmental impacts over the full life cycle, (2) to contribute to a better understanding on how to decouple economic growth from environmental degradation." (Resource Panel, 2020).

This means that the total mass of the upper 3 km of the continental crust is $2.7 \times 10^9 \times 459 \times 10^6 = 1238 \times 10^{15}$ metric tons. We will use the rounded amount of 120×10^{10} Mt. Suppose that the upper crustal abundance (UCA) of an element is 1 ppm. Then the extractable global resources (EGR) of that element in the upper 3 km of the Earth's crust are 0.01% of $120 \times 10^{10} \times 10^{-6} = 120$ Mt. In formula: EGR $= 120 \times$ UCA. EGR is expressed in Mt and UCA in ppm. In logarithmic form the formula is $\log(EGR) = \log(UCA) + 2.08$. Table 2.2 presents the amount of extractable global resources calculated on this basis.

The relation between EGR and UCA is graphically presented in Fig. 2.5. EGR includes past production.

The approach that the extractable amount of a resource is related to the upper crustal abundance of that resource is also used in *life cycle assessment*. One of the impact categories considered in *life cycle assessment* is abiotic resource depletion. Guinée and Heijungs (1995) defined the abiotic depletion potential (ADP) of a resource as the ratio of the annual production and the square of the ultimate reserve for the resource divided by the same ratio for a reference resource, which is antimony. The ultimate reserve for the resource is based on the crustal abundance of the resource multiplying the crustal content concentrations by the mass of the part of the crust considered. In this way the ADP expresses the geological scarcity of a resource compared to the geological scarcity of antimony. Recently, Van Oers et al. (2020) updated the ADPs using 2015 production data, derived from the US Geological Survey, and crustal content concentrations derived from Rudnick and Gao (2014).

Table 2.2 Materials, crustal abundance (CA), and extractable global resources (EGR) according to the approach of UNEP (2011).

Material	Symbol	UCA (ppm)[a]	EGR (0.01% of the amount of the material in the upper 3 km of the Earth's continental crust), 120 × UCA, rounded (Mt)
Aluminum	Al	81,000	10,000,000
Antimony	Sb	0.2	20
Arsenic	As	1.7	200
Barium	Ba	450	50,000
Beryllium	Be	2.6	300
Bismuth	Bi	0.052	6
Boron	B	13	2000
Cadmium	Cd	0.35	40
Chromium	Cr	110	10,000
Cobalt	Co	24	3000
Copper	Cu	50	6000
Gallium	Ga	17	2000
Germanium	Ge	1.6	200
Gold	Au	0.0025	0.3
Indium	In	0.24	30
Iron	Fe	48,000	6,000,000
Lead	Pb	14	2000
Lithium	Li	18	2000
Magnesium	Mg	23,000	2,000,000
Manganese	Mn	890	100,000
Mercury	Hg	0.067	8
Molybdenum	Mo	1.3	200
Nickel	Ni	68	8000
Niobium	Nb	17	2000
Platinum group metals	PGM total	0.023	3
Rare earth elements	REE total	170	20,000
Rhenium	Re	0.0010	0.1
Selenium	Se	0.050	6
Silver	Ag	0.069	8
Strontium	Sr	350	40,000
Tantalum	Ta	1.5	200
Tin	Sn	2.9	300
Titanium	Ti	4900	600,000
Tungsten	W	1.5	200
Vanadium	V	130	20,000
Zinc	Zn	72	9000

[a]The scientific literature provides different values for the crustal abundance. We have taken the average value of seven sources: McLennan (upper crustal abundance 2001), Darling (2020), Barbalace (2007), Webelements (2007), Jefferson Lab (2007), Wedepohl (1995), and Rudnick and Fountain (1995).

2.3 Estimation of total global deposits by extrapolation of US data

In 2000, the US Geological Survey published an assessment of *undiscovered* deposits of gold, silver, copper, lead, and zinc in the United States (US Geological Survey, 2000). The assessment was carried out by 19 teams and comprised the continental crust in the United States to a depth of 1 km. The results are presented in Table 2.3.

In the United States, the average ratio of the estimated total deposits of the five aforementioned raw materials (undiscovered plus identified resources) and the amount of their resources identified in 1998 was 3.14. We will apply this ratio to obtain an indicative estimate of the total global deposits of other mineral resources. For gold, silver, copper, lead, and zinc we will use the ratios mentioned in Table 2.3. For extrapolation to the global scale we will use 2018 data (US Geological Survey, 2018) on globally identified resources as the point of departure instead of the 1998 data, because we assume that—at least in 1998—the Earth's crust in the parts of the world outside the United States had been less explored than the crust in the United States.

If the identified resources of an element are not provided by the US Geological Survey we estimate them by multiplying the US Geological Survey reserve base (2009) by a factor of 2.8, which is the average ratio of identified resources (2018) to reserve base (2009) of the elements, of which both data are available.

The aforementioned assessment of available resources was to a depth of 1 km in the Earth's crust. We have assumed that future advances in mining

Table 2.3 Estimates of identified and undiscovered resources and past production of gold, silver, copper, lead, and zinc in the United States in 1998 to a depth of 1 km (US Geological Survey, 2000).

Category	Gold tons	Silver tons	Copper kilotons	Lead kilotons	Zinc kilotons	Average
Undiscovered	18,000	460,000	290,000	85,000	210,000	
Identified	15,000	160,000	260,000	51,000	55,000	
Past production	12,000	170,000	91,000	41,000	44,000	
Ratio (identified + undiscovered)/ identified	2.20	3.88	2.12	2.67	4.82	3.14

technology will make it economically feasible to mine all minerals to a depth of 3 km, comparable to the depth to which gold is mined nowadays. Hence, we have multiplied the results by a factor of 3 to reflect the amount of resources to a depth of 3 km in the Earth's crust as compared to the amount of resources to a depth of 1 km in the Earth's crust. Table 2.4 presents the results of this approach. Via regression analysis, using the method of the least squares, we have determined the relationship between the total global deposits (TGD) as presented in Table 2.4 and the upper crustal abundance (UCA) presented in Table 2.2. The relation is as follows: $\log(\text{TGD}) = 0.81 \log(\text{UCA}) + 1.86$ with $R^2 = 0.81$. The function is included as line 3 in Fig. 2.5.

2.4 The tectonic diffusion model

Instead of estimating the amount of available resources on the basis of the crustal abundance of the elements or on the basis of extrapolation of existing data regarding identified resources, Kesler and Wilkinson (2008) used the so-called tectonic diffusion model approach. This approach supposes that deposits of a certain type (e.g., porphyry copper deposits) that were formed during a certain period in geological history (e.g., the Phanerozoic eon) will move up and down at random in the crust. Arriving at the surface, crust erosion will destroy and dissipate the deposit. This results in a skewed bell-shaped distribution of deposits over the depth of the crust. The side of the bell near the crust's surface has partly disappeared. The model calculates the age frequency distribution of the deposits over the depth of the crust. After iterative calibration, the model results in a picture in which the modeled age frequency distribution of deposits near the surface corresponds with the real age frequency distribution of known deposits near the surface of the crust. For porphyry copper, Kesler and Wilkinson's best model results in 574 near-surface deposits with an age frequency distribution very similar to that of the known population of deposits. According to the model, the number of near-surface deposits (574) is only 1.2% of the total number of deposits in the crust. These known deposits contain 1.9 billion tons of extractable copper. Assuming that the grade tonnage distribution of porphyry copper deposits is the same in the rest of the crust, the total amount of porphyry copper deposits in the crust would be $100/1.2 \times 1.9 = 160$ billion tons.

Because porphyry copper is estimated at about 60% of all copper in the crust (US Geological Survey, 2014), total copper in ore deposits in the crust

Table 2.4 Indicative estimate of total global deposits. *IR*, identified resources; *na*, not available; *RB*, reserve base.

Column	RB (US Geological Survey, 2009) (Mt)	IR (US Geological Survey, 2018) (Mt)	IR/RB (column 2/column 1)	Calculated identified resources (column 1 × 2.8) (Mt)[a]	Ratio between assessed total deposits and identified resources (US Geological Survey, 2000)	Estimate for total global deposits to a depth of 3 km (3 × column 2 or 4 × column 5) (Mt)[b] (Rounded)
	1	2	3	4	5	6
Aluminum	na	26,000				200,000
Antimony	4.3	na		12		100
Arsenic	na	na				RB and IR na
Barium	520	na		1500		10,000
Beryllium	na	0.10				0.9
Bismuth	0.68	na		1.9		20
Boron	130	na		360		3000
Cadmium	1.2	na		3.4		30
Chromium	na	3700				35,000
Cobalt	13	25	1.9			200
Copper	1000	2100	2.1		2.12	10,000
Gallium	na	na				RB and IR na
Germanium	na	na				RB and IR na
Gold	0.10	na		0.28	2.20	2
Indium	0.016	na		0.045		0.4
Iron	160,000	230,000	1.4			2,000,000
Lead	170	2000	11.8		2.67	20,000
Lithium	11	53	4.8			500
Magnesium	2200	7200	3.3			70,000

Continued

Table 2.4 Indicative estimate of total global deposits. *IR*, identified resources; *na*, not available; *RB*, reserve base.—cont'd

Column	RB (US Geological Survey, 2009) (Mt)	IR (US Geological Survey, 2018) (Mt)	IR/RB (column 2/column 1)	Calculated identified resources (column 1 × 2.8) (Mt)[a]	Ratio between assessed total deposits and identified resources (US Geological Survey, 2000)	Estimate for total global deposits to a depth of 3 km (3 × column 2 or 4 × column 5) (Mt)[b] (Rounded)
	1	2	3	4	5	6
Manganese	4000	na		11,000		100,000
Mercury	0.24	0.60	2.5			6
Molybdenum	19	20	1			200
Nickel	150	130	1			1000
Niobium	3	na		8.4		80
Platinum group metals	0.080	0.10	1.3			0.9
Rare earth elements	130	na		350		3000
Rhenium	0.010	0.011	1.1	0.48		0.1
Selenium	0.17	na		1.6		5
Silver	0.57	na			3.88	20
Strontium	12	1000				9000
Tantalum	0.18	na		0.50		5
Tin	11	na		31		300
Titanium	900	1200	1.3			10,000
Tungsten	6.3	na		18		200
Vanadium	38	63	1.7			600
Zinc	480	1900	5.6		4.82	30,000
Average			2.8		3.14	

[a]If no figure is presented in column 3, the average (2.8) is used.
[b]If no figure is presented in column 5, the average (3.14) is used.

could be about 300 billion tons of copper according to this approach. Most of this amount will not be extractable in practice. According to the model of Wilkinson and Kesler, the upper kilometer in the crust contains about 6.5 billion tons of porphyry copper, the next kilometer about 8.5 billion tons, and the third kilometer about 9.5 billion tons (see Fig. 2.6). If we assume that the nonporphyry copper deposits are similarly distributed, the total amount of copper in the upper kilometer of the crust is $100/60 \times 6.5 = 11$ billion tons, in the next kilometer 15 billion tons, and in the third kilometer 17 billion tons. Hence, according to this approach, the upper 3 km of the Earth's crust contains about 40 billion tons of copper in enriched copper deposits (rounded). If the model is applied to the United States, deposits in the upper kilometer of the United States contain 700 Mt of copper. According to an assessment made in 2000 of undiscovered deposits of gold, silver, copper, lead, and zinc in the United States, deposits in the upper kilometer in the United States contain 640 Mt of copper

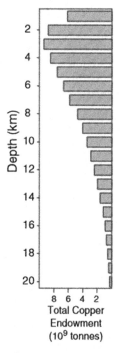

Figure 2.6 Vertical distribution of total copper endowments in porphyry copper deposits over 1 km crustal depth intervals, according to the tectonic diffusion model. *(Derived from Kesler, S.E., Wilkinson, B.H., March 2008, Earth's copper resources estimated from tectonic diffusion of porphyry copper deposits, Geology, 255–258.)*

(including past production), which is quite close to the estimate of 700 Mt according to the tectonic diffusion approach. See Table 2.3.

However, the supposed amount of 40 billion tons of extractable copper in the upper 3 km of the continental Earth's crust is about four times as much as the maximum estimated by the two other approaches, which seems quite optimistic. Wilkinson and Kesler (2010) applied their model to gold also and found that the total amount of gold deposits in the crust could be estimated to be 700 kt. This result does not fundamentally deviate from the two other approaches: extractable global gold resources of 300 kt in the upper 3 km (Table 2.2) and total global gold deposits of 2 Mt (Table 2.4). We have found no publications about the application of the tectonic diffusion model to the occurrence of other mineral resources.

2.5 Deeper mining

According to the results of the tectonic diffusion model, the occurrence of copper deposits does not depend very much of the depth. There is first a slight increase and then a decrease again per km depth (see Fig. 2.6). We suppose that this will not be very different for other minerals. This means that mining to 3 km deep could produce roughly three times more resources than mining to 1 km depth.

However, the temperature in the Earth's crust rises with increasing depth in the Earth's interior. This happens according to a certain geothermal gradient. According to Fridleifsson et al. (2008), away from tectonic plate boundaries, the geothermal gradient is about 25–30°C/km of depth near the surface in most of the world. This results in the temperatures shown in Table 2.5.

Table 2.5 Approximate temperatures at various depths below sea level in the continental Earth's crust.

Depth below sea level (m)	Temperature (°C)
1000	40
2000	60
3000	90
4000	120
5000	150

Derived from Fridleifsson, I.B., Bertani, R., Huenges, E., Lund, J.W., Ragnarsson, A., Rybach, L., 2008. The possible role and contribution of geothermal energy to the mitigation of climate change. In: Hohmeyer, O., Trittin, T., (Eds.), Proceedings of the IPPC Scoping Meeting on Renewable Energy Sources, Luebeck, Germany, 20–15 January, 2008: 59–80.

Mining at greater depths is therefore difficult and extremely costly. Conditions for the miners at such depths are very difficult because of the high temperature. Water, which is widely used in mining, must be cooled. So far, mines are seldom deeper than 1000 m.

The seven deepest mines in the world are gold mines in South Africa. They are almost 4 km deep. However, the elevation of the Earth's surface at the mining sites is about 1500 m. Therefore, the deepest shafts are about 2500 m below sea level. The temperature of the rocks below is more than 60°C. The mines are cooled with slurry ice to keep the temperature below 30°C (Wikipedia, 2018).

The eighth deepest mine in the world is the Kidd copper-zinc mine in Timmins, Ontario (Canada), which is about 3 km deep (below sea level!); the ninth deepest is again a gold mine in South Africa and the tenth is the Creighton nickel mine in Canada, which is 2.5 km below sea level. For the time being, a mine depth of 3 km below sea level seems to be the limit for profitable mining, although currently there are not many mines deeper than 1 km.

Bardi and Lavacchi (2009) state: "What creates the peak is not the resource that has run out; rather, what runs out is the financial capital needed to extract or to produce it."

2.6 Resources in seawater

In the previous sections we have only considered resources in the Earth's continental crust. However, seawater contains large amounts of resources, but in low concentrations. Of the dissolved salts, nearly 99.8 wt.% is made up by six elements: chlorine, sodium, magnesium, sulfur, calcium, and potassium (Rankin, 2011). Mineral production from seawater is practically limited to salts of these six elements (Bardi, 2008). Except for situations like in the Dead Sea, with a high salt concentration, mineral resources other than the six materials mentioned above are not produced from seawater because the high dilution rate makes their extraction prohibitively expensive, despite their enormous quantity in seawater. Table 2.6 presents the chemical composition of seawater.

Table 2.7 provides the concentration of nine other elements in seawater.

In the future, lithium might be commercially extractable from seawater, once the extraction of lithium resources from the continental Earth becomes more costly than extracting lithium from seawater (i.e., five times higher than current extraction costs according to Yaksic and Tilton, 2009).

Table 2.6 Major ions in "standard mean ocean water" (with a salinity of 35 g dissolved matter per kilogram seawater).

Ion	Symbol	Concentration (ppm)	Concentration (%)	% of dissolved matter
Chloride	Cl^-	19,353	1.9	55.29
Sodium	Na^+	1076	1.1	30.74
Magnesium	Mg_2^{2-}	1292	0.13	3.69
Sulfate	SO_4^{2-}	2712	0.27	7.75
Calcium	Ca^{2+}	412	0.04	1.18
Potassium	K^+	399	0.04	1.14
Total		34,928	3.5	99.8%

Derived from Chemical composition of seawater; Salinity and the major constituents, https://www. soest.hawaii.edu/oceanography/courses/OCN623/Spring%202015/Salinity2015web.pdf, retrieved on 31-7-2018.

Table 2.7 Concentration of nine elements in seawater (Bardi, 2008).

Element	Concentration (ppb)
Lithium	178
Barium	21
Molybdenum	10
Nickel	8
Zinc	5
Iron	3.4
Copper	0.9
Chromium	0.02
Lead	0.003

2.7 Resources on the sea floor

Mineral resources occur in three manners on the sea floor:
- Manganese nodules;
- Manganese crusts;
- Seafloor massive sulfide (SMS) deposits.

Manganese nodules are centimeter- to decimeter-sized lumps of manganese and iron oxides that occur on much of the ocean floors at depths of about 5500 m below sea level (Scott, 2001). Manganese constitutes 20%–25% of the higher grade nodules (Scott, 2001). The richer deposits contain an average of about 2.4% Cu + Ni + Co. According to Hein et al. (2013), seafloor nodules in the Clarion-Clipperton Fe–Mn Nodule Zone contain more thallium, manganese, tellurium, nickel, cobalt, and yttrium than the entire global terrestrial reserve base for these metals. According to the same authors, the nodules in the Clarion–Clipperton (CC) zone also contain significant amounts of copper, molybdenum, tungsten, lithium, niobium, and rare earth oxides. According to Scott (2001), as a copper resource, seafloor nodules contain about 10% of the amount of known land reserves of copper. According to the US Geological Survey, cited by Hannington et al. (2011), the global amount of copper in nodules is estimated at 700 million tons. However, only a small part of the Fe—Mn nodules in the CC zone can be considered as potentially of economic interest (Glasby, 2000; Hein et al., 2013).

Manganese crusts form centimeter- to decimeter-thick pavements of manganese and iron oxides on the flanks of sea mounts at water depths of 1000—2500 m. The manganese crust with the highest economic interest is the Prime Fe—Mn Crust Zone (PCZ) in the Central Pacific. According to Hein et al. (2013), the PCZ contains significantly more thallium, tellurium, cobalt, and yttrium than the entire terrestrial reserve base. Other metals occurring in significant amounts in the PCZ are bismuth, rare earth oxides, niobium, and tungsten (Hein et al., 2013). According to Hein et al. (2013), only a limited portion of these amounts can be economically recovered.

Polymetallic sulfides are produced by seafloor hot springs at water depths of 10 to 3500 m (Scott, 2001). Most of these seafloor massive sulfide (SMS) deposits are small. But there are some that are similar in size to land deposits (Scott, 2001). According to Hoagland et al. (2010), only about one-third of the 327 active and inactive hydrothermal sites that had been documented prior to 2010 are known to host significant SMS mineralization. SMS may contain substantial grades of copper, zinc, lead, silver, and gold (Scott, 2001; Hein et al., 2013). Using data from 10,000 km of ridge with SMS deposits, Hannington et al. (2011) came to the conclusion that the total accumulation of copper and zinc in these areas is on the order of 30 million tons. Although this is not insignificant, it is a small quantity compared to the extractable amounts of copper and zinc in the continental Earth's crust and in Fe—Mn nodules.

None of these three types of occurrences on the sea floor is being exploited thus far. There have been attempts, but they failed. Nevertheless, in January 2013, exploration contracts had been signed or were pending signature for about 1.8 million km^2 of seabed (Hein et al., 2013). About 45% of the contracted area concerned SMS deposits. The remainder mainly concerned Fe—Mn nodules (Hein et al., 2013). The total amount of resources on the sea floor is not expected to be high enough to offer a fundamental long-term solution for the exhaustion of resources in the

Earth's continental crust (Glasby, 2000; Hein et al., 2013; Sharma, 2011; Hannington et al., 2011). However, exploitation of some of the largest SMS deposits may be economically attractive (Scott, 2001). According to Sharma (2011), the total estimated cost of a single deep-sea mining venture for exploitation of Fe–Mn nodules would cost on the order of 10 billion USD, whereas the total gross-in-place value over 20 years would be on the order of 20–40 billion USD. This calculation does not take into consideration any unforeseen risks or failures. The mining circumstances are extreme: low temperatures (1–2°C, see Fig. 2.7), high pressure (500 bars at 5000 m depth), total darkness, cross-cutting currents at different water levels, uneven microtopography, variable seafloor characteristics, and heterogeneous nodule sizes and compositions (Sharma, 2011).

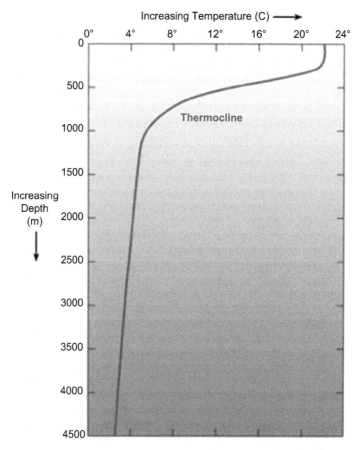

Figure 2.7 Water temperature–ocean depth profile. *(Derived from (Windows2universe, 2020).*

The potential environmental impact of deep-sea mining will be controversial. The main impact will be benthic disturbance and resuspension of fine sediment at the mining site. Hydrothermal vents host ecosystems with species adapted in special ways to the vent environment (Van Dover, 2011). However, Scott (2001) believes that deep-sea mining may be less environmentally harmful than land-based mining because there will be no piles of waste and no landscape deterioration. In Scott's view, the resuspension of material does not fundamentally differ from the ongoing suspension of material from seafloor hot springs.

2.8 Extraterrestrial resources

As the nearest neighbor to the Earth, the Moon is the most interesting celestial body for exploring the presence of exploitable mineral resources. Although the Moon possesses abundant raw materials "it is difficult to identify any single lunar resource that will be sufficiently valuable to drive a lunar resource extraction industry on its own" (Crawford, 2015). Crawford contends that the exploitability of a raw material on the Moon depends on whether the costs of such exploitation will be lower than the costs of extracting these materials on the Earth.

According to Eilingsfield (2016), the current cost of space transportation is on an order of magnitude of EUR 100,000 per kilogram payload. This is four orders of magnitude (10,000 times) more than aircraft transport, five orders of magnitude (100,000 times) more than transport by road, rail, and river barge, and six orders of magnitude (a million times) more than transport by container ship. Eilingsfield (2016) estimates that the cost of space transport would have to be three orders of magnitude lower for it to become close to the cost of terrestrial air transportation. Nagatomo and Collins (1997) state that "a two-order-of-magnitude reduction of launch costs will open the market of space tourism and power from space which is potentially large enough to make space activities profitable."

Clearly, large-scale mining activities on the Moon, in an atmosphere without oxygen and water, will be a challenge not only technologically, but also financially. No quantitative estimates are available for the costs of such Moon mining.

The July 2018 market price of platinum was about EUR 23,000 per kilogram. Let us assume that costs for extraction and concentration of platinum on the Moon would be comparable to the 2018 market price, which is a very optimistic assumption. Only if the current space transportation costs of EUR 100,000 per kilogram payload were to decrease by two orders of magnitude to EUR 1000, would the costs of space transportation approach an acceptable level. Current space transportation costs would only become surmountable if the platinum market price were to at least quadruple, assuming that the extraction and concentration costs of platinum on the Moon are not much higher than on Earth. Clearly, lunar exploitation of other industrial metals with a much lower market price, such as zinc, copper, and chromium, is very far from reality.

Near-Earth asteroids are also being considered in view of their potential interest for supplying mineral resources to the Earth. However, "the costs and risks of a private initiative to develop the systems needed for successful asteroid mining are simply too high and ... government participation ... is required ..." (Blair, 2000). However, the costs of exploiting mineral resources on near-Earth asteroids other than the Moon will presumably be higher than the costs of exploiting resources on the Moon.

The conclusion is that extraterrestrial exploitation may be an option in the future for only some selected mineral resources with a very high market price, such as platinum group metals, gold, and rhenium, and only if space transportation costs decrease by two or three orders of magnitude and the resources of such minerals on Earth become exhausted.

2.9 Conclusion

Although the ocean floor contains considerable amounts of mineral resources in manganese nodules, hydrothermal deposits, and ferromanganese crusts, large-scale extraction of these resources is not yet profitable. Additionally, it is risky from an environmental point of view. Although the extractable quantities are not yet well mapped, the amounts do not seem to be so high that the overall picture of available resources will change fundamentally (Glasby, 2000; Hein et al., 2013; Sharma, 2011; Hannington et al., 2011).

Although the resources in seawater and the extraterrestrial resources are also not negligible, they will not fundamentally change the estimate of available resources.

Hence, to estimate the ultimately available resources we will make use of the results of the two different generic approaches presented in Sections 2.2 and 2.3. Obviously, the results of the two approaches are not precise and only give order of magnitude estimates of the ultimately available resources. Nevertheless, the results will enable us to compare resources with respect to their relative scarcity and to get an indicative estimate of the exhaustion periods. In Table 2.8 we compare the results of the two approaches for estimating the available mineral resources (columns 1 and 2). We select the highest result and assume that this is the ultimately available amount of resources extractable from the Earth's crust (column 3).

Table 2.8 Estimate of the ultimately available resources.

Element	Extractable global resources (from Table 2.2) (Mt)	Total global deposits (from Table 2.4) (Mt)	Estimate of the ultimately available resources (rounded, Mt) [a]
Column	(1)	(2)	(3)
Aluminum	10,000,000	200,000	10,000,000
Antimony	20	100	100
Arsenic	200	Na	200
Barium	50,000	10,000	50,000
Beryllium	300	0.9	300
Bismuth	6	20	20
Boron	2000	3000	3000
Cadmium	40	30	40
Chromium	10,000	35,000	35,000
Cobalt	3000	200	3000
Copper	6000	10,000	10,000
Gallium	2000	Na	2000
Germanium	200	Na	200
Gold	0.3	2	2
Indium	30	0.4	30
Iron	6,000,000	2,000,000	6,000,000
Lead	2000	20,000	20,000
Lithium	2000	500	2000
Magnesium	3,000,000	70,000	3,000,000
Manganese	100,000	100,000	100,000
Mercury	8	6	8
Molybdenum	200	200	200
Nickel	8000	1000	8000
Niobium	2000	80	2000
Platinum group metals[b]	3	0.9	3
Rare earth elements[c]	20,000	3000	20,000
Rhenium	0.1	0.1	0.1
Selenium	6	5	6

Continued

Table 2.8 Estimate of the ultimately available resources.—cont'd

Element	Extractable global resources (from Table 2.2) (Mt)	Total global deposits (from Table 2.4) (Mt)	Estimate of the ultimately available resources (rounded, Mt) [a]
Column	(1)	(2)	(3)
Silver	8	20	20
Strontium	40,000	9000	40,000
Tantalum	200	5	200
Tin	300	300	300
Titanium	600,000	10,000	600,000
Tungsten	200	200	200
Vanadium	20,000	600	20,000
Zinc	9000	30,000	30,000

[a]The highest of the results in column 1 and column 2.
[b]Platinum group metals are: ruthenium, rhodium, palladium, osmium, iridium, and platinum.
[c]Rare earth elements are scandium, yttrium, lanthanum, cerium, praseodymium, neodymium, samarium, europium, gadolinium, terbium, dysprosium, hosmium, erbium, thulium, ytterbium, lutetium, and promethium.

The figures in Table 2.8 can be used for setting priorities regarding scarce mineral resources and for getting an indication of the period of time within which a resource may be exhausted, even with very optimistic assumptions. The results are not a precise absolute estimate of the ultimately available resources of each of the mentioned elements in the Earth's crust. In reality, the ultimate economical availability of resources in the Earth's crust may be much less than indicated in Table 2.8, but probably not much more.

In Sections 7.2 to 7.14, which cover the sustainable production of individual scarce resources in more detail, we will consider both the lower and higher availability estimates.

References

Allwood, J.M., Ashby, M.F., Gutowski, T.G., Worrell, E., 2011. Material efficiency: a white paper. Resour. Conserv. Recycl. 55, 362–381.

Barbalace, K., 2007. Periodic Table of Elements. http://environmentalchemistry.com/yogi/periodic. Retrieved 4 14, 2007.

Bardi, 2008. Mining the oceans: can we extract minerals from seawater?. In: Posted by Ugo Bardi on September 22, 2008 in the Oil Drum: Europe, Topic Supply/Production.

Bardi, U., Lavacchi, A., 2009. A simple interpretation of Hubbert's model of resource exploitation. Energies 2, 646–661.

Blair, B.R., May 5, 2000. The Role of Near-Earth Asteroids in Long-Term Platinum Supply, EB535 Metal Economics. Colorado School of Mines.

Crawford, I.A., February 8, 2015. Lunar resources: a review. Prog. Phys. Geogr. https://doi.org/10.1177/0309133314567585.

Darling, D., 2020. Terrestrial Abundance of elements. Europe's # 1 Data Room. http://www.daviddarling.info/encyclopedia/E/elterr.html. Retrieved on July 2 , 2020.

Eilingsfield, F., 2016. Halfway to anywhere: long term trends in space transportation cost. In: Presentation at the 2016 International Training Symposium of the International Cost Estimating and Analysis Association (ICEAA), Bristol, England, 19 October 2016.

Erickson, R.L., 1973. Crustal occurrence of elements, mineral reserves and resources. In: US Geological Survey Professional Paper 820, pp. 21—25.

Fridleifsson, I.B., Bertani, R., Huenges, E., Lund, J.W., Ragnarsson, A., Rybach, L., 2008. The possible role and contribution of geothermal energy to the mitigation of climate change. In: Hohmeyer, O., Trittin, T. (Eds.), Proceedings of the IPPC Scoping Meeting on Renewable Energy Sources, Luebeck, Germany, 20—15 January, 2008, pp. 59—80.

Glasby, G.P., July 28, 2000. Lessons learned from deep-sea mining. Science 289 (5479), 551—553.

Graedel, T.E., Nassar, N.T., 2015. The criticality of metals: a perspective for geologists. In: Jenkin, G.R.T., Lusty, P.A.J., McDonald, I., Smith, M.P., Boyce, A.J., Wilkinson, J.J. (Eds.), Ore Deposits in an Evolving Earth, vol. 393. Geological Society, London, Special Publications, pp. 291—302.

Guinée, J.B., Heijungs, R., 1995. A proposal for the definition of resource equivalency factors for use in product life-cycle assessment. Environ. Toxicol. Chem. 14, 917—925.

Hannington, M., Jamieson, J., Monecke, T., Petersen, S., Beaulieu, S., December 2011. The abundance of seafloor massive sulfide deposits. Geology 1155—1158.

Hein, J.R., Mizell, K., Koschinsky, A., Conrad, T.A., 2013. Deep-ocean mineral deposits as a source of critical metals for high- and green-technology applications: comparison with land-based resources. Ore Geol. Rev. 51, 1—14.

Hoagland, P., Beaulieu, S., Tivey, M.A., Eggert, R.G., German, C., Glowka, L., Kin, J., 2010. Deep-sea mining of seafloor massive sulfides. Mar. Pol. 34, 728—732.

Jefferson Lab, 2007. It's Elemental, the Periodic Table of Elements. http://education.jlab.org/itselemental/index.html, 4 29, 2007.

Jowitt, S.M., Mudd, G.M., 2014. Global Gold: Current Resources, Dominant Deposit Types and Implications. Proc. Gold14@Kalgoorlie, Kalgoorlie, Western Australia, pp. 56—58.

Kesler, S.E., Wilkinson, B.H., March 2008. Earth's copper resources estimated from tectonic diffusion of porphyry copper deposits. Geology 255—258.

McLennan, S.M., April 20, 2001. Relationships between the trace element composition of sedimentary rocks and upper continental crust. G-cubed 2.

Mohr, S.H., Mudd, G., Giurco, D., 2012. Lithium resources and production: critical assessment and global projections. Minerals 2, 65—84.

Mookherjee, A., Panigrahi, D., 1994. Reserve Base in relation to crustal abundance of metals — another look. J. Geochem. Explor. 51, 1—9.

Mudd, G.M., 2007. Global trends in gold mining: towards quantifying environmental and resource sustainability? Resour. Pol. 32, 42—56.

Mudd, G.M., Jowitt, S.M., 2014. A detailed assessment of global nickel resource trends and endowments. Econ. Geol. 109, 1813—1841.

Mudd, G.M., Jowitt, S.M., 2018. Growing global copper resources, reserves and production; discovery is not the only control on supply. Econ. Geol. 113 (6), 1235—1267.

Mudd, G.M., Weng, Z., Jowitt, S., Turnbull, I.D., Graedel, T.E., 2013. Quantifying the recoverable resources of by-product metals: the case of Cobalt. Ore Geol. Rev. 55, 87—98.

Mudd, G.M., Jowitt, S.M., Werner, T.T., 2017. The world's lead-zinc mineral resources: scarcity, data, issues and opportunities. Ore Geol. Rev. 80, 1160—1190.

Mudd, G.M., Jowitt, S.M., Werner, T.T., 2018. Global platinum group element resources, reserves and mining — a critical assessment. Sci. Total Environ. 622—623, 614—625.

Nagatomo, M., Collins, P., 1997. A common cost target of space transportation for space tourism and space energy development. In: Presented at the 7th International Space Conference of Pacific Basin Societies, vol 96. AAS paper no 97-460, AAS, pp. 617—630.

Nishiyama, T., Adachi, T., 1995. Resource depletion calculated by the ratio of the reserve plus cumulative consumption to crustal abundance for gold. Nat. Resour. Res. 4 (3), 253—261.

Phillips, W.G.B., 1977. Statistical estimation of global mineral resources. Resour. Pol. 3 (4), 268—280.

Rankin, W.J., 2011. Minerals, Metals and Sustainability, Meeting Future Material Needs. CRC Press.

Resource Panel, 2020. Derived from Web-Site of UNEP International Resource Panel. www.resourcepanel.org. Retrieved on 01 23, 2020.

Rudnick, R.L., Fountain, D.M., 1995. Nature and composition of the continental crust, a lower crustal perspective. Rev. Geophys. 267—309.

Rudnick, R., Gao, S., 2014. Composition of the continental crust. In: Holland, H., Turekian, K. (Eds.), Treatise on Geochemistry, second ed. Elsevier, Oxfords, pp. 1—51.

Scott, S.D., 2001. Deep ocean mining. Geosci. Can. 28 (2), 87—96.

Sharma, R., 2011. Deep-sea mining: economic, technical, technological and environmental considerations for sustainable development. Mar. Technol. Soc. J. 45 (5), 28—41.

Singer, D.A., 2017. Future copper resources. Ore Geol. Rev. 86, 271—279.

Skinner, B.J., 1976. A second iron age ahead? Am. Sci. 64, 158—169.

Skinner, B.J., 2001. Long run availability of minerals. In: Keynote Talk at a Workshop "Exploring the Resource Base", Resources for the Future, Washington DC, April 22—23, 2001.

Sverdrup, H.U., 2016. Modelling global extraction, supply, price and depletion of the extractable geological resources with the lithium model. Resour. Conserv. Recycl. 114, 112—129.

Sverdrup, H.U., Ragnarsdottir, K.V., 2016. A system dynamics model for platinum group metal supply, market price, depletion of extractable amounts, ore grade, recycling and stocks in use. Resour. Conserv. Recycl. 114, 130—152.

Sverdrup, H., Koca, D., Ragnarsdottir, K.V., 2014a. Investigating the sustainability of the global silver supply, reserves, stocks in society and market price using different approaches. Resour. Conserv. Recycl. 83, 121—140.

Sverdrup, H.U., Ragnarsdottir, K.V., Koca, D., 2014b. On modelling the global copper mining rates, market supply, copper price and the end of copper reserves. Resour. Conserv. Recycl. 87, 158—174.

Sverdrup, H.U., Ragnarsdottir, K.V., Koca, D., 2015. Aluminium for the future: modelling the global production, market supply, demand and price and long term development of the global reserves. Resour. Conserv. Recycl. 103, 139—154.

Tilton, J.E., 2001. Depletion and the long-run availability of mineral commodities. In: Workshop on the Long-Run Availability of Mineral Resources Sponsored by the Mining, Minerals and Sustainable Development Project and Resources for the Future in Washington DC, April 22—23, 2001.

Tilton, J.E., 2003. On borrowed time? Assessing the threat of mineral depletion. Miner. Energy - Raw Mater. Rep. 18 (1), 33–42.

UNEP International Resource Panel on Sustainable Resource Management, April 6, 2011. Estimating Long-Run Geological Stocks of Metals, Working Group on Geological Stocks of Metals, Working Paper.

US Geological Survey, 2000. 1998 Assessment of undiscovered deposits of gold, silver, copper, lead and zinc in the United States. Circular 1178.

US Geological Survey, 2017. Mineral Commodity Summaries 2017, Appendix C.

US Geological Survey, January 2009. Mineral Commodity Summaries.

US Geological Survey, January 2014. Estimate of Undiscovered Copper Resources of the World, 2013, Fact Sheet 2014-3004.

US Geological Survey, January 2018. Mineral Commodity Summaries.

Van Dover, C.L., 2011. Mining seafloor massive sulphides and biodiversity: what is at risk? ICES J. Mar. Sci. 68 (2), 341–348.

Van Oers, L., Huinée, J.B., Heijungs, R., 2020. Abiotic resource depletion potentials (ADPs) for elements revisited — updating ultimate reserve estimates and introducing time series for production data. Int. J. Life Cycle Assess. 25, 294–308.

Van Vuuren, D., Strengers, B., de Vries, H., 1999. Long term perspectives on world metal use- a system-dynamics model. Resour. Pol. 25 (4), 239–255.

Webelements, 2007. Abundance in Earth's Crust. http://www.webelements.com/webelements/properties/tekst/image-flah/abund-crust.html. March 9, 2007.

Wedepohl, K.H., 1995. The composition of the continental crust. Geochimica et Cosmochimica 59 (7), 1217–1232.

Weng, Z., Jowitt, S.M., Mudd, G.M., Hague, N., 2015. A detailed assessment of global rare earth resources: opportunities and challenges. Econ. Geol. 110, 1925–1952.

Wikipedia, 2018. List of Deepest Mines. In: Wikipedia. https://en.wikipedia.org/wiki/List_of_deepest_mines. Retrieved 7 16, 2018.

Wilkinson, B.H., Kesler, S.E., 2010. Tectonic-diffusion estimates of orogenic gold resources. Econ. Geol. 105, 1321–1334.

Wikipedia, 2020. Gangue, n.d., in: Wikipedia, https://en.wikipedia.org/wiki/Gangue; Derived from http://www.mindat.org/photo-229273.html, Ralph Bottril, Retrieved 12-7-2020

Windows2universe, 2020. Temperature of Ocean Water, n.d. In: Wikipedia, derived from https://www.windows2universe.org/earth/Water/temp.html. Retrieved 7 2, 2020.

Yaksic, A., Tilton, J.E., 2009. Using the cumulative availability curve to assess the threat of mineral depletion. Resour. Pol. 34, 185–194.

CHAPTER 3

Limits to the extractability of mineral resources[1]

3.1 Introduction

Whether we will run out of mineral resources is not a hypothetical question. The only situation in which humanity would not run out of mineral resources would be if mineral resources were unlimited; in other words, if they were available in a sufficient amount for thousands of years, regardless of future economic and population growth. Chapter 2 of this book provided an indication of the amounts of ultimately available resources. These figures enable us to calculate, indicatively, how soon a mineral resource may be exhausted. By exhaustion of a mineral resource, we mean a lack of profitability or feasibility of further extraction of that resource due to financial, environmental, energetic, climate change, waste generation, water use, or social impact factors, or a combination of these.

The exhaustion calculation is made in Section 3.2, answering the question whether mineral resources will be sufficiently available for many centuries to come — whatever development takes place — or whether humanity needs to take action to prevent a resource from being exhausted at relatively short notice. We go on to discuss Hubbert's peak theory, which is an alternative approach for predicting the depletion of mineral resources, in Section 3.3. As the extractability of a mineral resource is limited by the amount of energy needed for the extraction and its environmental impact, this is the subject of Section 3.4. We then discuss the environmental impacts of mining and using a mineral resource in Section 3.5.

Many mineral resources are produced as a by-product of the production of another mineral resource. These materials are called companion materials,

---

[1] This section is partly based on the publication of Henckens MLCM, Driessen PPJ, Worrell E, Metal scarcity and sustainability, analyzing the necessity to reduce the extraction of scarce metals, Resources Conservation and Recycling, 93, 2014, 1–8.

Governance of the world's mineral resources
ISBN 978-0-12-823886-8
https://doi.org/10.1016/B978-0-12-823886-8.00014-2

and in Section 3.6 we discuss the availability and exhaustion of these mineral resources.

In this chapter extraction is synonymous to primary production. By production we mean primary production.

3.2 Mineral resource exhaustion

3.2.1 Scarcity

In Chapter 2 of this book, we demonstrated that there is a strong relationship between the extractable amount of a raw material and the crustal abundance of the raw material in the Earth's crust. Hence, an element with a relatively low crustal abundance will also be expected to have a relatively low extractable quantity (see Fig. 2.5). The higher the annual extraction rate of a mineral resource compared to its crustal abundance, the higher the scarcity of that resource will be. The other main factor that determines exhaustion is the annual production growth of a resource. Moreover, the scarcity of a mineral resource compared to the scarcity of another resource may change as its extraction rate increases (or decreases) faster (or slower) than the extraction rate of other mineral resources. This may be caused by: (1) a higher (or lower) demand for that mineral resource due to new applications (or disappearing applications), (2) the discovery of new deposits of that resource, or (3) technological developments that make extraction of that mineral resource feasible to a lower grade. For the past, these developments are reflected in the production rates. As a proxy for future production growth (or decrease) for the coming decades, we use the average production growth or decrease of a mineral resource during the 35-year period between 1980 and 2015. For the further future, we expect mineral resource production growth to decouple from economic growth. This is analyzed in two scenarios in Section 3.2.2.

3.2.2 Two scenarios

The time period within which a resource may be exhausted depends on the temporal development of the extraction rate of that resource. We consider two schematic scenarios to estimate the development in annual mineral resource production:

Scenario 1: the annual production increase of mineral resources between 2015 and 2050 is equal to the average annual production increase between 1980 and 2015; the annual production increase between 2050 and 2100 is half of the assumed annual 2015–2050 production increase; no further production increase is assumed after 2100.

> Scenario 2: the annual production increase of mineral resources between 2015 and 2100 is equal to the average annual production increase between 1980 and 2015; the annual production increase between 2100 and 2200 is half of the assumed annual 2015–2100 production increase; no further production increase is assumed after 2200.

The two growth scenarios are presented schematically in Fig. 3.1.

In both scenarios, the production growth of mineral resources is assumed to slow down at a certain point and finally stop. In scenario 1, growth starts to slow in 2050 and stops in 2100. In scenario 2, growth starts to slow in 2100 and stops in 2200. These scenarios are not hypothetical, as explained below.

Growth in the global production and consumption of mineral resources is determined by a combination of population growth and gross domestic product (GDP) per capita increase. The increase in raw material use is partly offset by increasing material efficiency and recycling. According to the current expectations of the UN, the world population will reach about 11 billion by the end of the 21st century and will not grow much more after that (United Nations, 2017). However, GDP per capita will probably not stop growing, and there is a positive relationship between GDP and metal consumption. Therefore, the wealthier a country, the higher its metal use per capita (Graedel and Cao, 2010). However, above a certain GDP, material consumption per capita does not necessarily continue to increase with GDP (Halada et al., 2008; United Nations Environmental Programme, 2011). In fact, as these studies indicate, material consumption may decouple from GDP growth, starting from a per capita GDP of about

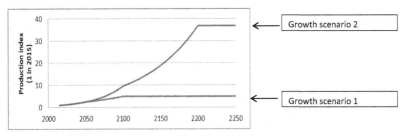

Figure 3.1 The two growth scenarios between 2015 and 2300. These are based on an annual production increase of raw materials of 2.7% between 1980 and 2015. The production is indexed at 1 in 2015.

US$10,000 (in 1980 US dollars) (Halada et al., 2008). UNEP (United Nations Environmental Programme, 2011) has developed a "freeze and catching up" scenario that shows that stabilization of raw material use would be possible from 2050, assuming a 3% annual growth in primary material use globally up to 2050. This growth would be composed of zero growth in raw material use in the industrialized world and a higher (>3%) growth in raw material use in the developing world. In this scenario, the developing world has caught up with the industrialized world in 2050 and, from that year on, primary material use could be globally decoupled from further GDP growth. A "freeze" scenario for the industrialized world is described by various sources (Eurostat, 2002; Weisz et al., 2006; NIES/MOE, 2007; Roglich et al., 2008), and the "catching up" scenario for the rest of the world is considered in a number of other studies (Giljum, 2002; Gonzalez-Martinez and Schandl, 2008; Chen and Qiao, 2001; Perez-Rincon, 2006; Russi et al., 2008; OECD, 2008).

Our growth scenarios are more pessimistic (or optimistic, depending on the perspective you take) than the UNEP "freeze and catching up" scenario. In scenario 1, overall production growth does not stop in 2050, but continues at half speed until it stops in 2100. In scenario 2, overall growth slows to half speed from 2100 and comes to a stop in 2200.

Table 3.1 presents the two scenarios for 36 elements.

Currently, the fastest production growth is in gallium, indium, lithium, the rare earth elements, niobium, and cobalt; all of which have an annual growth rate of more than 4%. Three elements had negative growth between 1980 and 2015: barium, beryllium, and mercury. The average production growth between 1980 and 2015 was 2.67%.

The purpose of the above schematic scenarios is not to correctly simulate real developments in the use of each mineral resource over a longer period of time, but to enable an indicative calculation of the time period within which exhaustion of a particular resource may become reality. The actual development of mineral extraction will probably follow a more S-curve-shaped type of path.

3.2.3 Estimated exhaustion periods

By combining the data in Table 3.1 and Table 2.8, we can calculate within how many years after 2015 mineral resources will be exhausted. Because of the uncertainties and assumptions, all calculated numbers are rounded to single-figure numbers (see Table 3.2). The elements are ranked in order of

Table 3.1 Two scenarios for the future production increase of raw materials.

	Annual production increase between 1980 and 2015. Derived from USGS (2017a,b)	Scenario 1, stabilization of production rates from 2100			Scenario 2, stabilization of production rates from 2200		
		Assumed production increase between 2015 and 2050	Assumed production increase between 2050 and 2100	Assumed production increase after 2100	Assumed production increase between 2015 and 2100	Assumed production increase between 2100 and 2200	Assumed production increase after 2200
Aluminum	3.8%	3.8%	1.9%	0.0%	3.8%	1.9%	0.0%
Antimony	2.2%	2.2%	1.1%	0.0%	2.2%	1.1%	0.0%
Arsenic	0.5%	0.5%	0.2%	0.0%	0.5%	0.2%	0.0%
Barium	−0.1%	−0.1%	−0.04%	0.0%	−0.1%	−0.04%	0.0%
Beryllium	−1.4%	−1.4%	−0.7%	0.0%	−1.4%	−0.7%	0.0%
Bismuth	3.8%	3.8%	1.9%	0.0%	3.8%	1.9%	0.0%
Boron	4.1%	4.1%	2.0%	0.0%	4.1%	2.0%	0.0%
Cadmium	0.7%	0.7%	0.4%	0.0%	0.7%	0.4%	0.0%
Chromium	3.5%	3.5%	1.7%	0.0%	3.5%	1.7%	0.0%
Cobalt	4.1%	4.1%	2.0%	0.0%	4.1%	2.0%	0.0%
Copper	2.8%	2.8%	1.4%	0.0%	2.8%	1.4%	0.0%
Gallium	9.8%	9.8%	4.9%	0.0%	9.8%	4.9%	0.0%
Germanium	0.9%	0.9%	0.5%	0.0%	0.9%	0.5%	0.0%
Gold	2.7%	2.7%	1.4%	0.0%	2.7%	1.4%	0.0%
Indium	8.1%	8.1%	4.1%	0.0%	8.1%	4.1%	0.0%
Iron	2.7%	2.7%	1.3%	0.0%	2.7%	1.3%	0.0%
Lead	1.0%	1.0%	0.5%	0.0%	1.0%	0.5%	0.0%
Lithium	5.4%	5.4%	2.7%	0.0%	5.4%	2.7%	0.0%
Magnesium	2.6%	2.6%	1.3%	0.0%	2.6%	1.3%	0.0%

Continued

Table 3.1 Two scenarios for the future production increase of raw materials.—cont'd

	Annual production increase between 1980 and 2015. Derived from USGS (2017a,b)	Scenario 1, stabilization of production rates from 2100			Scenario 2, stabilization of production rates from 2200		
		Assumed production increase between 2015 and 2050	Assumed production increase between 2050 and 2100	Assumed production increase after 2100	Assumed production increase between 2015 and 2100	Assumed production increase between 2100 and 2200	Assumed production increase after 2200
Manganese	1.7%	1.7%	0.9%	0.0%	1.7%	0.9%	0.0%
Mercury	−2.1%	−2.1%	−1.0%	0.0%	−2.1%	−1.0%	0.0%
Molybdenum	2.2%	2.2%	1.1%	0.0%	2.2%	1.1%	0.0%
Nickel	3.1%	3.1%	1.6%	0.0%	3.1%	1.6%	0.0%
Niobium	4.2%	4.2%	2.1%	0.0%	4.2%	2.1%	0.0%
Platinum group metals	2.3%	2.3%	1.2%	0.0%	2.3%	1.2%	0.0%
Rare earth elements	4.6%	4.6%	2.3%	0.0%	4.6%	2.3%	0.0%
Rhenium (linear)[a]	1.1 kt/year	1.1 kt/year	0.6 kt/year	0 kt/year	1.1 kt/year	0.6 kt/year	0 kt/year
Selenium	1.6%	1.6%	0.8%	0.0%	1.6%	0.8%	0.0%
Silver	2.5%	2.5%	1.2%	0.0%	2.5%	1.2%	0.0%
Strontium	3.8%	3.8%	1.9%	0.0%	3.8%	1.9%	0.0%
Tantalum	2.2%	2.2%	1.1%	0.0%	2.2%	1.1%	0.0%
Tin (linear)[a]	1.5 kt/year	1.5 kt/year	0.8 kt/year	0 kt/year	1.5 kt/year	0.8 kt/year	0 kt/year
Titanium	1.8%	1.8%	0.9%	0.0%	1.8%	0.9%	0.0%
Tungsten	1.6%	1.6%	0.8%	0.0%	1.6%	0.8%	0.0%
Vanadium	2.2%	2.2%	1.1%	0.0%	2.2%	1.1%	0.0%
Zinc	2.2%	2.2%	1.1%	0.0%	2.2%	1.1%	0.0%

[a]Increase in rhenium and tin production is linear rather than exponential.

Table 3.2 Exhaustion periods of elements at different scenarios (years after 2015).

	Production in 2015 (kt)	Indicative Estimate of the Ultimately available Resources (Table 2.8) (Mt)	Indicative exhaustion periods (years after 2015)		
			Growth scenario 1	Growth scenario 2	Average
Copper	19,100	10,000	100	100	100
Antimony	142	100	200	100	150
Gold	3.1	2	200	100	150
Boron	1,638	3,000	200	100	150
Silver	25.1	20	200	100	150
Bismuth	13.3	20	200	100	150
Molybdenum	235	200	300	100	200
Indium	0.77	30	400	100	250
Chromium	9,360	35,000	500	200	350
Zinc	12,800	30,000	600	200	400
Nickel	2,280	8,000	600	300	450
Tungsten	89.4	200	900	300	600
Tin	291	300	800	600	700
Rhenium	0.049	0.1	900	500	700
Selenium	2.2	6	1,000	400	700
Cadmium	23.2	40	1,000	700	850
Iron	1,400,000	6,000,000	2,000	200	1,100
Cobalt	126	3,000	2,000	200	1,100
PGM group metals (1)	0.473	3	2,000	400	1,200
Manganese	17,500	100,000	2,000	700	1,350
Lead	4,950	20,000	2,000	1,000	1,500
Niobium	64.3	2,000	3,000	300	1,700
Lithium	32.5	2,000	3,000	200	1,600
Arsenic	27.6	200	5,000	4,000	4,500
Gallium	0.469	2,000	10,000	200	5,100
Rare Earth Elements (2)	109.2	20,000	10,000	600	5,300
Strontium	354	40,000	10,000	800	5,400
Aluminum	57,500	10,000,000	20,000	1,000	10,500
Titanium	6,120	600,000	40,000	8,000	24,000
Tantalum	1.17	200	50,000	9,000	29,500
Vanadium	77.8	20,000	70,000	10,000	40,000
Magnesium	9,005	3,000,000	70,000	10,000	40,000
Germanium	0.16	200	700,000	400,000	550,000

Scarce
Moderately scarce
Not scarce

(1) Platinum group metals are ruthenium, rhodium, palladium, osmium, iridium, and platinum
(2) Rare earth elements are scandium, yttrium, lanthanum, cerium, praseodymium, neodymium, samarium, europium, gadolinium, terbium, dysprosium, holmium, erbium, thulium, ytterbium, lutetium, and promethium

increasing exhaustion period according to the average of the two scenarios. For the purpose of grouping the considered elements according to scarcity, we have categorized the elements as scarce, moderately scarce, and non-scarce elements.

Considering the results presented in Table 3.2, the most important observation is that there are eight elements that may be exhausted within 100 years. These are copper, gold, bismuth, silver, boron, antimony, molybdenum, and indium in scenario 2. We define this group of elements as *scarce* elements.

There is another group of five elements that may be exhausted within an average of 350 to 700 years: zinc, chromium, tungsten, tin, and nickel. We define this group as *moderately scarce* elements. The remaining elements are defined as *non-scarce* elements. This group includes the rare earth elements, the platinum group metals, iron, aluminum, cobalt, and lead.

Another observation is that the outcome of the calculations is fairly sensitive to the production growth. For instance, for the elements with rapid production growth: gallium, lithium, the rare earth elements, niobium, and cobalt (currently all above 4% annually), the exhaustion period in scenario 2 is much shorter than the exhaustion period in scenario 1. For elements with negative production growth, such as barium, beryllium, and mercury, the outcome of scenario 2 is higher than that of scenario 1.

It should be noted that the assumptions made are generic and not precise. This is justified because the main goal of the calculations is to determine the relative scarcity of resources and to obtain an indicative estimate of the exhaustion periods. In reality, the growth in the extraction rate of certain mineral resources (or groups of mineral resources) may be considerably higher (or lower) and/or longer (or shorter) than that calculated in Table 3.2. This may apply in particular to mineral resources that are essential for the energy transition from fossil-based to fossil-free, such as some rare earth elements (Kleijn, 2012; Achzet et al., 2011).

Note too that the data in Table 3.2 is based on current recycling rates and on the quantities extracted annually from the Earth's crust. This is not equal to the annual consumption figures, which are higher due to end-of-life recycling.

3.3 Peak minerals

The assumptions made in this book to estimate the exhaustion of mineral resources are fairly schematic: a continuation of the historic annual extraction growth until 2050 or 2100, a halving of this growth for 50 or 100 years, and stabilization of the extraction rate from 2100 or 2200 until the resource is exhausted. From that moment on, extraction is zero. The estimates of the ultimately available resources are also schematic. According to the peak theory, however, the growth and decline of the global extraction of a mineral resource will follow a more bell-shaped pathway.

3.3.1 Peak oil

In the 1950s, oil geoscientist Hubbert (1956a,b, 1971) developed a model to predict oil production in the United States. He did this through empirical research into the production rates of individual oil wells in the United States. Before the start of oil extraction from a well, the amount of oil extracted per unit of time is zero and, once the well is depleted, it becomes zero again. Hubbert demonstrated that the well production over time is a more or less bell-shaped curve with a maximum ("peak") somewhere in the middle.

According to Calvo et al. (2017), the Hubbert curve can be described by the following formula:

$$P(t) = \frac{R}{b_0\sqrt{2\pi}} e^{-\frac{1}{2}\left(\frac{t-t_0}{b_0}\right)^2}$$

$P(t)$ = the production in a certain year;
R = total available resource;
t_0 and b_0 are constants.

The maximum of the function, the peak, is at time t_0. At that moment, the production is:

$$P(t_0) = \frac{R}{b_0\sqrt{2\pi}}$$

When combining multiple oil sources, the sum curve will remain approximately bell-shaped, as old sources run empty while new sources are at the beginning of their production. Eventually, however, any new source will also run out of oil and the sum curve will also drop to zero. The area under the curve represents the amount of extractable oil.

However, reality deviates from the ideal curve. Calvo et al. (2017) give as possible reasons for this: new discoveries, new insights regarding the impact of the use of the resource on the environment and the climate, geopolitical factors, and substitution of the resource by another resource. Hubbert curves are based on a *business-as-usual* development, so without artificially braking or accelerating the exploitation of a resource.

Taking into account the uncertainty of how much new oil may be discovered in the future or, perhaps more relevant, the extent to which oil consumption will be limited due to climate impacts, the future shape of the global oil production curve is uncertain. However, what is certain is that, at some point, it will go down to zero again; for example, when all oil wells are exhausted or when mankind decides to stop oil extraction to address climate effects. The Hubbert curve is not necessarily beautifully Gaussian, but it will have a "peak" somewhere in time. That point (or period) is called "peak oil". If the total amount of oil is known in a particular source, or in a specific region or continent, the Hubbert curve for that particular source, region or continent can be determined on the basis of past annual production data. If there is more uncertainty about the ultimately recoverable oil resources, the curve will also be more uncertain.

Technological improvements can lead to the accelerated extraction and use of oil, which means that "peak oil" comes faster. Technological improvements may, however, also extend the exploitation of oil wells, so that more oil can be extracted from the same source, or so that an equal amount of oil can be extracted from a smaller source. Although technology development makes the final form of Hubbert curves uncertain, the curves are nevertheless considered to be fairly robust for oil (Bardi, 2009b; Kaufmann and Shiers, 2008; Pesaran and Samiei, 1995; all cited by May et al., 2011). According to Hubbert, cited by Pesaran and Samiei (1995), Hubbert curves can only be used with sufficient certainty for prediction purposes if the cumulative production has exceeded one-third of the ultimately recoverable oil resources. This means that this technique can be applied nationally, for example in Australia, and the United States or regionally, for example in Europe where the geological data are fairly complete, but not yet globally, because of the uncertainty about the ultimately recoverable oil resources. Obviously, Hubbert curves can be constructed based on scenarios. For example, Bartlett (2000) analyzed the impact of doubling the amount of available oil. He came to the conclusion that such an assumption would lead to a postponement of the oil peak by only about 15 years. By considering various scenarios, an impression can be obtained with regard to the future availability of oil.

Because exploration results in new discoveries, as long as exploration continues it is not certain when the final peak will ultimately be on a global scale. However, the peak has been reached as far as new discoveries of oil are concerned, as the amount of newly discovered oil (discovery rate) has been decreasing since the mid-1960s (Bardi, 2009a). "Peak production follows peak discovery … and …. production curves mirror discovery curves" (May et al., 2011; see Fig. 3.2).

3.3.2 Peak approach for non-fossil mineral resources

Hubbert's theory relates to oil, but can the theory also be applied to other mineral resources (May et al., 2011)? Other than oil, the crustal abundance of elements is enormous. However, it is not economically, environmentally, socially, or politically possible to completely extract this stock from the Earth's crust, so that only the enriched occurrences, the ores, are exploited. This book therefore concerns the exhaustion of ores of minerals, not the depletion of minerals in the Earth's crust. It is plausible that the production of non-fuel mineral resources will also show a peak somewhere in time. Like the oil discovery rate, the discovery rate of non-fossil mineral resources is decreasing (Lambert, 2001; Bleischwitz et al., 2009). However, the form of the peak will depend on whether it concerns major or minor elements (see Section 2.1). As the enrichment factor of a substance is higher in an ore, that substance, with respect to its occurrence, is more equivalent to oil, meaning that the substance can be found at clearly limited locations

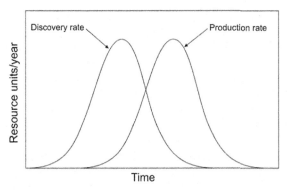

Figure 3.2 Simplified model of the behavior of production and discovery rates (Engelhardt, 1975).

just like oil. Major elements such as iron and aluminum occur more or less everywhere in higher or lower concentrations and are less comparable to oil with regard to their exhaustibility.

For major elements like iron and aluminum, large extra amounts of resources will become economically extractable if technology development enables the extraction of ores that are only slightly lower graded than those already extracted. For minor elements, technology development will not lead directly to much more available resources, but to a relatively easier extraction, and therefore possibly to the accelerated exhaustion of the same relatively high-graded ores. If the discovery rate of a substance, the ultimately available amount of the resource, and its demand are known, then, a Hubbert curve can in theory be drawn up. For example, Cordell et al. (2009), Glaister and Mudd (2009), Mudd and Ward (2008), Mudd (2007a,b), Sverdrup et al. (2013), and Sverdrup et al. (2015) derive Hubbert curves from the assumed amount of ultimately available resources of a given substance and the development of the production volume of that substance so far. In individual countries, where much is known about resources and where the peak production of certain non-fossil mineral resources has already been reached, the Hubbert model can represent reality fairly well. This is less clear for other resources. See Fig. 3.3 for the extraction of gold and copper in Canada.

Calvo et al. (2017) estimated the Hubbert peak of 47 different mineral commodities based on available information about the amount of extractable resources (see Fig. 3.4).

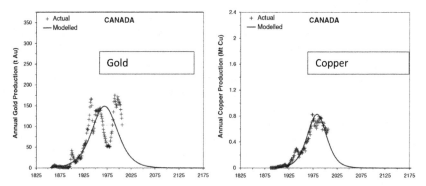

Figure 3.3 Examples of modeled Hubbert peak curves compared to historical production numbers. *(Derived from May, D., Prior, T., Cordell, D., Giurco, D., 2011. Peak minerals: theoretical foundations and practical application. Nat. Resour. Res. 21 (1), 43–60 (2012).)*

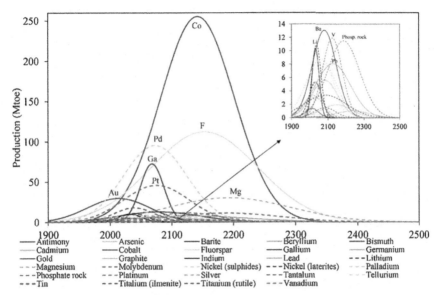

Figure 3.4 Hubbert peaks of selected minerals, as provided by Calvo et al. (2017). The amount of resources has been derived from USGS (2015) data: identified resources. Where these were not available, the 2009 reserve base data were used. The production of the various metals is expressed as an amount of exergy in million tons of oil equivalents. Exergy represents the minimum energy costs involved in producing a mineral resource with a specific chemical composition and grade from common rock containing the average abundance of the resource in the continental Earth's crust (see Section 3.4).

When interpreting the Hubbert curves, it must be borne in mind that the use of resources per capita will not increase forever, but that there is some kind of ceiling. It is estimated that this ceiling is relatively close to the current consumption level of an average resident of an industrialized country (Halada et al., 2008). This upper limit is about four times higher than the consumption of the average world citizen (see Section 5.4.3 and Table 5.3). This means that the peaks in a business-as-usual scenario may become flat at a level that is about four times higher than the current global use. It is illustrative and relevant to realize that the position of the peak is not very sensitive to the amount of resources. Based on a case study for lithium, Calvo et al. (2017) show that doubling the amount of resources in the case of lithium leads to a delay of the peak by only about 20 years (from 2060 to 2078; see Fig. 3.5).

The curves in Fig. 3.5 only include lithium resources in the continental Earth's crust. Lithium can however also be extracted from seawater at costs that are four to five times higher than the current costs (Yaksic and Tilton, 2009; see Fig. 6.2).

Some authors have doubts regarding the applicability of Hubbert curves to non-fossil mineral resources. Crowson (2011), for example, is of the opinion that Hubbert peak models for non-fuel mineral resources lead to uncertain results due to unknown and changing input variables. Hence, according to this author, peak modeling is not suitable for the estimation of future production trends, although he recognizes that extractable mineral resources are finite. Nevertheless, the value of the Hubbert peak approach is that it offers the possibility to apply different scenarios to available resources and future demand development, and therefore offers a tool for making calculations for policymaking. As knowledge about the extractable amount

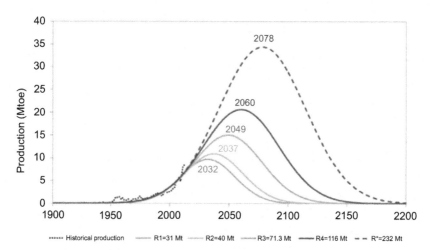

Figure 3.5 Hubbert peaks applied to lithium. The four lower curves (R1, R2, R3, R4) represent resource estimates made by Kesler et al. (2012), USGS (2016), Mohr et al. (2012), and Sverdrup (2016), all cited by Calvo et al. (2017). The curve with a peak in 2078 (R*) was calculated by assuming a doubling of the most optimistic estimate. *(Derived from Calvo, G., Valero, A., Valero, A., 2017. Assessing maximum production peak and resource availability of non-fuel mineral resources: analyzing the influence of extractable global resources. Resour. Conserv. Recycl. 125, 208–217.)*

of a mineral resource increases, the outcomes of the model will become more precise.

3.4 Energy use as a limiting factor for the extractability of mineral resources

The lower the grade of a mineral resource and the deeper the mine, the more energy is required for extraction of the resource. Once a certain amount of energy needed for the extraction of a unit of raw material is exceeded, extraction is no longer profitable. Mineral resources have been "processed" by nature from lower concentrations in the average Earth's crust to higher concentrations in ores. Therefore, mining companies do not start at zero, but make use of nature's beneficial action over a long period of time: the formation of enriched deposits of minerals.

Fig. 3.6 shows the natural and man-made processes in combination.

The energy that nature has used to transform a material from its appearance in the average Earth's crust into an ore with an enriched concentration of that material is called exergy. Exergy represents "the minimum energy costs involved in producing a mineral resource with a specific chemical composition and concentration of common materials in the environment" (Valero et al., 2009). Or, in other words: "The exergy of a mineral resource is the energy required by the given available technology to return a resource from the dispersed state of the reference environment into the conditions in which it was delivered by the ecosystems" (Valero

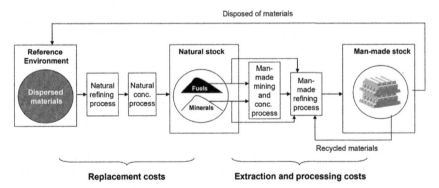

Figure 3.6 Processes involved in the production of a raw material. *(Derived from Valero A., Valero, A., Mudd, G.M., 2009. Exergy: A Useful Indicator for the Sustainability of Mineral Resources and Mining. https://www.researchgate.net/publication/268002763.)*

et al., 2010). Exergy is expressed in TOE (tons of oil equivalents). The exergy approach makes it possible to take into account all of the facets that make a natural resource valuable (quantity, chemical composition, and concentration) in a single indicator. This therefore allows the total share of mineral resources in the natural capital of a country (and of the world) to be expressed in TOE.

Conventional economics only takes the energy into account that is necessary for the extraction, further concentration, and processing of a mineral resource. However, it would be fairer and more balanced to also take the embedded exergy into account, which expresses the decrease in natural capital.

The lower the concentration of a mineral resource in an ore, or the lower the exergy of a mineral resource, the higher the amount of energy needed for extraction and concentration. To produce the same amount of metal, a larger amount of ore is needed. With halving of the grade, about double the amount of energy is required for the extraction of a raw material. This has been visualized for copper by Harmsen et al. (2013) (see Fig. 3.7).

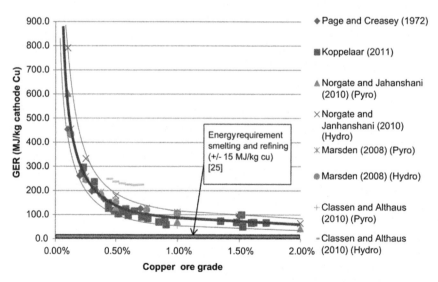

Figure 3.7 Gross energy requirement for different copper ore grades. *(Derived from Harmsen et al. (2013). These authors derived the data from Page and Creasy (1975), Koppelaar (2011), Norgate and Jahanshani (2010), Marsden (2008), and Classen and Althaus (2007).)*

The energy requirements for smelting and refining are minor in comparison to the energy requirements for mining and beneficiation, which is the removal of the gangue from the ore to produce a concentrate and tailings. According to Harmsen et al. (2013), about 15 MJ/kg Cu is used for smelting and refining in the pyro-metallurgical processing of copper. For copper ore grades between 0.5% and 1.0%, this is about 15% of the total gross energy requirement (GER), while 85% is used for mining and beneficiation. At lower ore grades, beneficiation will require relatively more energy. Deeper mining also requires more energy. According to Harmsen et al. (2013), the extra GER due to deeper mining for copper ore is 2.9 MJ/kg Cu per 100 m depth increase.

It is no longer useful to extract a material if the amount of energy needed for its extraction (energy investment) falls below the energy return. Take wind turbines as an example, where copper is needed in the turbines and in the electricity network connected to them. Switching from conventional electricity generation to electricity generation with wind turbines therefore requires extra copper, both for the turbines themselves and for the transport of the electricity produced. More electricity production requires more copper. As soon as copper extraction costs more energy than can be produced by the extracted copper, copper extraction for electricity production no longer makes sense. The energy return on energy investment (EROEI) has then become less than one. The GER for the extraction of a metal is therefore an important scarcity indicator. Harmsen et al. (2013) estimated the EROEI for wind turbines and the related transmission system to be an order of 20, also in scenarios with a high energy demand. This means that, from the point of view of the EROEI, there is still room for the extraction of lower graded copper occurrences.

3.5 Environmental impact as a limiting factor

According to Bardi and Lavacchi (2009), "what creates a peak is not that the resource has run out; rather, what runs out is the … capital needed to extract or produce it." To the extent that costs of mining, including the societal costs of greenhouse gas production, water use, waste generation, and landscape deterioration are internalized, extraction costs may become prohibitively high. Fig. 3.8 shows for copper and gold how CO_2 emissions and energy costs increase as their ore grade decreases.

The ultimate extractability of a mineral resource is limited to a larger extent by economic, environmental, climatic, and social restrictions than

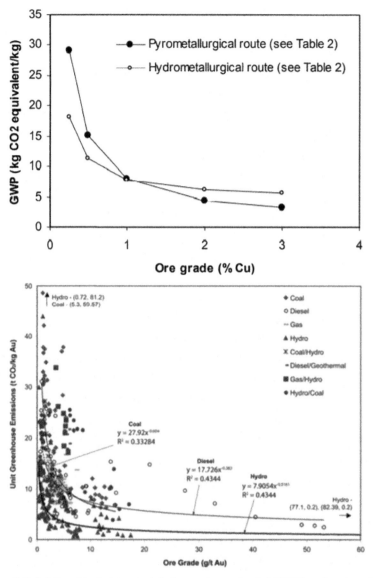

Figure 3.8 Relationship between global warming potential (GWP) and ore grade for copper (Norgate and Rankin, 2002) and between unit greenhouse emissions and ore grade for gold (Prior et al., 2012).

by the mere availability of the resource. For more information on this issue, please refer to the papers of Mudd et al. (2007, 2008, 2009, 2010, 2017a,b.

3.6 Carriers and companions

Mineral deposits are always a complex mixture of various compounds (e.g., sulfides, oxides) of a blend of different elements and other materials (e.g., silicates, carbonates, sulfates). Depending on the monetary value of each of the different elements, some are considered to be carrier (or host) materials and others to be companion materials (or by-products). The concentration of the companion materials is so low that their deposits are not economically viable on their own. Depending on the economic viability of the companion materials, they may be recovered as by-product during the processing of the host material. Fig. 3.9 shows the relationship between carrier materials and different co-elements.

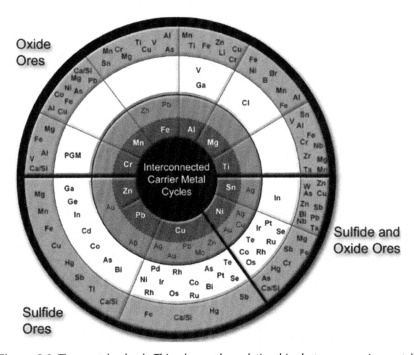

Figure 3.9 The metal wheel. This shows the relationship between carrier metals (dark blue: dark gray in printed version), co-elements that have their own production infrastructure (light blue: gray in printed version), co-elements that have no, or a limited, own production infrastructure (mostly highly valuable, high-tech metals, white), and co-elements that end up in residues or as emission (green: light gray in printed version) (Verhoef et al., 2004).

About 60% of 62 investigated metals have a companionality of more than 50%, meaning that more than 50% of the metal is produced as a by-product (see Fig. 3.10). This concerns antimony, arsenic, bismuth, cadmium, cobalt, gallium, germanium, hafnium, indium, iridium, most of the lanthanides group, osmium, palladium, rhenium, rhodium, ruthenium, scandium, selenium, silver, tellurium, thallium, vanadium, and zirconium. Of these materials, antimony, bismuth, indium, and silver are the relatively scarcest.

The availability of a companion material depends on the demand for the host material. This means that adjusting the production of companion materials to demand is more difficult than adjusting the production of carrier materials. The price movements of companion materials are particular. If demand for a companion material is lower than supply, the price may decrease, unless the producers agree on a fixed price. After all, companion materials are extracted whether they are wanted or not. If demand is higher than supply, the price of the companion metal will increase without suppliers increasing the production because, if they were to do so, they would produce a surplus of the host material, leading to overall lower profits. The lack of incentive at the suppliers' side to increase the

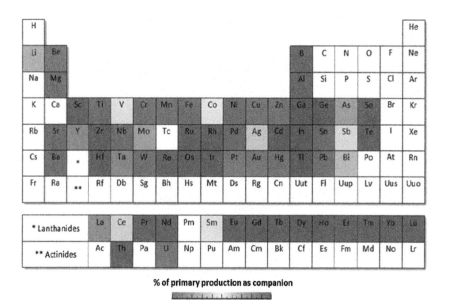

% of primary production as companion

0 10 20 30 40 50 60 70 80 90 100

Figure 3.10 The periodic table of companionality on a global basis in 2008 (Nassar et al., 2015).

production of companion materials can be explained by the relatively low added value of companion materials compared to the added value of the carrier material (Fu et al., 2018). Therefore, the recovery efficiency of by-products is often relatively low, due to the relatively low added value of the by-product compared to the value of the main product. Based on this, we would expect companion materials to have structurally and significantly higher price volatility than carrier materials, but this is not the case (see Chapter 6).

The short-term price movements of mineral resources are determined to a large extent by external factors, such as geopolitical developments and general economic developments. Only when the mining of a companion material becomes more rewarding than the mining of the carrier material will the mining of the companion material become leading compared to the former carrier material. This happens when the total value of the companion material mined becomes higher than the value of the carrier material, despite the fact that the carrier material may have a much larger volume. However, usually, even very high prices for companion materials have only a minor impact on decisions to open or close a mine (Werner et al., 2017). If there is a surplus of companion materials, they are generally included in mine tailings and slags. The recovery efficiency of the companion material decreases, but the advantage is that this creates a buffer of by-product for the future.

There are therefore two buffer mechanisms by which the production of companion materials can be regulated: (1) increasing (or decreasing) the recovery efficiency, and (2) extracting tailings and slags. However, the reaction to price increases will probably remain slow, because the demand for the main product remains leading.

3.7 Conclusions

Eight mineral resources are considered to be scarce: antimony, bismuth, boron, copper, gold, indium, molybdenum, and silver. This means that the geological resources of these elements may be exhausted within a century, if no saving measures are taken. It should be stressed that we have used the following three optimistic assumptions to draw this conclusion: (1) for all resources, mining will be economically viable until three kilometers deep in the continental Earth's crust; (2) as a point of departure for the amount of ultimately available resources we have taken the highest result of two generic approaches; and (3) production growth ceases in 2100 or 2200.

Although the total occurrence of elements in the Earth's crust is enormous, and quasi-unlimited, most of this amount cannot be profitably extracted. The exhaustion of a mineral resource is limited to a greater extent by financial, environmental, energy, and social restrictions than by the mere availability of the resource.

References

Achzet, B., Reller, A., Zepf, V., Rennie, C.B.P., Ashfield, M., Simons, J., University of Augsburg, 2011. ON Communication, Materials Critical to the Energy Industry. An introduction.

Bardi, U., 2009a. The four stages of a new idea. Energy 34, 323–326.

Bardi, U., 2009b. The mineral economy: a model for the shape of oil production curves. Energy Pol. 34 (3), 323–326.

Bardi, U., Lavacchi, A., 2009. A simple interpretation of Hubbert's model of resource exploitation. Energies 2, 646–661.

Bartlett, A.A., 2000. An analysis of US and world oil production patterns using Hubbert-style curves. Math. Geol. 32 (1), 1–17.

Bleischitz, R., Welfens, P.J., Zhang, Z., 2009. Sustainable Growth and Resource Productivity, : Economic and Global Policy Issue. Greenleaf Publishing Limited, Sheffield.

Calvo, G., Valero, A., Valero Antonio, 2017. Assessing maximum production peak and resource availability of non-fuel mineral resources: analyzing the influence of extractable global resources. Resour. Conserv. Recycl. 125, 208–217.

Chen, X., Qiao, L., 2001. A preliminary material output analysis of China. Popul. Environ. 23 (1), 117–126.

Classen, M., Althaus, H., 2007. Life Cycle Inventories of Metals (Ecoinvent 2.0) 2007, Part III, Copper, Molybdenum & Telluride and Silver Containing Byproducts, Data v2.0; 2007.

Cordell, D., Drangert, J.O., White, S., 2009. The story of phosphorus: global food security and food for thought. Global Environ. Change 19 (2), 292–305.

Crowson, P.C.F., 2011. Mineral reserves and future minerals availability. Miner. Econ. 24, 1–6.

Engelhardt, W., 1975. Resources and reserves of metals. Resour. Pol. 1 (4), 186–191.

Eurostat, 2002. Materials Use in the European Union, 1980-2000, Indicators and Analysis, Office for Official Publications of the European Communities, 2002.

Fu, X., Polli, A., Olivetti, E., 2018. High-resolution insight into materials criticality. In: Quantifying Risk for by-product Metals From Primary Production, Research and Analysis, Yale University. https://doi.org/10.1111/jiec.12757.

Giljum, S., 2002. Trade, material flows and economic development in the South: the example of Chile. In: Pater Presented at the 7th Biennial Conference of the International Society for Ecological Economics, Sousse Tunisia.

Glaister, B.J., Mudd, G.M., 2009. Platinum mining and sustainability – understanding the environmental costs of future technologies. In: Sustainable Development Indicators in the Minerals Industry (SDIMI) Conference, Gold Coast, QLD, pp. 259–272.

Gonzalez-Martinez, A.C., Schandl, H., 2008. The biophysical perspective of a middle income economy: material flows in Mexico. Ecol. Econ. 68 (1–2), 317–327.

Graedel, C., Cao, J., 2010. Metal spectra as indicators of development. Proc. Natl. Acad. Sci. U.S.A. 107 (49), 20905–20910.

Halada, K., Shimada, M., Ijima, K., 2008. Decoupling status of metal consumption from economic growth. Mater. Trans. 49 (3), 411–418.

Harmsen, J.H.M., Roes, A.L., Patel, M.K., 2013. The impact of copper scarcity on the efficiency of 2050 global renewable energy scenarios. Energy 50, 62–73.

Hubbert, M.K., 1956a. Nuclear energy and the fossil fuels. Drill. Prod. Pract. 95, 1–40.

Hubbert, M.K., 1956b. Nuclear Energy and the fossil fuels. In: Presented before the Spring Meeting of the Southern District, American Petroleum Institute, Plaza Hotel, San Antonio, Texas, March, 1956.

Hubbert, M.K., 1971. The energy resources of the earth. Sci. Am. 225 (3), 1–61.

Kaufmann, R.K., Shiers, L.D., 2008. Alternatives to conventional crude oil: when, how quickly, and market driven? Ecol. Econ. 67 (3), 405–411.

Kesler, S.E., Gruber, P.W., Medina, P.A., Keoleian, G.A., Everson, M.P., Wallington, T.J., 2012. Global lithium resources: relative importance of pegmatite, brine and other deposits. Ore Geol. Rev. 48, 55–69.

Kleijn, R., 2012. Materials and Energy: A Story of Linkages. University of Leiden.

Koppelaar, R.H.E.M., January 2011. Modeling Mineral Cost Developments, the Case of Copper, Iron Ore, Tungsten and Phosphorus. Wageningen University.

Lambert, I.B., 2001. Mining and sustainable development: considerations for minerals supply. Nat. Resour. Forum 25 (4), 275–284.

Marsden, J., 2008. Energy efficiency and copper hydrometallurgy. In: Proceedings of the 6th International Symposium Phoenix: Hydrometallurgy, August 2008, pp. 29–42.

May, D., Prior, T., Cordell, D., Giurco, D., 2011. Peak minerals: theoretical foundations and practical application. Nat. Resour. Res. 21 (1), 43–60, 2012.

Mohr, S.H., Mudd, G., Giurco, D., 2012. Lithium resources and production: critical assessment and global projections. Minerals 2, 65–84.

Mudd, G.M., 2007a. Global trends in gold mining: towards quantifying environmental and resource sustainability? Resour. Pol. 32, 42–56.

Mudd, G.M., 2007b. The Sustainability of Mining in Australia: Key Production Trends and Their Environmental Implications for the Future, Research Report No RR5. Department of Civil Engineering, Monash University and Mineral policy Institute. Revised April 2009.

Mudd, G.M., 2009. Historical trends in base metal mining: back casting to understand the sustainability of mining. In: Proc. 48th Annual Conference of Metallurgists, Canadian Metallurgical Society, Sudbury, Ontario, Canada, August 2009.

Mudd, G.M., 2010. The environmental sustainability of mining in Australia: key megatrends and looming constraints. Resour. Pol. 35, 98–115.

Mudd, G.M., Ward, J.D., 2008. Will sustainability constraints cause "peak minerals"?. In: 3rd International Conference on Sustainability Engineering & Science: Blueprints for Sustainable Infrastructure, Auckland, New Zealand, 9–12 December 2008.

Mudd, G.M., Jowitt, S.M., Werner, T.T., 2017a. The world's lead-zinc mineral resources: scarcity, data, issues and opportunities. Ore Geol. Rev. 80, 1160–1190.

Mudd, G.M., Jowitt, S.M., Werner, T.T., 2017b. The world's by-product and critical metal resources part I: uncertainties, current reporting practices, implications and grounds for optimism. Ore Geol. Rev. 86, 924–938.

Nassar, N.T., Graedel, T.E., Harper, E.M., April 3, 2015. By-product metals are technologically essential but have problematic supply. Sci. Adv. 1, e1400180.

NIES/MOE, 2007. Material Flow Account for Japan. National Institute of Environmental Studies on behalf of the Ministry of the Environment.

Norgate, T., Jahanshahi, S., 2010. Low grade ores —smelt, leach or concentrate? Miner. Eng. 23, 65–73.

Norgate, T.E., Rankin, W.J., 2002. The role of metals in sustainable development. In: Proceedings of "Green Processing 2002 — International Conference on Sustainable Processing of Minerals", AusIMM, CSIRO and AMEEF, Cairns, QLD, May 2002, pp. 49–55.

OECD, 2008. Measuring Material Flows and Resource Productivity. Synthesis Report, OECD, Paris.

Page, N.J., Creasy, S.C., 1975. Ore grade, metal production and energy. J. Res. U.S. Geol. Surv. 3 (1), 9–13.

Perez-Rincon, M.A., 2006. Colombian international trade from a physical perspective: towards an ecological "Prebisch thesis". Ecol. Econ. 59 (4), 519–529.

Pesaran, M.H., Samiei, H., 1995. Forecasting ultimate resource recovery. Int. J. Forecast. 11 (4), 543–555.

Prior, T., Giurco, D., Mudd, G., Mason, L., Behrisch, J., 2012. Resource depletion, peak minerals and the implications for sustainable resource management. Global Environ. Change 22, 577–587.

Roglich, D., Cassara, A., Wernick, I., Miranda, M., 2008. Material Flows in the United States: A Physical Accounting of the US Industrial Economy. WRI Report.

Russi, D., Gonzalez-Martinez, A.C., Silva-Macher, J.C., Giljum, S., Martinez-Alier, J., Vallejo, M.C., 2008. Material flows in Latin America : a comparative analysis of Chile, Ecuador, Mexico and Peru (1980-2000). J. Ind. Ecol. 12 (5–6), 704–720.

Sverdrup, H.U., 2016. Modeling global extraction, supply, price and depletion of the extractable geological resources with the (LITHIUM) model. Resour. Conserv. Recycl. 114, 112–129.

Sverdrup, H.U., Koca, D., Ragnarsdottir, K.V., 2013. Peak metals, minerals, energy wealth, food and population: urgent policy considerations for a sustainable society. J. Environ. Sci. Eng. B2 189–222.

Sverdrup, H.U., Ragnarsdottir, K.V., Koca, D., 2015. An assessment of metal supply sustainability as an input to policy: security of supply extraction rates, stocks-in-use, recycling and risk of scarcity. J. Clean. Prod. https://doi.org/10.1016/j.jclepro.2015.06.085.

UNEP, 2011. International Resource Panel, Decoupling Natural Resource Use and Environmental Impacts from Economic Growth.

United Nations, 2017. World Population Prospects, the 2017 Revision.

US Geological Survey, January 2015. Mineral Commodity Summaries.

US Geological Survey, 2016. Mineral Commodity Summaries 2016, Lithium.

US Geological Survey, January 2017a. Mineral Commodity Summaries.

US Geological Survey, 2017b. Historical Statistics for Minerals and Mineral Commodities in the USA.

Valero, A., Valero, A., Mudd, G.M., 2009. Exergy: A Useful Indicator for the Sustainability of Mineral Resources and Mining. https://www.researchgate.net/publication/268002763.

Valero, A., Valero, A., Martínez, A., 2010. Inventory of the exergy resources on earth including its mineral capital. Energy 35, 989–995.

Verhoef, E., Dijkema, G., Reuter, M.A., 2004. Process knowledge, system dynamics and metal ecology. J. Ind. Ecol. 8, 23–43.

Weisz, H., Krausmann, F., Amann, C., Eisenmenger, N., Erb, K.-H., Hubacek, K., Fischer-Kowalski, M., 2006. The physical economy of the European Union: cross-country comparison and determinants of material consumption. Ecol. Econ. 58 (4), 676–698.

Werner, T.T., Mudd, G.M., Jowitt, S.M., 2017. The world's by-product and critical metal resources, Part III: a comprehensive assessment of indium. Ore Geol. Rev. https://doi.org/10.1016/j.oregeorev.2017.01.015.

Yaksic, A., Tilton, J.E., 2009. Using the cumulative availability curve to assess the threat of mineral depletion. Resour. Pol. 34, 185–194.

CHAPTER 4

Critical raw materials

4.1 Introduction

Critical raw materials are not identical to geologically scarce mineral resources.

The criticality of a raw material reflects the vulnerability of a system (a country, the world, the EU, a company, or an industrial sector) to the risk that supplies of that raw material might be disrupted. The European Union (EU) considers a raw material to be critical if it has a high economic importance to the EU combined with a high supply risk. Criticality has two dimensions: risk of supply disruption and vulnerability of the system using the raw material. Different systems have different vulnerabilities to supply disruption of a specific raw material. Hence, there is no such thing as an unambiguous global list of critical raw materials. The more important a raw material is for the economy of a country, the more vulnerable that country is to the nonavailability of that raw material. This depends on the specific needs of the country. Obviously, countries producing a raw material are less vulnerable to nonavailability of that material than nonproducing countries. Also countries with a large production of modern electronics, for example, need more specific rare earth elements than countries without such a production. Vulnerability is moderated by the extent to which a material is more easily substituted and recycled.

Physical exhaustion is a supply risk in the long term. Short-term (but temporary) supply risks are caused by, for instance, accidents in important mines, strikes, wars, political instability and weak governance in production countries, boycott actions, or export limitations with a political or economic background.

This chapter presents an overview of different approaches to the concept of criticality of raw materials including the different supply risk and vulnerability indicators. We present criticality lists of different countries and we compare criticality with the scarcity of elements as presented in Table 3.2.

Governance of the world's mineral resources
ISBN 978-0-12-823886-8
https://doi.org/10.1016/B978-0-12-823886-8.00023-3

4.2 Different approaches

Erdman and Graedel (2011) and Schrijvers et al. (2020) present reviews of different approaches to the criticality of nonfuel minerals. The goal of all approaches is to identify critical raw materials. However, the scope of the approaches differs widely with regard to the considered system (national economies, EU economy, a societal sector (such as defense), a single company, clean technology sector). The considered time horizons differ also, but most of the studies have a time horizon of less than 10 years. Most approaches result in a criticality matrix or a criticality index. Mostly, the dimensions of the criticality matrices are supply risk and vulnerability (importance for the system or impact on the system), but three-dimensional matrices have also been developed (Nassar et al., 2012). The more easily a raw material is substituted or recycled, the less the impact of a supply disruption will be. Different criticality assessments will result in different classifications, whether or not a raw material is critical for a system. There is a rather high sensitivity of the results to minor methodological changes.

4.3 Supply risk and vulnerability indicators

Different supply risk and vulnerability indicators are presented in Table 4.1. The indicators are explained in the Sections 4.3.1 and 4.3.2. Note that not every indicator mentioned in the table is included in each approach.

4.3.1 Supply risk indicators

- Global reserves and/or geological scarcity

    Global reserves of raw materials are based on data of the United States Geological Survey (USGS). Geological scarcity is usually based as well on USGS data: the reserve base, identified resources, or identified plus undiscovered resources. For the definitions, we refer to Chapter 2. The indicator is the ratio between reserves (or resources) and annual global production. Some approaches, like the EU approach, do not include reserves or geological scarcity as an indicator for raw material criticality.

- Global supply concentration

    Usually the so-called Herfindahl–Hirschman index is used to take the supply concentration into account. This index measures the size of the production of a raw material in a country in relation to the global production of the raw material.

Table 4.1 Supply risk and vulnerability indicators.

Supply risk indicators	References	Vulnerability indicators	References
Global reserves (medium term)	a,b	Economic importance	a,b
Geological scarcity (long term)	a,b	Substitutability	a,b,e
Global supply concentration	a,b	Recyclability	f
Reserves distribution	c	Innovation potential	a,b
Companion metal fraction	a,b	Percentage of the population using the raw material	a,b
Environmental implications	a	Import reliance ratio	a,b
Governance quality in production countries	a,b		
Human Development Index of production countries	a,b		
Political stability in production country	a,b		
Price volatility	d		

[a] Erdmann and Graedel (2012).
[b] Graedel et al. (2012).
[c] British Geological Survey (2015).
[d] Bastein and Rietveld (2015).
[e] Nassar et al. (2012).
[f] Roelich et al. (2012).

- Reserves distribution

    Supply risk is higher to the extent that the reserves are concentrated in one or only a few countries. The indicator can be expressed as the proportion of the reserves in the three countries with the highest reserves compared to the global reserves. The higher the indicator is, the larger the supply risk is.
- Companion metal fraction

    Many technologies rely on raw materials, which are a by-product (companion material) of the mining of so-called host products or carrier materials. Companion materials are considered to be a supply risk because their production depends on the production of their carriers rather than on their demand from the market. The monetary value of the fraction of a raw material, which is produced as a by-product, compared to the monetary value of the main product is an indicator for the supply risk. The supply risk is less to the extent that the recovery efficiency of the by-product can be increased without much effort.

- Environmental implications

  This concerns the environmental impact of the extraction and production of the raw materials, for example, the greenhouse warming potential, the toxicity of the process and the produced waste, the potential impact on health and safety of the miners, and the material/water use per kilogram of raw material produced. These are factors which may hamper or even disrupt the production of the raw material. The indicator may be based on the score for a cradle to gate life cycle assessment (from unmined ore to the manufacturing front gate). See Graedel et al. (2012).

- Governance quality in production countries

  As an indicator for the governance quality, the World Governance Index (WGI) published by the World Bank (2020) can be used. The WGI includes voice and accountability, political stability, government effectiveness, regulatory quality, rule of law, and control of corruption. To the extent that the governance quality in a production country is lower, the risk of supply disruption is supposedly higher.

- Human Development Index of production countries

  The Human Development Index (HDI) is developed (and adapted) by the United Nations Development Programme (UNDP, 2018). The index is composed of life expectancy, education, and national per capita income indicators. The UNDP publishes reports on an annual basis with the Human Development Indices of countries. To the extent that the HDI of a production country is lower, it can be supposed that the supply risk by such a country is higher.

- Price volatility

  In a way, the price volatility of a raw material mirrors all the other supply risks. However, in some approaches (Bastein and Rietveld, 2015) price volatility is also included as a separate indicator. A system (nation, company, sector) can be vulnerable to price volatility to the extent that sudden price increases cannot be passed on to customers and they have an impact on the competitiveness of the system. The influence depends strongly on the cost of the raw material as compared to the cost of the final product in which it is included.

4.3.2 Vulnerability indicators

- Economic importance

    There are several approaches to determining economic importance. Here, we present the approach of the EU. For the assessment of the economic importance of a raw material, the EU (2020) combines the proportion of a material associated with industrial mega sectors at an EU level with the mega sectors' proportional gross value added to the EU's gross domestic product. The two proportions (percentages) are multiplied, providing an indicator for the economic importance of the raw material for the EU economy.

- Substitutability

    It is assumed that the supply risk of a raw material is lower to the extent that the raw material can be more easily substituted for another raw material. There are no quantitative data on the substitutability of materials. Substitutability is based upon qualitative or semiquantitative data. See, for example, the report on critical raw materials for the EU (2020). Substitutability is composed of substitute performance, substitute scarcity, environment, health, and safety (EHS) impacts of the substitute compared to the EHS impacts of the original and the price of the substitute as compared to the price of the original (Graedel et al., 2012).

- Recyclability

    Increased recycling is assumed to reduce the supply risk. Recyclability can be quantified by the current end-of-life recycling rate of a raw material. There is also a study which considers the potential recyclability of a raw material by taking into consideration aspects such as the dissipative usage share, physical and chemical limitations to recycling, price incentives for recycling, and the availability of appropriate technologies and infrastructure (Buchert et al., 2009).

- Innovation potential

    For innovative nations it will be easier to cope with a supply restriction of a raw material than for less innovative nations. As an indicator, the Global Innovation Index which is published annually by Cornell University, INSEAD, and the World Intellectual Property Organization (2019), can be used.

- Percentage of the population using the raw material

    To the extent that a larger part of the population benefits from a specific raw material, the impact of a supply restriction is greater. The indicator is the percentage of the population using the raw material.

- Import reliance ratio

    To the extent that a nation has its own geological reserves of a raw material, it is less dependent on other nations with respect to the import of that raw material. The indicator is the ratio of the net import and the apparent domestic consumption of the raw material.

4.4 Criticality lists

For the determination of the criticality of a raw material, different approaches attribute different weights to the individual indicators. The indicator weights may differ per system considered (EU, individual country, industrial sector, company, societal sector). Eventually, it is a political, financial, or economic choice as to how much weight is attributed to each of the indicators in a given system. An overview of indicator weights in different criticality approaches is given by Achzet and Helbig (2013).

Table 4.2 provides an overview of lists of critical raw materials in 10 countries (Great Britain, Germany, France, United States of America, Japan, Netherlands, China, Canada, Australia, and South Korea) and in the European Union. Moreover, Table 4.2 shows the occurrence of raw materials in a recent overview of 36 different criticality lists (Schrijvers et al., 2020). The five materials with the highest occurrence in the 36 considered criticality lists are cobalt, gallium, indium, lithium and Rare Earth Elements.

The different criticality lists differ quite a lot from each other. The reason is that vulnerabilities of countries and sectors to supply disruption differ a lot per raw material. However, the number of lists in which a specific raw material occurs could be considered as a measure to the criticality of that material on a global scale.

In Fig. 4.1 and Table 4.3 we compare the criticality according to the EU list of critical raw materials (EU, 2020) with the scarcity of materials according to Table 3.2. Fig. 4.1 and Table 4.3 clarify that there is not a strong relation between criticality and scarcity. Only five elements (19% of the elements considered) are both critical and scarce. It concerns antimony, bismuth, boron, indium and tungsten. Eight elements (31% of the elements considered) is scarce or moderately scarce according to Table 3.2, but is not considered critical according to the EU list of critical raw materials. Thirteen elements (50% of the elements considered) are critical according to the EU list of critical raw materials (EU, 2020), but not scarce according to Table 3.2.

Table 4.2 Criticality lists.

Raw material	EU	GB (top 22)	De	F	USA	Jp	Nl	CN	CA	AU	KR	% of 36 different criticality lists, in which the material occurs
References	a	b	c	d	e	f	f	f	f	f	f	g h
Aluminum + bauxite								X	X			65%
Antimony	X	X	X				X	X				50%
Arsenic		X					X				X	25%
Barium + baryte	X	X										36%
Beryllium	X	X					X					47%
Bismuth	X	X	X				X					36%
Boron	X											36%
Cadmium		X					X					33%
Carbon		X										
Chromium			X									53%
Cobalt	X	X		X	X	X	X				X	75%
Coking coal	X											
Copper									X			61%
Fluor	X											36%
Gallium	X	X	X	X	X		X				X	75%
Germanium	X	X	X	X			X					53%
Gold							X		X			44%
Graphite	X											42%
Hafnium	X											
Helium	X											
Indium	X	X	X	X	X		X				X	75%
Iron								X	X			50%
Lead							X		X			39%
Lithium		X		X	X		X			X	X	72%
Magnesium + magnesite	X	X										55%
Manganese						X					X	53%
Mercury							X	X				28%
Molybdenum		X					X	X	X	X		X	58%
Natural rubber	X											
Nickel							X	X		X		69%
Niobium	X			X	X		X					58%

Continued

Table 4.2 Criticality lists.—cont'd

Raw material	EU	GB (top 22)	De	F	USA	Jp	NI	CN	CA	AU	KR	% of 36 different criticality lists, in which the material occurs
Phosphorus	X											25%
Platinum Group Metals	X	X	X				X					60%
Rare Earth Elements	X	X	X	X	X		X	X		X	X	70%
Rhenium		X	X	X			X					50%
Scandium	X						X					
Selenium				X			X					47%
Silicon metal	X											25%
Silver		X	X				X		X			50%
Strontium		X					X					28%
Tantalum	X	X		X			X			X	X	56%
Tellurium					X		X					61%
Thallium							X					
Tin			X				X	X				53%
Titanium							X				X	53%
Tungsten	X	X	X			X	X	X			X	53%
Vanadium	X	X				X	X	X		X	X	53%
Yttrium							X					
Zinc							X	X				44%
Zirconium							X					39%

[a] EU (2020); Dozolme (2020).
[b] British Geological Survey (2015).
[c] KfW (2011).
[d] Dozolme (2020).
[e] US Department of Energy (2011).
[f] DEFRA (2012).
[g] Schrijvers et al. (2020).
[h] In case of groups of elements/materials, such as PGMs and REEs, the occurrences overlap in different criticality lists. For example the different REEs occur all together 156 times in the 36 criticality lists and the PGMs have 63 occurrences. Details have not been provided in the article. Therefore, the figures for the groups of materials in this column are a rounded approximation.

The conclusion is that mitigation policies with respect to critical raw materials do not necessarily match with mitigation policies regarding geologically scarce resources. Critical raw materials on the one hand, and geologically scarce mineral resources on the other, require different policies.

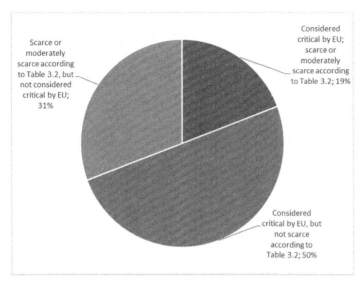

Figure 4.1 *Comparison of criticality according to the EU list of critical raw materials (2020) with scarcity according to Table 3.2.*

Table 4.3 Comparison of criticality and scarcity. Critical raw materials, which have not been considered in Table 3.2 are not included.

Critical according to the EU list of critical raw materials (EU, 2020) **and** scarce or moderately scarce according to Table 3.2: 19% of the raw materials considered	Antimony, Bismuth, Boron, Indium, Tungsten
Critical according to the EU list of critical raw materials (EU, 2020), **but not** scarce according to Table 3.2: 50% of the raw materials considered	Aluminum, Cobalt, Gallium, Germanium, Lithium, Magnesium, Nickel, Platinum Group Metals, Rare Earth Elements, Strontium, Tantalum, Titanium, Vanadium
Scarce or moderately scarce according to Table 3.2, **but not** critical according to the EU list of critical raw materials (EU, 2020): 31% of the raw materials considered	Chromium, Copper, Gold, Molybdenum, Nickel, Silver, Tin, Zinc

Another conclusion is that there is a risk that policies for mitigating the criticality of a specific raw material by substitution may lead to the situation that this critical raw material is substituted for a geologically scarce raw material; for example, substitution of critical cobalt by the geologically scarce nickel in lithium-ion batteries and the critical tungsten in tungsten carbide by the geologically scarcer molybdenum.

References

Achzet, B., Helbig, C., 2013. How to evaluate raw material supply risks — an overview. Resour. Pol. 38, 435—447.

Bastein, T., Rietveld, E., 2015. Materials in the Dutch economy. A vulnerability analysis. In: TNO Report 2015 R11613 for the Ministry of Economic Affairs of the Netherlands, 1 December 2015. TNO.

British Geological Survey, Risk list, 2015. Current Supply Risk for Chemical Elements or Element Groups Which Are of Economic Value.

Buchert, M., Schüler, D., Bleher, D., 2009. Critical Metals for Future Sustainable Technologies and Their Recycling Potential. Öko Institute, United Nations Environmental Programme, Nairobi, Kenia.

Cornell University, 2019. INSEAD, World Intellectual Property Organization. Global Innovation Index.

DEFRA, March 2012. A Review of National Resource Strategies and Research. Department for Environment. Food and Rural Affairs. www.defra.gov.uk.

Dozolme, P., 2020. Securing Metals Sourcing: The emergence of a New French Policy? http://mining.about.com/od/UnderseaExploration/a/Securing-Metals-Sourcing-The Emergence-Of-A-New-French-Policy.htm. accessed 09-01-2020.

Erdman, L.K., Graedel, T.E., 2011. The criticality of non-fuel minerals: a review of major approaches and analyses, Environment. Sci. Technol. 45 (18), 7620—7630.

EU, 2020. Study on the EU's List of Critical Raw Materials (2020), Final Report, Directorate-General for Internal Market, Industry, Entrepreneurship and SMEs.

Graedel, T.E., Barr, R., Chandler, C., Chase, T., Choi, J., Christoffersen, L., Friedlander, E., Henly, C., Jun, C., Nassar, N.T., Schechner, D., Warren, S., Yang, M., Zhu, C., 2012. Methodology of metal criticality determination. Environ. Sci. Technol. 46, 1063—1070.

KfW, 2011. Lack of Raw Materials Endangers Future Viability of German Economy. http://www.kfw.de/kfw/en/KfW_Group/Press/Latetst_News_/PressArchiv/2011/ 20111110_54444.jsp.

Nassar, N.T., Barr, R., Browning, M., Diao, Z., Friedlander, E., Harper, E.M., Henly, C., Kavlak, G., Kwatra, S., Jun, C., Warren, S., Yang, M.Y., Graedel, T.E., 2012. Criticality of the geological copper family. Environ. Sci. Technol. 46, 1071—1078.

Roelich, K., Purnell, P., Steinberger, J., Dawson, D., Bisch, J., February 27, 2012. Undermining Infrastructure: Avoiding the Scarcity Trap. University of Leeds.

Schrijvers, D., Hool, A., Blengini, G.A., Chen, W.Q., Dewulf, J., Eggert, R., Van Ellen, L., Gauss, R., Goddin, J., Habib, K., Hagelüken, C., Hirohata, A., Hofmann_Amtenbrink, M., Kosmol, J., Le Gleuher, M., Grohol, M., Ku, A., Lee, M.-H., Liu, G., Nansai, K., Nuss, P., Peck, D., Reller, A., Sonnemann, G., Tercero, L., Thorenz, A., Wäger, P.A., 2020. A review of methods and data to determine raw material criticality. Resour. Conserv. Recycl. 155, 1—17. https://doi.org/104617.

UNDP, 2018. Human Development Report 2018 — Human Development Indices and Indicators.

US Department of Energy, 2011. Critical Materials Strategy for the US.

World bank, 2020. http://info.worldbank.org/governance/wgi/. (Accessed 9 January 2020).

CHAPTER 5

Mineral resources ethics[1]

5.1 Introduction

In the previous sections, we showed that we cannot assume that all mineral resources will always be easily and abundantly available, although there are big differences between the availability of different mineral resources. In this section, we examine how humanity could deal with future scarcity of mineral resources. What is the current generation's responsibility for the availability of raw materials for future generations, if any? If so, how far-reaching is this responsibility and how long does it extend? We must also see these questions in the context of a significant part of the current world population having much less access to raw materials than the wealthiest part. Another relevant question relating to the future availability of raw materials concerns the potential of ongoing technological development. Will technological development enable ever-lower concentrations of raw materials to be extracted economically and environmentally responsible for centuries to come? Views on these questions range from "*après nous le déluge*" and "*technological development will always enable access to sufficient resources*" to "*we need to be committed to responsible stewardship of mineral resources*" and "*better err on the side of caution.*"

Below, we will first present different ethical views on the issue of mineral resources availability for future generations. Then we examine whether normative principles in existing international agreements justify internationally agreed policy measures on the conservation and sustainable use of geologically scarce mineral resources. Based on these normative principles, we will then propose a definition of sustainable use of raw materials and show the implications of different sustainability ambitions.

---

[1] This chapter is based on the publication of Henckens, M.L.C.M, Ryngaert, C.M.J., Driessen, P.P.J., Worrell, E., 2018. Normative principles and the sustainable use of geologically scarce mineral resources, Res. Pol. 59, 351–359.

Governance of the world's mineral resources
ISBN 978-0-12-823886-8
https://doi.org/10.1016/B978-0-12-823886-8.00002-6

5.2 Different ethical views

5.2.1 Pessimists versus optimists

In Chapters 2 and 3 it was shown that the absolute quantity of the elements in the earth's crust is enormous and quasi-inexhaustible. Moreover, unlike fossil fuels, elements do not disappear or change when used: they remain available and may be recycled. Hence, theoretically, there are no limits with respect to the use of mineral resources. In practice, however, financial and environmental reasons impose certain limits on the extraction of mineral resources. These practical limits shift over time as a result of technological development and the willingness to pay for a resource.

It is also important to take into account that minor elements are supposedly bimodally distributed in the earth's crust (Fig. 2.2 in Chapter 2), and thus that in a number of cases, there is a rather sharp transition from occurrences of a relatively high concentration to occurrences of much lower grades. As a result, once the higher grades of a resource are depleted, extraction becomes ever more expensive. It is not the availability of minerals as such that is the problem, but rather the costs involved in transforming these minerals into mineral resources. Extraction costs include the so-called external costs of the impacts of extraction on the environment (waste generation, water pollution, air pollution, landscape deterioration) and on climate change.

At this point, views start to diverge. The *resource optimists* (so-called by Tilton, 2003) strongly believe that humanity will be able to cope with the effects of depletion of the higher-grade occurrences of a mineral resource. The optimists contend that when demand outpaces supply, the price for the resource will rise and so the pressure to find substitutes or alternatives for the mineral will increase. When the price for a mineral commodity is rising, society has to consider what to give up in order to obtain an additional ton of that scarce commodity. The economic pressure to find substitutes for the material and to recycle it to a greater extent will increase concomitantly with the price. To the extent that extraction costs increase and substitutes, which do not contain the scarce resource, will become more achievable, technological innovations will become more feasible. Hence, according to the resource optimists, the market will automatically solve the scarcity problem. The optimists trust that humankind will timely find a clever solution for replacing scarce resources by substitutes, as has always been the case thus far. The resource optimists support the so-called opportunity cost paradigm.

The *resource pessimists* (so-called by Tilton, 2003) support the fixed stock paradigm: the earth is finite and so the amount of mineral resources is also finite. Hence, after a certain point has been passed, the costs of further extraction of a mineral resource will become prohibitively high. Although recycling technology continues to improve, a 100% circular economy is impossible: inevitably, a small part of a resource will always be dissipated. Dissipation happens both in the usage phase (for instance, when an element is used as fertilizer or in detergents) and in the end-of-life stage because recycling is never 100%. The resource pessimists are of the opinion that humanity should not deliberately deprive future generations of relatively abundant and reasonably priced resources without taking adequate measures to reduce the danger of potential future scarcity as much as possible.

5.2.2 Different sustainability paradigms

The most influential definition of sustainability was formulated in 1987 by the so-called Brundtland Commission in the report "Our common future": "*Sustainable development is the kind of development that meets the needs of the present without compromising the ability of future generations to meet their own needs*" (World Commission on Environment and Development, 1987). Since then, governments, politicians, scientists, and representatives from industry have been trying to make the Brundtland sustainability definition operational. In the 2 years after the Brundtland sustainability definition was published, about 140 variously modified definitions of sustainable development emerged, and by 2007, some 300 different elaborations of the concept of sustainability and sustainable development existed in the domain of environmental management and associated disciplines (Johnston et al., 2007). This demonstrates that the Brundtland sustainability definition has been interpreted in many different, even contradictory, ways.

The concept of sustainability has been hardly or not concretely operationalized for the extraction of primary resources. We found only one single concrete approach. According to this approach, the extraction rate of a mineral resource can be deemed to be sustainable if a world population of nine billion people can be provided sufficiently for 50 years (Graedel and Klee, 2002). But this can hardly be interpreted as a long-term goal for the sustainable extraction of mineral resources. It covers only two generations, whereas a sustainable approach must be able to continue forever or at least for a long time.

Two main conceptions of sustainability can be identified: a weak one and a strong one (Hansson, 2010). According to the *weak sustainability* concept, elements of sustainable development are interchangeable as long as economic development and welfare as a whole do not diminish. In this vision, maintaining sustainability is a matter of assuring that total capital (human plus natural) does not diminish. We may pass on less environmental resources to coming generations as long as we pass on more human-made capital instead. Or in Hansson's (2010, p 275) words: *"If we hand over to coming generations new technologies that reduce their needs of natural resources, then according to this view we can deplete more resources now and yet comply with the precepts of sustainability."* According to Van Den Bergh (2010), adoption of the weak concept of sustainability for mineral resources is not a problem as long as the environmental externalities are fully taken into account.

In contrast, the *strong sustainability* concept in its pure form sees human-made and natural capital as different categories that are not interchangeable and that must be preserved separately. In the most extreme version of the concept of strong sustainability, every species and resource must be preserved since it cannot be replaced. Further extraction of exhaustible resources should therefore no longer be possible.

The strong concept is widely considered as not practicable, whereas the weak notion of sustainability has been criticized as being too lax because it enables depletion of resources provided that this is compensated for by improvements in other fields—for instance, better health care Hansson (2010). Therefore, Ayres et al. (2001) suggest a compromise: the strong concept of sustainability should focus on critical ecosystems and on environmental assets that cannot be replaced by anything else, while the weak sustainability concept should apply to mineral resources. Ultimately, it appears that the interpretation of the Sustainable Development Principle and the connected obligations of States are evolving over time (Barral, 2012).

5.3 Applicable normative principles

5.3.1 International agreements on natural resources

For *the short term*, in view of future crisis situations, some governments have decided to strategically stockpile certain raw materials. In European Union (EU) Directive 2009/119/EC, EU Member States are obliged to maintain minimum stocks of crude oil and/or petroleum products (European Union, 2009). After the end of World War I, the government of the United States started considering whether to maintain strategic stocks of a number of materials. The first strategic stockpiling in the United States was established in 1938 (Chappel, 2016). Many governments (including the EU, Germany, France, Finland, the Netherlands, the United States, Canada, Japan, Korea, and Taiwan) have prepared strategies based on the so-called criticality studies of the short-term availability of materials critical for the economy (Department for Environment, Food, and Rural Affairs, 2012). The background of these strategies is principally geopolitical: the fear that monopolist producers might deliberately stop or reduce the production and delivery of certain materials essential for the economy, for political or financial reasons. See Chapter 4 of this book.

However, for *the long term*, no government has so far taken action, nor has an internationally binding agreement limiting the extraction of geologically scarce mineral resources ever materialized. According to Brilha et al. (2018), the abiotic natural resources are persistently neglected in international and national politics. Nevertheless, some international declarations and charters have directly addressed depletion of mineral resources. Thus, the Declaration of the United Nations Conference on the Human Environment in Stockholm (1972) provides in Principle 5 that *"[t]he non-renewable resources of the earth must be employed in such a way as to guard against the danger of their future exhaustion and to ensure that benefits from such employment are shared by all mankind."* The UN World Charter for Nature (1982), for its part, provides in Principle II (d) that *"[n]on-renewable resources which are consumed as they are used shall be exploited with restraint, taking into account their abundance, the rational possibilities of converting them for consumption, and the compatibility of their exploitation with the functioning of natural systems."* Finally, Principle II.5.f of the UNESCO Earth Charter (2000) calls on the international community to *"manage the extraction and use of non-renewable resources such as minerals and fossil fuels in ways that minimize depletion and cause no serious environmental damage."* The 25-9-2015 Resolution adopted by the UN General Assembly on the 2030 Agenda for Sustainable Development includes Sustainable Development Goal 12.2: *"by 2030, achieve the sustainable management and efficient use of natural resources."*

While these declarations have not yet paved the way for the adoption of binding international agreements on the depletion of geologically scarce mineral resources, the normative principles they contain remain relevant in the context of this Chapter because they might inform future customary and treaty law specifically aimed at regulating mineral resources depletion.

Additionally, there are some international agreements that address the *exploitation* (although not the *scarcity* or *depletion*) of mineral resources, namely the United Nations Convention on the Law of the Sea (UNCLOS, 1982), the Protocol on Environmental Protection to the Antarctic Treaty (1991), and the Agreement Governing the Activities of States on the Moon and other Celestial Bodies (1979). However, these agreements are limited to areas outside national jurisdiction. There have also been international agreements on the conservation and protection of specific *renewable resources* such as biodiversity, endangered fish species, wild flora and fauna, and tropical timber: the Convention for the Conservation of Antarctic Seals (1972), the Convention on the Conservation of Migratory Species of Wild Animals (1979), the Convention on Biological Diversity (1992), the Convention on the Conservation and Management of the Highly Migratory Fish Stocks in the Western and Central Pacific Ocean (2000), and the International Tropical Timber Agreement (2006).

In the next section, we will investigate which normative principles contained in existing international environmental agreements could be relevant as a foundation for international policy measures on the conservation and sustainable use of geologically scarce mineral resources.

5.3.2 Inventory of normative principles in existing agreements and their relevance for mineral resources depletion

The main source for the relevant existing international environmental agreements presented here is the International Environmental Agreements database of the University of Oregon (Mitchell, 2016). This database comprises over 1100 multilateral and 1500 bilateral agreements. We specifically focus on agreements with broad international support and on normative principles repeatedly used in such broadly supported agreements. Such principles may be considered to have broad international support too. The 29 multilateral agreements selected and shown in Table 5.1 have been signed since 1960 onwards, have a global or quasi-global scope, and have been signed and ratified by a substantial number of relevant countries.

Table 5.1 does not include agreements with a regional scope, except agreements on oceans and on the Antarctic and Arctic regions. We have not selected multilateral agreements on energy, radioactive material, weaponry and other military issues, creation of institutions, financing, patents, occupational health, training, data confidentiality, communication, information management and public participation, disasters and emergency situations, sustainable housing, research and monitoring, meteorology, liability, industrial safety, human health, related to tobacco, compliance

Table 5.1 Goal-oriented normative principles in multilateral international environmental agreements with global or quasi-global scope signed since 1960.

No	Year	Agreement	Conservation and/or Sustainable Use of Resources	Protection of wild flora and fauna, environment and nature	Precautionary Principle	Intergenerational Equity	Sustainable Development
1	1971	Convention on Wetlands of International Importance Especially as Waterfowl Habitat		X			
2	1972	Convention for the Conservation of Antarctic Seals	X				
3	1972	Convention for the Prevention of Marine Pollution by Dumping from Ships and Aircraft		X			
4	1972	Convention on the Prevention of Marine Pollution by Dumping of Wastes and Other Matter		X			
5	1972	Convention for the Protection of the World Cultural and Natural Heritage				X	
6	1973	Agreement on Conservation of Polar Bears		X			

Continued

Table 5.1 Goal-oriented normative principles in multilateral international environmental agreements with global or quasi-global scope signed since 1960.—cont'd

No	Year	Agreement	Conservation and/or Sustainable Use of Resources	Protection of wild flora and fauna, environment and nature	Precautionary Principle	Intergenerational Equity	Sustainable Development
7	1973	Convention on International Trade in Endangered Species of Wild Fauna and Flora			X		
8	1973	International Convention for the Prevention of Pollution from Ships		X			
9	1974	Convention on the Prevention of Marine Pollution from Land-Based Sources		X			
10	1976	Convention on Conservation of Nature in the South Pacific		X		X	X
11	1979	Convention on Long-Range Transboundary Air Pollution		X			
12	1979	Convention on the Conservation of Migratory Species of Wild Animals		X		X	

Continued

#	Year	Convention					
13	1980	Convention on the Conservation of Antarctic Marine Living Resources		X			
14	1982	United Nations Convention on the Law of the Sea	X	X	X		
15	1985	Convention for the Protection of the Ozone Layer (Vienna Convention)			X		
16	1985	Convention on the Control of Transboundary Movements of Hazardous Wastes And their Disposal		X			
17	1992	Convention on Biological Diversity	X				
18	1992	United Nations Framework Convention on Climate Change	X		X	X	X
19	1994	Agreement Relating to the Implementation of Part XI of the United Nations Convention on the Law of the Sea		X			

Table 5.1 Goal-oriented normative principles in multilateral international environmental agreements with global or quasi-global scope signed since 1960.—cont'd

No	Year	Agreement	Conservation and/or Sustainable Use of Resources	Protection of wild flora and fauna, environment and nature	Precautionary Principle	Intergenerational Equity	Sustainable Development
20	1994	1994 WTO Agreement on the Application of Sanitary and Phytosanitary Measures			X		
21	1995	Agreement for the Implementation of the Law of the Sea Convention Relating to the Conservation and Management of Straddling Fish Stocks and Highly Migratory Fish Stocks	X		X		X
22	2000	Convention on the Conservation and Management of the Highly Migratory Fish Stocks of the Western and Central Pacific Ocean	X	X	X	X	

23	2001	International Convention on the Control of Harmful Anti-fouling Systems on Ships		X	X		
24	2001	International Treaty on Plant Genetic Resources for food And Agriculture	X			X	
25	2001	Convention on Persistent Organic Pollutants		X	X		
26	2004	International Convention for the Control and Management of Ships' Ballast Water and Sediments		X	X		
27	2006	International Tropical Timber Agreement	X				X
28	2009	Agreement on Port State Measures to Prevent, Deter and Illuminate Illegal, Unreported and Unregulated Fishing					X
29	2013	Minamata Convention on Mercury		X			

and enforcement, cultural heritage, and transport. The remaining selection consists of the 29 international environmental agreements in Table 5.1. We have made an inventory of the normative principles included in the preambles of the 29 selected agreements. In the inventory, we have distinguished (1) normative principles that directly pertain to solving an environmental problem and (2) normative principles that pertain to the acceptability of the agreement for the signatories. We have mentioned the first category: goal-oriented principles. From the inventory, five principles are considered to be goal-oriented principles:

- The Principle of Conservation and/or Sustainable Use of Resources (in 8 of the 29 agreements)
- The Principle of Protection of Wild Flora and Fauna, Environment and Nature (in 18 of the 29 agreements)
- The Precautionary Principle (in 10 of the 29 agreements)
- The Intergenerational Equity Principle (in 6 of the 29 agreements)
- The Sustainable Development Principle (in 5 of the 29 agreements).

Table 5.1 indicates for every agreement which goal-oriented normative principles are included in that agreement. These principles could also be the normative building blocks for international policy measures for achieving sustainable extraction of mineral resources.

The normative principles that are not directly related to the goal of the agreement but that pertain to the acceptability of the agreement for the signatories are related to aspects such as fairness, burden-sharing, and assignment of responsibility. These normative principles are the Sovereignty over Natural Resources Principle, the Intragenerational Equity Principle, the Principle of Priority for the Special Situation and Needs of Developing Countries, the Principle of Common but Differentiated Responsibilities in Accordance with Capabilities, the Principle of Equitable Contribution to Achieving the Goal of a Convention, the Principle that Activities may not Cause Damage to the Environment of other States, and the Polluter Pays Principle (Kiss and Shelton, 2004). These normative principles are relevant as boundary conditions in an agreement on the conservation and sustainable use of mineral resources, thereby determining the architecture of such an agreement.

While the five goal-oriented normative principles buttress the analyzed international agreements, they are not necessarily all *relevant in* the framework of international policy measures on geologically scarce mineral resources. Notably, the Principle of Protection of Wild Flora and Fauna, Environment and Nature seems less relevant, as scarce mineral resources have no value of their own without further utilization, certainly when compared to endangered species and beautiful landscapes. As far as we know, the mineral ores in the earth do not have a specifically important role as habitat. Their stocks, outside ores, are very large so that absolute depletion of a mineral is not at stake, and unlike landscapes, mineral ores are not considered to have a specific beauty. Accordingly, the Principle of Protection of Wild Flora and Fauna, Environment, and Nature is not considered relevant for international policy measures on the conservation and sustainable use of geologically scarce mineral resources.

The Principle of Sustainable Development is not a stand-alone principle. It is an integrative principle that includes other principles, notably the Sustainable Use of Resources, Intragenerational and Intergenerational Equity, and the Precautionary Principle. Hence, in the next subsections, the three remaining normative goal-oriented principles are applied to the problem at hand. It concerns the Principle of Conservation and Sustainable Use of Resources, the Precautionary Principle and the Intergenerational Equity Principle.

5.3.3 The Principle of Conservation and/or Sustainable Use of Resources

The very title of the Principle of Conservation and/or Sustainable Use of Resources makes obvious that this normative principle is relevant for the conservation and sustainable use of mineral resources. Up to now, the Principle of Conservation and Sustainable Use of Resources in existing international environmental agreements has concerned specific individual or types of *renewable* resources, such as the conservation and sustainable use of fish, biodiversity, and tropical timber. These resources are only renewable on the condition that their use is in balance with their natural recovery. Irreversible, global, and short-term disappearance was feared unless urgent action was taken at international level, as their consumption rate was exceeding their recovery rate. The conservation and sustainable use of *nonrenewable* resources (such as mineral resources), however, has only been included in nonbinding declarations and charters. Therefore, let us compare the gravity of the extinction of endangered biotic resources and species with the gravity of the depletion of geologically scarce mineral resources.

The gravity of an (environmental) problem is characterized by three main dimensions: size, seriousness, and urgency. Having analyzed the preambles of the 29 selected international agreements for the presence of sentences relating to these dimensions of gravity, we conclude that the main

Table 5.2 Elements for assessing the gravity of a resource scarcity problem.

Elements of gravity of a resource scarcity problem	Subelements
Potential size of the problem	- The number of countries affected (spatial dimension) - The proportion of the resource that is endangered (volume dimension) - The extent that future generations are affected (temporal dimension)
Potential seriousness of the problem	The extent of - (Ir)reversibility - The impact on human life, directly or indirectly - The impact on health, safety, and survival of the living environment (animals, plants, natural cycles and equilibria, ecosystems, natural tipping points, food chains, biodiversity, and habitats) - The impact on the uniqueness of the endangered resource - The impact on economy and welfare
Potential urgency of the problem	- Time span available to redress the developments in order to prevent the problem from becoming too grave to be adequately solved

dimensions of gravity are composed of nine subelements (Table 5.2). We assume that certain thresholds (or a combination of thresholds) must be exceeded before the subelements are grave enough to be deemed relevant. It is difficult to design a generally accepted quantitative measuring stick composed from the subelements in Table 5.2 that could be used to assess a mineral resources scarcity problem and that would enable a mathematical determination of whether the problem is grave enough to merit action. However, we consider the gravity of depletion of geologically scarce mineral resources to be comparable with the gravity of the problem of extinction of endangered biotic resources and species for at least four of the nine gravity subelements in Table 5.2: the number of countries affected (all countries); the extent that future generations are affected (for eight resources possibly starting within 100 years); the irreversibility (ores can only be extracted once); and the impact on economy and welfare. The potential impact of depletion of mineral resources on economy and welfare is substantial because, after depletion of the enriched deposits (ores), the extraction of the resource from the earth's crust is only possible at much higher costs. Recyclability is never 100%. A perfectly circular economy is not possible. New applications of the mineral resource will be hampered by its increased costs, and the flexibility and degrees of freedom of future generations will be less when the resource is running out. Our conclusion is that the normative Principle of Conservation and Sustainable Use of Resources justifies international policy measures to achieve a sustainable use of scarce mineral resources.

5.3.4 The precautionary principle

In this section, we investigate whether the precautionary principle would justify international policy measures regarding the conservation and sustainable use of geologically scarce mineral resources. According to many authors (i.e., Sirinskiene, 2009; Sandin et al., 2002; Tickner et al., 2003; Tickner and Kriebel, 2006; Sachs, 2011), there is sufficient state practice and *opinio iuris* to support the contention that the Precautionary Principle has already crystallized into a rule of general customary international law. Nevertheless, some scholars and politicians, especially in the United States, consider the Precautionary Principle to be incoherent, internally inconsistent, and having a paralyzing effect on industrial and economic development. They therefore strongly oppose application of the Precautionary Principle, especially of stronger versions (Sunstein, 2005). Nonetheless, the Precautionary Principle has been broadly adopted as a cornerstone of international and national declarations, agreements, and regulations on the environment, whether binding or non-binding.

Criteria for invocation of the Precautionary Principle that are shared in most Precautionary Principle definitions are the threat of serious or irreversible damage (Rio Declaration, 1992), the potentially dangerous effects (European Commission, 2000), the fact that scientific evaluation of the risk cannot be determined with sufficient certainty (European Commission, 2000), and the reasonable foreseeability of damage falling short of conclusive scientific proof (International Law Association, 2014).

The irreversible damage resulting from the depletion of mineral resources that we foresee is economic: once ores have been exhausted, the extraction of mineral resources will become 10–1000 times more expensive (Steen and Borg, 2002). Thus, the access to resources that are currently easily available may become much more difficult for future generations. This economic damage can be considered serious because it concerns all humanity and because future generations will be deprived of the current generation's possibilities to access certain mineral resources, which will encompass substantial economic costs. The damage is irreversible because ores will be definitively depleted. In light of our considerations in Chapters 2 and 3, our conclusion is that there is a reasonable foreseeability of depletion of the ores of the geologically scarcest mineral resources within a century, although there is debate on the seriousness of the consequences.

We acknowledge that the *scope* of the Precautionary Principle is generally limited to the protection of the environment and human, animal, and plant health. This includes the protection and conservation of biotic resources such as biodiversity. Trouwborst (2007, p. 190) writes that *"the Precautionary Principle has, from the outset, been an environmental principle."* However, non-environmental problems need not necessarily be excluded from the scope of a broadly defined Precautionary Principle. In international humanitarian law, for instance, the principle informs the constant care States shall take in the conduct of military operations, to spare the civilian population, civilians, and civilian objects (Article 57 Additional Protocol I to the Geneva Conventions, 1977). Precaution could similarly inform economic decisions. The European Commission (2000) stresses in a Communication, in rather general terms, that *"the Precautionary Principle goes beyond the problems associated with a short- or medium-term approach to risks. It also concerns the longer run and the well-being of future generations."* And in the 2002 EU Regulation on state aid to the coal industry (EU, 2002; Preamble, para 7), the EU extended the scope of the Precautionary Principle beyond the environment only, in the context of broadly defined energy security: *"Strengthening the Union's energy security, which underpins the general Precautionary Principle, therefore justifies the maintenance of coal-producing capability supported by State aid."* From these instruments, it could be inferred that at least the EU is willing to include the well-being of future generations in general, including economic security, into the Precautionary Principle.

Economic theories concerned with future uncertainty do in any event rely on a version of the Precautionary Principle (see, for instance, the publications of Gollier and Treich (2003), Gollier et al. (2001), Gollier (2010a,b), and Farrow and Hayakawa (2002)). According to these theories, irreversible developments (e.g., the depletion of mineral ores), as well as uncertainty and lack of knowledge about the consequences and size of future risks (such as the risks of climate change), lead to a loss of flexibility for future generations and therefore have a cost. Gollier (2010a,b) contends, in this respect, that uncertain future costs must be discounted to a net present value using lower discount rates to the extent that decisions relate to a more distant future and the uncertainty is bigger. Whereas in economic theory, precaution in principle informs market participants' decisions, Chapter 6 will demonstrate that the market may not take sufficient precautions in the face of future depletion of mineral ores. *Regulatory precautions to prevent economic loss as a result of depletion may have to be taken instead. An economically flavored Precautionary Principle may be appropriate in this case.* Nevertheless, it must be admitted that there is no international consensus on the scope and invoking criteria of the Precautionary Principle or on the potential seriousness of the consequences of resources depletion. The conclusion is therefore that the applicability of the Precautionary Principle to the problem of depletion of geologically scarce mineral resources may not be sufficiently unambiguous to lend itself to being the basis of international policy measures on this issue.

5.3.5 The Intergenerational Equity Principle

In this section, we investigate whether the Intergenerational Equity Principle would justify international policy measures on the conservation and sustainable use of geologically scarce mineral resources. The Principle of Intergenerational Equity is included in many international conventions, including the Rio Declaration on Environment and Development (1992) and the United Nations Framework Convention on Climate Change (1992). Intergenerational equity embodies care for future generations. It means that the current generation is merely "borrowing" the earth from future generations.

Intergenerational equity is a legal principle and means that future generations may have a legitimate expectation of equitable access to planetary resources (International Law Association, 2014). According to Padilla (2002, p 81), *"we should recognize and protect the future generations' right to enjoy at least the same capacity of economic and ecological resources that present generations enjoy."* According to Shelton (2007, p 643), *"[t]hose living have received a heritage from their forebears in which they have beneficial rights of use that are limited by the interests and needs of future generations. This limitation requires each generation to maintain the corpus of trust and pass it on in no worse condition than it was received."* It is a matter of justice that an intergenerational community gives a voice to voiceless future generations. In this respect, as Agius et al. (1998, p 11) have pointed out, *"future generations are similar to those that our society has declared legally incompetent."* It is undeniable that previous and current generations have been or are still irretrievably and inevitably depleting mineral resources. As a result, the options for future generations are gradually being constrained and their flexibility is dwindling. In what state the current generation leaves the earth for future generations will ultimately be a political decision, though mandated by the legal Principle of Intergenerational Equity.

For *non-renewable* resources such as mineral resources, an equilibrium in which the resource does not decrease in quantity and quality would imply zero use. This would deprive current and future generations of the resource, however. In this sense, the sustainable extraction of mineral resources is an oxymoron. Anyhow, given the expected depletion, within a relatively short period of time, of a number of mineral resources at current extraction rates, the needs and rights of future generations compel the current generation to deal with mineral resources in general as economically (or sustainably) as possible, especially if the mineral resources are geologically scarce. The conclusion is that the Intergenerational Equity Principle justifies international policy measures for the conservation and sustainable use of geologically scarce mineral resources.

5.4 Sustainability goals

In this section, we examine how the normative principles discussed in the previous section can be made operational. Which measures must humanity take to ensure that sufficient affordable mineral resources remain available for an acceptable period of time for future generations, at a service level, which is enjoyed in developed countries in 2020? A definition of the conservation and sustainable use of mineral resources includes the following elements: (1) the period of time during which humanity wants sufficient resources to be available for the world population (called the sustainability ambition or sustainability period or time horizon), (2) the future affordability of the mineral resources, (3) intragenerational equity to assure that all world citizens have the same access to scarce resources, if possible at a service level, which is the same as the one in developed countries in 2020, and (4) the world population during that period. We will discuss these elements in the context of the applicable normative principles as presented in the previous sections. Thereafter, we will propose a definition for the sustainable production of mineral resources and discuss its implications.

5.4.1 The time horizon

An activity is sustainable if it can be continued forever. This means that the combination "sustainability period" and "sustainable extraction" is a contradiction in terms. Theoretically, sustainable extraction of an ore is impossible, since ores are not renewable and will eventually be exhausted. We therefore need to find a suitable proxy for the concept of "sustainability" related to the use of mineral resources. We must try to find an approach that is practically feasible and acceptable for the current world community, but that also satisfies the long-term objective of meeting the needs of future generations. A period of 50 years with guaranteed supply of mineral resources as proposed by Graedel and Klee (2002), so just two generations ahead, is—in our view—insufficient to be considered sustainable. In our view, the minimum period with a sufficient supply of mineral resources should cover at least several centuries. Ultimately, however, the minimum length of the sustainability period (the sustainability ambition) is a matter of political choice. The short-term needs of the current generation, especially of the poor, should be in balance with the long-term needs of the world population. We will consider sustainability periods of 200, 500, and 1000 years. A shorter period, for instance of 100 years, would be, in our opinion, too short to simulate sustainability and to satisfy the long-term goal of ensuring that future generations have a sufficient and affordable supply of scarce resources. A sustainability period of 100 years would allow just a few generations to deplete a substantial part of the extractable resources, which might make it more difficult for future generations to adapt. However, a period two orders of magnitude longer, 10,000 years, seems unnecessarily long.

5.4.2 Affordability

As explained in Chapter 2, the total amount of mineral resources in the earth's crust is enormous. Only the enriched deposits (ores) of a mineral in the upper part of the earth's crust (seldom deeper than about 1 km) are currently being extracted. Extraction of occurrences that are of a lower grade or that are located deeper is technically possible but is costlier. In the light of the Principle of Intergenerational Equity, we believe that the use of a mineral resource can only be defined as sustainable as long as the market price of that resource does not rise sharply due to geological scarcity. A much higher market price of a mineral resource in the future may make today's "normal" applications unaffordable for future generations. Our grandchildren would be justified in accusing us of having plundered the treasures of the earth, if we allow the real price of a resource to rise structurally by more than an order of magnitude (factor 10) compared to the price that we pay ourselves.

5.4.3 Intragenerational equity

Intragenerational equity needs to be distinguished from *intergenerational* equity. Intragenerational equity does not limit the *use* of resources as such, but it governs the *distribution* of resources and the distribution of the costs and benefits between people and peoples of the same generation. There is surely a tension between intragenerational and intergenerational equity: the wish to distribute resources more equitably over the current generation could imply that there is pressure to use more resources than justified on the basis of intergenerational equity. It is hardly justifiable to impose austerity on the current poor for the sake of the future rich. Now and in the future, intragenerational equity will be an important precondition to be able to reach agreement on international policy measures on the conservation and sustainable use of geologically scarce mineral resources. Geologically scarce mineral resources need to become and stay attainable for all countries and people, including for people and countries that previously did not have abundant access to these resources. In our view, this is a prerequisite for obtaining the cooperation of all countries for achieving global conservation and sustainable use of geologically scarce mineral resources. Intragenerational and intergenerational equity are complementary.

The World Commission on Environment and Development (1987) reconciles the Intergenerational and Intragenerational Principles in its definition of sustainable development. Hence, the normative principle of *intragenerational equity* means that—in principle—all countries have the right to equal access to the available mineral resources. One part of humanity may not deprive another part of humanity from resources. This starting point leads to the conclusion that a production rate of a raw material can only be labeled as sustainable if all countries have the right to a fair share of that material per capita. The current unequal distribution of mineral resources consumption per capita over the world is not acceptable as a starting point in a definition of sustainable use of mineral resources. At present, the average consumption of raw materials per inhabitant differs greatly per country, depending on the stage of its economic development (Fig. 5.1 with copper as example). Per capita, industrialized countries consume much more mineral resources than non-industrialized countries.

Fig. 5.1 shows that the per capita consumption of mineral resources in industrialized regions like Europe and North America is stable and that the consumption level per capita in industrializing countries like South Korea and China is rapidly increasing. The consumption of mineral resources in a

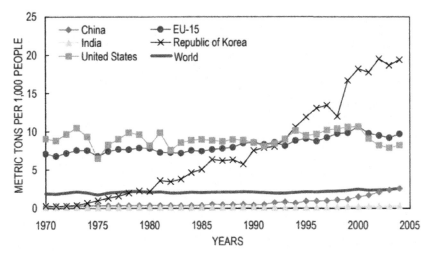

Figure 5.1 Copper consumption per capita by country (US Geological Survey, 2008).

country increases with growing GDP per capita. However, from a certain level of per capita income, the consumption of many mineral resources per capita may decouple partly or fully from GDP growth or even decline again (Halada et al., 2008; Bleischwitz et al., 2018). The explanation is that in wealthier countries, infrastructure (electric, water, transport) has been completed and only needs maintenance and the inhabitants spend relatively more money on immaterial things, such as culture, travel, sport, health, and education, which require less material per unit of GDP per capita than tangible objects do, such as electric infrastructure and washing machines. In a schematic way, Fig. 5.2 depicts the different ways of decoupling above a certain GDP per capita per year. This is derived from data of Halada et al. (2008). In Fig. 5.2,

- picture (A) represents the consumption development of steel, aluminum, antimony, nickel, silver, molybdenum, and palladium in relation to GDP per capita
- picture (B) represents the consumption development of copper in relation to GDP per capita
- picture (C) represents the consumption development of gold, tungsten, tin, zinc, chromium, and manganese in relation to GDP per capita
- picture (D) represents the consumption development of rare-earth elements, platinum, gallium, cobalt, lithium, and indium in relation to GDP per capita.

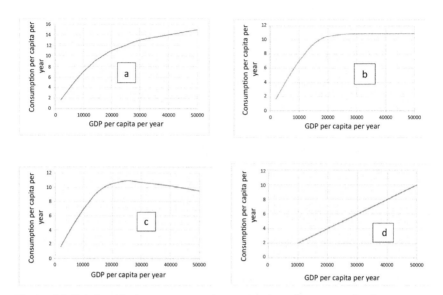

Figure 5.2 Relationship between annual consumption of a commodity in a country depending on the GDP per capita per year in that country (Halada et al., 2008). (A) partial decoupling; (B) full decoupling; (C) slow decrease after fast increase; (D) relative recent start of consumption at higher GDP per capita per year. The horizontal axis is in US Dollars (2008).

In Japan, decoupling started at a GDP of approximately USD 15,000–20,000 per capita per year (purchasing power 2008) (Halada et al., 2008). Bleischwitz et al. (2018) came to similar conclusions.

In 2020, the per capita consumption of resources in developed countries was about four times higher than the world average consumption. We base this on the assumption that the level of consumption of resources is related to the GDP per capita. In 2020, the per capita GDP in high-income countries was about four times as high as the global average GDP per capita (Worldbank, 2020). The assumption that, in 2020, the consumption of mineral resources in developed countries must have been about four times as high as the global average consumption of mineral resources per capita is supported by observations of Halada et al. (2008), Bleischwitz et al. (2018), and Pan Pacific Copper (2019). Based on this assumption, Table 5.3 calculates the approximate consumption level per capita of the 13 scarcest resources in developed countries in 2020.

The factor 4 gap between the average global consumption of resources per capita and the consumption in developed countries per capita could be bridged by about 2050, if United Nations Environment Programme

Table 5.3 Supposed consumption level of the 13 scarcest resources in developed countries in 2020.

Column	1 Average annual global extraction in the period 2015–19 (t) (a)	2 Ratio consumption/ extraction (b)	3 Average annual global consumption between 2015 and 2019 (t) (c)	4 Average global consumption per capita per year in the period 2015–19(g) (d)	5 Supposed consumption level per capita per year in developed countries in 2020 (g) (e)	6 Supposed consumption level per capita per year in developed countries in 2020, rounded (g)
Antimony	146,715	1.2	176,058	23	94	90
Bismuth	16,460	1.0	16,460	2	8.8	9
Boron	1,142,474	1.0	1,142,474	152	609	600
Chromium	11,860,916	1.2	14,233,099	1,898	7,591	8,000
Copper	19,860,000	1.3	25,818,000	3,442	13,769	10,000
Gold	3,208	1.2	3,850	0.5	2.1	2
Indium	733	1.0	733	0.1	0.39	0.4
Molybdenum	279,600	1.1	307,560	41	164	200
Nickel	2,314,000	1.3	3,008,200	401	1,604	2,000
Silver	26,300	1.1	28,930	3.9	15	20
Tin	304,000	1.2	364,800	49	195	200
Tungsten	85,140	1.3	110,682	15	59	60
Zinc	12,680,000	1.2	15,216,000	2,029	8,115	8,000

(a)Derived from data of US Geological Survey.
(b)Derived from flow schemes in Sections 7.2–7.14.
(c)The product of column 1 x column 2.
(d)The average world population between 2015 and 2019 was 7.5 billion.
(e)This is based on the difference between GDP per capita in developed countries and the GDP per capita worldwide (factor 4).

(UNEP)'s *"freeze and catch-up"* scenario (UNEP, 2011) is realized. In this UNEP scenario, the industrialized countries *freeze* raw material consumption at the current level, while developing countries build up to the same average per capita consumption as the industrialized countries by 2050 (*catch up*). In this way, intragenerational equity could be achieved, theoretically. This would be possible with a global growth rate of 3% annually. Anyhow, for a number of raw materials, the consumption level in rich countries has already reached the kind of saturation point as observed by Halada et al. (2008) in Japan as shown in Fig. 5.2.

5.4.4 World population

The world population trend is monitored by the United Nations, which in its most recent report (2017) forecasts that the world population will have increased to about 11 billion at the end of the 21st century and will stabilize from then on. The longer-term trend in world population is highly uncertain. The number may continue to increase, or it may decrease, as is now happening in many developed countries. In our definition of the sustainable use of mineral resources, we will assume a long-term stabilization of the world population at 10 billion people.

5.5 The sustainable extraction rate of mineral resources

Based on the above four elements we propose the following operational definition for the sustainable extraction rate of a mineral resource:

> The extraction of a mineral resource is sustainable if a world population of 10 billion can be provided with that resource for a period of at least 200/500/1000 years in such a way that during that period every country can enjoy the same service level of that resource as enjoyed by developed countries in 2020, for an affordable price.

Though the duration of the sustainability period is a matter of political choice, it should be at least 200 years in our opinion and preferably 1000 years, if feasible.

In Table 5.4, we present the sustainable extraction rates of the 13 scarcest mineral resources and we compare the results with the use of primary resources of these materials in developed countries in 2020.

Table 5.4 Sustainable extraction rates compared to the use of primary resources in developed countries in 2020.

Column	(1)	(2)			(3)	(4)	(5)
	Ultimately available resources[a] (Mt)	Sustainable extraction rate per capita[b] (g/capita/year) Sustainability ambition (years)			Estimated consumption in developed countries in 2020[c] (g/capita/year)	Estimated ratio consumption/extraction in 2020[d]	Estimated use of primary resources in developed countries in 2020[e] (g/capita/person)
Element		200	500	1000			
Antimony	100	50	20	10	90	1.2	80
Bismuth	20	10	4	2	9	1	9
Boron	3,000	1,500	600	300	600	1	600
Chromium	35,000	18,000	7,000	4,000	8,000	1.2	6,000
Copper	10,000	5,000	2,000	1,000	10,000	1.3	10,000
Gold	2	1	0.4	0.2	2	1.2	2
Indium	30	15	6	3	0.4	1	0.4
Molybdenum	200	100	40	20	200	1.1	100
Nickel	8,000	4,000	1,600	800	2,000	1.3	1,000
Silver	20	10	4	2	20	1.1	10
Tin	300	150	60	30	200	1.2	200
Tungsten	200	100	40	20	60	1.3	50
Zinc	30,000	15,000	6,000	3,000	8,000	1.2	7,000

[a]Table 2.8.
[b]Column (1) divided by 10 billion divided by the sustainability ambition (200, 500, or 1000 years).
[c]Table 5.3.
[d]Derived from flow schemes in Sections 7.2–7.14.
[e]Column (3) divided by column (4) and rounded.

This leads to the following conclusions:

- With a sustainability ambition of 200 years, the per capita use in developed countries of the primary resources of antimony, copper, gold, and tin exceeds the sustainable extraction rate or is equal to the sustainable extraction rate (molybdenum and silver).
- With a sustainability ambition of 500 years, this is also the case for bismuth, boron, tungsten, and zinc.
- With a sustainability ambition of 1000 years, this is additionally the case for chromium and nickel.
- Indium use is currently sustainable. But the annual production growth of indium (8.1%) between 1980 and 2015 is so high that it is likely that the sustainable extraction threshold will be reached within a few decades.

Notes

(1) Column 4 of Table 5.4 assumes that the current buildup of the in-use stocks of resources will continue. However, to the extent that this buildup decreases in the future, the ration consumption/extraction (column (4)) will increase and the estimated use of primary resources (column (5)) will decrease.

(2) Table 5.4 supposes that no future saving measures, such as substitution and more recycling, will be taken.

(3) Table 5.4 supposes that the consumption level in developed countries in 2020 will not increase any further. However, it may not be excluded that, in reality, the consumption of several of the 13 considered elements in developed countries continues growing for some time and may only falter by 2050 or 2100. According to the Intragenerational Equity Principle, poorer nations will have the right to catch up the same consumption level as developed countries. Note also that the estimates of the ultimately available resources are quite optimistic.

In Chapter 7, we will analyze for the 13 scarcest mineral resources (antimony, bismuth, boron, chromium, copper, gold, indium, molybdenum, nickel, silver, tin, tungsten. and zinc) at which conditions a sustainable extraction rate of these resources can be combined with increasing their global service level to the service level in developed countries in 2020.

References

Agius, E., Busuttil, S., Kim, T.-C., Yazaki, K., 1998. Future generations & international law. In: Proceedings of the International Experts' Meeting Held by the Future Generations Programme at the Foundation for International Studies, University of Malta, in Collaboration with the Foundation for International Environmental Law and Development (FIELD), University of London; the Future Generations Alliance Foundation, Japan; and the Ministry for the Environment, Malta. Earthscan Publications Ltd, London.

Ayres, R.U., van den Bergh, J.C.J.M., Gowdy, J.M., 2001. Strong versus weak sustainability: economics, natural sciences, and "consilience". Environ. Ethics 23 (2), 155−168.

Barral, V., 2012. Sustainable development in international law: nature and operation of an evolutive legal norm. Eur. J. Int. Law 23 (2), 377−400.

Bleischwitz, R., Nechifor, V., Winning, M., Huang, B., Geng, Y., 2018. Extrapolation or saturation − revisiting growth patterns, development stages and decoupling. Glob. Environ. Chang. 48, 86−96.

Brilha, J., Gray, M., Pereira, D.I., Pereira, P., Geodiversity, 2018. An integrative review as a contribution to the sustainable management of the whole of nature. Environ. Sci. Pol. 86, 19−28.

Chappel CG, December 12, 2016. www.dla.mil/Portals/104/Documents/Strategic Material/DNSC%20History.pdf, 12-12-2016.

Department for Environment, Food and Rural Affairs, UK, 2012. A Review of National Resource Strategies. www.defra.gov.uk.

European Commission, 2000. Communication from the Commission on the Precautionary Principle, COM, 2000.

European Union, September 14, 2009. Council Directive 2009/119/EC, Imposing an Obligation on Member States to Maintain Minimum Stocks of Crude Oil and/or Petroleum Products.

Farrow, S., Hayakawa, H., 2002. Investing in safety. An analytical precautionary principle. J. Saf. Res. 33, 165−174.

Gollier, C., 2010a. Debating about the discount rate: the basic economic ingredients. Perspekt. Wirtsch. 11 (s1), 38−55.

Gollier, C., 2010b. Ecological discounting. J. Econ. Theor. 145, 812−829.

Gollier, C., Treich, N., 2003. Decision making under scientific uncertainty: the economics of the precautionary principle. J. Risk Uncertain. 27 (1), 77−103.

Gollier, C., Moldovanu, B., Ellingsen, T., 2001. Should we beware of the precautionary principle? Econ. Pol. 16 (33), 301−327.

Graedel, T.E., Klee, R.J., 2002. Getting serious about sustainability. Environ. Sci. Technol. 36 (4), 523−529.

Halada, K., Shimada, M., Ijima, K., 2008. Decoupling status of metal consumption from economic growth. Mater. Trans. 49 (3), 411−418.

Hansson, S.O., 2010. Technology and the notion of sustainability. Technol. Soc. 32, 274−279.

International Law Association, 2014. In: Legal Principles Relating to Climate Change, Washington Conference.

Johnston, P., Everard, M., Santillo, D., Robert, K.H., 2007. Reclaiming the definition of sustainability. Env. Sci. Pollut. Res. 14 (1), 60−66.

Kiss, A., Shelton, D., 2004. International Environmental Law. Transnational Publishers Inc., Ardsley, NY, USA.

Mitchell, R.B., 2002−2016. International Environmental Agreements Database Project (Version 2014.3). Available at: http://iea.uoregon.edu/Date (Accessed 2 May 2016).

Padilla, E., 2002. Intergenerational equity and sustainability. Ecol. Econ. 41, 69−83.

Pan Pacific Copper, 2019. https://www.metalbulletin.com/events/download.ashx/document/speaker/7570/a0ID000000X0kAJMAZ/Presentation, Downloaded 30-1-2019.

Rio Declaration, 1992. In: United Nations Conference on Environment and Development, Rio de Janeiro.

Sachs, N.M., 2011. Rescuing the Strong Precautionary Principle from its Critics. University of Illinois Review, pp. 1285–1338.

Sandin, P., Peterson, M., Hansson, S.O., Rudén, C., Juthe, A., 2002. Five charges against the precautionary principle. J. Risk Res. S (4), 287–299.

Shelton, D., 2007. Equity. In: Bodansky, D., Brunnée, J., Hey, E. (Eds.), The Oxford Handbook of International Environmental Law. Oxford University Press.

Sirinskiene, A., 2009. The status of the precautionary principle: moving towards a rule of customary law. Jurisprudence 4 (118), 349–364.

Steen, B., Borg, G., 2002. An estimation of the cost of sustainable production of metal concentrates from the earth's crust. Ecol. Econ. 42, 401–413.

Sunstein, C.R., 2005. Laws of Fear, beyond the Precautionary Principle. Cambridge University Press.

Tickner, J.A., Kriebel, D., 2006. The role of science and precaution in environmental and public health policy. In: Elisabeth, F., Judith, J., René von, S. (Eds.), Implementing the Precautionary Principle, Perspectives and Prospects, ELGAR,2006, Section 3, pp. 42–62.

Tickner, J.A., Kriebel, D., Wright, S., 2003. A compass for health: rethinking precaution and its role in science and public health. Int. J. Epidemiol. 32, 489–492.

Tilton, J.E., 2003. On borrowed time? Assessing the threat of mineral depletion. Miner. Energy - Raw Mater. Rep. 18 (1), 33–42.

Trouwborst, A., 2007. The precautionary principle in general international law: combating the Babylonian confusion. Reciel 16 (2), 185–195.

UNEP International Resource Panel, 2011. Decoupling Natural Resource Use and Environmental Impacts from Economic Growth. UNEP.

United Nations, 1992. Framework Convention on Climate Change. Preamble.

United Nations, 2017. World Population Prospects, the 2017 Revision.

US Geological Survey, 2008. The Global Flows of Metals and Minerals, Open File Report 2008–1355.

Van Den Bergh, J.C.J.M., 2010. Externality or sustainable economics. Ecol. Econ. 69, 2047–2052.

World Bank, 2020. https://www.data.worldbank.org/indicator/NY.GDP.PCAP.KD.

World Commission on Environment and Development (Brundtland Commission), 1987. Our Common Future.

CHAPTER 6

The price mechanism[1]

6.1 Introduction

In this Chapter, we examine whether the price mechanism of the free market system leads to sufficient conservation of geologically scarce mineral resources for future generations timely and automatically. This question was already asked in 1979 by Dasgupta and Heal (1979), but it has never been unambiguously answered and remains relevant today. According to Seyhan et al. (2012) and De Bruyn et al. (2009), the price of a mineral resource is only a limited indicator for geological scarcity. According to Farley and Constanza (2002), markets are not efficient mechanisms for allocating scarce resources. Famous in this context is the wager between the economist Julian Simon and the environmentalist Paul Ehrlich made in 1980, on the price development of commodities (Worstall, 2013). Ehrlich expected prices to increase because of growing demand combined with limited supply. Simon argued that more people mean more brains and better methods of extraction, combined with a more efficient use of primary materials. Although Simon won the bet for the 10-year period considered (1980−90), it is not certain that the outcome can be extrapolated to any period in the (far) future

In Section 6.2, we explore how prices of minerals might react to increasing geological scarcity in general. We will formulate a hypothesis on the correlation between price trend and geological scarcity.

Then, in Section 6.3, we study trends in market prices of mineral resources for the period between 1900 and 2013. We compare the real market price development of mineral resources with the price development that can be expected on the basis of the hypothesis.

---

[1] This chapter is based on the publication of Henckens, M.L.C.M., Van Ierland, E.C., Driessen, P.P.J., Worrell, E., 2016. Geological scarcity, market price trends and future generations, Resour. Pol. 49, 102−111.

Governance of the world's mineral resources
ISBN 978-0-12-823886-8
https://doi.org/10.1016/B978-0-12-823886-8.00018-X

The potential impact of geological scarcity of a mineral resource is not only interesting with regard to the price development but also with regard to the prices of mineral resources as such. In Section 6.4, we investigate whether mineral resources prices are correlated to geological scarcity.

Our conclusions are presented in Section 6.5.

6.2 Mineral resources prices and geological scarcity

In this section, we will formulate a hypothesis on the relation between the long-term price development of a mineral resource and its geological scarcity.

It should be noticed that historical prices for mineral commodities do not necessarily reflect the costs of extraction and processing only. Factors that may influence the market price are geopolitical circumstances, such as political instability in an important producing country, wars, accidents, or strikes. Monopolist or oligopolistic producers may influence the market price by increasing or decreasing the offer of the raw material on the world market. A strong increase of demand due to new applications can cause the supply of commodities to the market to lag behind the demand so that prices will increase. Another important circumstance in which the market price does not reflect the production costs is when a relatively large proportion of the resource is obtained as a by-product of the production of other mineral resources, as is the case, for instance, for rhenium, molybdenum, and cobalt. The largest part of these minerals extracted worldwide is a by-product of copper extraction (rhenium: 90%, cobalt: 70%, molybdenum: 60%) (Copper Alliance, 2015). The market prices of by-product minerals will partly be determined by the trade-off between the volume of by-product generated and the demand for this by-product.

On the longer term, the bottom price for a mineral is determined by the marginal extraction costs of that mineral. Cyclical variations of the market price are superposed on the extraction costs. Nevertheless, the market price will not decrease structurally below the level of the marginal costs for extraction and exploration because mine owners will not want to work at a loss. At that point, mines will be closed, as has happened to many mines in Europe and the United States. In the past century, ore grades declined, and mines became ever deeper and were located in ever remoter places. If technology had not developed, commodities would have been substantially more costly now than they are. However, technology has developed and has partly or wholly neutralized the cost increasing effects of depletion of the richest ores in easily accessible places as we will see in Section 6.3.

The cumulative supply curve (or cumulative availability curve) is a concept that reflects how the cumulative supply of a mineral could vary over *all time* with the extraction costs (Tilton and Skinner, 1987). For mineral commodities, supply at a certain price is fixed by the amount of the mineral that can be produced profitably at that price. A rising price permits the extraction of a lower grade mineral resource, deeper in the crust at a remoter place and with higher external costs, as far as these are included in the price. The higher the price that consumers are willing to pay for a mineral, the greater its possible cumulative extraction is.

The type of distribution of the resource in the earth's crust (unimodal or bimodal) determines the slope of the cumulative curve. Fig. 6.1, derived from Tilton and Skinner (1987), presents three model cumulative supply curves.

In Fig. 6.1A, a small price increase allows a large increase in cumulative supply and, inversely, a growing demand will only trigger a relatively small increase in price. This type of curve belongs typically to a mineral with the bell-shaped unimodal distribution of abundant minerals in the earth's crust, such as aluminum. If a mineral is bimodally distributed in the earth's crust, the cumulative supply curve will have the form shown in Fig. 6.1B or C. The steep part of the curves in (B) and (C) represents the so-called mineralogical barrier between the occurrence of a mineral in enriched ores and its occurrence in common rock. Costs may rise by a factor of 10 to 1000 in a relatively short period of time (Steen and Borg, 2002). Fig. 6.1 supposes a stable state of technology, so technological development over time is not included in Fig. 6.1.

For lithium, the cumulative availability curve has been determined (Yaksic and Tilton, 2009) (Fig. 6.2). According to these authors, the 2009 price of lithium carbonate is USD 6 per kg. The right-hand—flat—part of the curve represents the situation when lithium is extracted from seawater (an almost inexhaustible source of lithium). The costs would then increase until USD 16—22 per kg (USD 7—10 per pound). Such costs do not seem to be insurmountable for application of lithium in lithium batteries. It should be noted that only lithium, sodium, potassium, calcium, and chlorine are elements that can be extracted economically from seawater because of their relatively high abundance in seawater, as demonstrated by Bardi (2009).

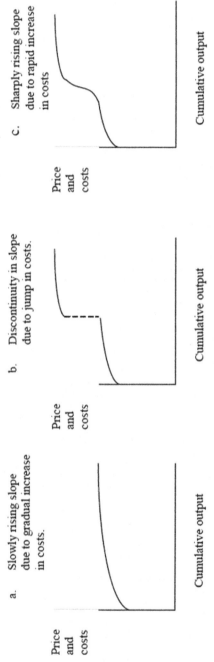

Figure 6.1 Illustrative cumulative supply curves (Tilton and Skinner, 1987).

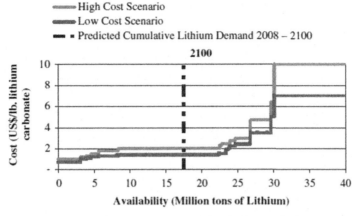

Figure 6.2 Cumulative availability curve for lithium (Yaksic and Tilton, 2009). On the vertical axis, the costs are expressed in USD per lb. of lithium. 1 lb. is 0.45359237 kg.

The maximum extraction costs of a mineral resource are determined by the costs of extraction of that resource from common rock or seawater. Once this is the case, scarcity no longer plays a role. The quantities of minerals available in common rock and seawater are very large compared to the amount of minerals available in ores, but it is expensive to extract them, although technically not impossible. The energy consumption for extracting copper from common rock is 10 times higher than that for extracting copper from copper ore (Skinner, 1976; Harmsen et al., 2013; Norgate and Jahanshahi, 2010). Table 6.1 compares the costs of extraction of minerals from common rock for a number of mineral resources with the current costs.

Table 6.1 implies that ultimately the extraction costs of a mineral will maximally reach the costs of extraction of the mineral from common rock

Table 6.1 Cost increase for the production of ore-like metal concentrates from common rock in a sustainable way, including the external costs, compared to the current price level (Steen and Borg, 2002).

	Cost increase compared to current price level		Cost increase compared to current price level
Cd	4,000—100,000	Ni	40
Co	30	Pb	700
Cr	20	Sn	200
Cu	90	W	20,000—200,000
Mn	10	Zn	50

and/or seawater. Taking the cost decreasing effect of technology development into consideration, the extraction costs may slowly decrease again, once a resource is being extracted from common rock or seawater. The start-up of extraction and processing of a new mineral resource may be relatively more costly because specific technology needs to be developed and because the scale is still small, for example, aluminum and chromium. But once the scale has become bigger and the technology more developed, the costs stabilize. The extra costs of lower grades and deeper and remoter mining have been neutralized by technology development. This means that the graph describing the development of resource extraction costs over time will probably be duck shaped, assuming that mineral extraction continues after ore exhaustion (Fig. 6.3).

The duck shape is applicable for both major and minor elements. In the case of major elements, the slope of the curve (during time period B) will be gentle and stretched out, whereas in the case of minor elements with a bimodal distribution in the earth's crust, the slope of the curve will be much steeper.

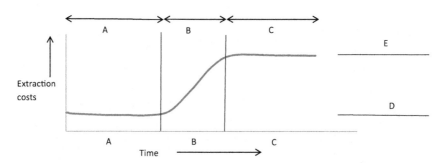

Figure 6.3 Expected trend in extraction cost of mineral commodities.
A: Period of relatively low extraction costs. This period corresponds with the exhaustion periods indicated in Table 3.3. Lower grades are mitigated by technology development.
B: Period of increasing extraction costs. This is the transition period between the extraction of minerals from ores and extraction of minerals from common rock and seawater.
C: Period of high extraction costs. Ores are exhausted. Minerals are extracted from common rock and seawater. Due to technological developments, the costs may decrease slightly.
D: Current level of extraction costs of commodities.
E: Ultimate cost level for extraction of commodity from common rock or from seawater. For a number of elements, the E/D ratio is presented in Table 6.1.

In practice, the high price level connected to the high level extraction costs at the right-hand side of Fig. 6.3 will probably not be attained for most minerals in most applications. Depending on the application, from a certain price level on, a substitute will replace the mineral and the so-called choke price will be attained. The choke price is the price level at which the demand for a commodity for a given application will fall to zero because an adequate substitute is available. Extraction will stop when the choke price of a mineral for its last application is reached. Prediction of the specific form of Fig. 6.3 for a specific element will be complex, for example, because data are lacking on the availability and distribution of mineral ores in the earth's crust. Moreover, knowledge on the availability and application of substitutes, which may delay exhaustion of ores, is limited.

An important question is how much time will elapse between leaving the low price level and reaching the high price level (the duration of the period B in Fig. 6.3). Will the market timely anticipate future scarcity and will prices start to rise appreciably a long time before scarcity of ore reserves is in sight, so well before the period A in Fig. 6.3 has ended? This seems unlikely. The price will probably follow the extraction costs until relatively close to or even within period B.

If scarcity increases rapidly and adequate substitutes are absent or few in number, the market price will probably rise relatively rapidly to a level sufficient to cover the higher costs of extraction or of developing suitable substitutes. However, even a quick price increase of a raw material may not have an immediate effect on demand. This depends on the share of the costs of the raw material in the total costs of its main product applications. A threefold price increase of a raw material may make an average end product no more than about 10% more expensive. This is based on the fact that raw material prices nowadays usually make up only a small percentage of the cost of an end product (5% according to De Bruyn et al., 2009). Only if prices of raw materials were to increase by a factor $10-100$, this would result in products that are in the order of one and a half to five times more expensive. This means that the eventual price increase of a raw material due to exhaustion of the ores in which it is contained does not necessarily lead to a proportional decrease in the demand for the raw material. The conclusion is that extraction of ores may continue at the same pace even when they are practically exhausted. This will certainly be the case if a proper substitute is available. In that case, mine owners will probably try to make their capital as profitable as possible and try to sell the remaining ore before it becomes "worthless." An example is the mass marketing of natural gas by the Dutch government in the years 1960 and 1970, in an attempt to maximize profit for Dutch society, before the introduction of nuclear energy would make this gas superfluous according to the concept of the time (de Pous, 1962)

We expect that, ultimately, the price of mineral resources is determined by geological scarcity and is duck shaped, as indicated in Fig. 6.3. The scarcer a mineral resource, the earlier the market price will start to increase.

6.3 Price trends and geological scarcity

To investigate whether geological scarcity is visible in the price development of a mineral resource, we selected the 35 minerals and groups of minerals presented in Table 3.3.

We have carried out a trend analysis of the market price development of each of the mineral resources of Table 3.3. The results are visualized in the Annex to this Chapter. We used the inflation correction figures provided by the United States Geological Survey (2017). US Geological Survey uses the official consumer price index provided by the Bureau of Labor Statistics.

The rate of increase or decrease of the real price of minerals over time is represented by the coefficients of the linear functions, which are presented in the annexed pictures. If the coefficient is positive, the long-term trend is a real price increase. If the coefficient is negative, the long-term trend is a real price decrease. The higher (or lower) the coefficient, the faster the price increase (or decrease).

The line that represents the trend in the price of a specific mineral over time is calculated by the least squares linear regression method. The quality of the fit is presented by R^2. The so-called P-value of the coefficient represents the probability that the value of the coefficient is determined by chance. If the P-value is smaller than or equal to .05, the slope of the linear function (the coefficient) is considered to represent the price trend of the mineral in question over time in a significant way. Table 6.2 presents the coefficients and the related P-values for all the minerals that we have investigated. We compare these coefficients with the geological scarcity of the mineral. To obtain a simple number for the degree of scarcity, we express the geological scarcity as the natural logarithm of 1,000,000 divided by the exhaustion time after 2015. For the exhaustion time, we have taken the average column of Table 3.3. This results in a scale from 0 to 10, subdivided in the same way as Table 3.3:

- Scarcity between 8.1 and 10: scarce
- Scarcity between 7.4 and 8: moderately scarce
- Scarcity lower than 7.4: not scarce

Table 6.2 Price trend/time coefficient and geological scarcity.

Mineral	Period considered	Price-trend / time coefficient	P-value of price trend / time coefficient	Significance of the price trend/ time coefficient[a]	Scarcity[b]	Scarcity class
Copper	1920-2013	0.0020	0.010	**	9.2	Scarce
Antimony	1900-2013	0.0010	0.050	*	8.8	
Gold	1900-2013	0.0202	5.3E-08	***	8.8	
Boron	1900-2013	-0.0066	0.0047	**	8.8	
Silver	1900-2013	0.0037	0.068	NS	8.8	
Bismuth	1900-2013	-0.010	1.1E-22	***	8.8	
Molybdenum	1912-2013	0.0022	0.45	NS	8.5	
Indium	1946-2013	-0.0057	0.11	NS	8.3	
Chromium	1900-2013	0.015	1.9E-13	***	8.0	
Zinc	1900-2013	-0.0036	0.0025	**	7.8	Moderately scarce
Nickel	1920-2013	0.0047	0.00002	***	7.7	
Tungsten	1900-2013	-0.000092	0.95	NS	7.4	
Tin	1900-2013	0.0024	0.027	*	7.7	
Rhenium	1980-2013	0.0062	0.11	NS	7.3	Not scarce
Selenium	1920-2013	-0.0012	0.383	NS	7.3	
Cadmium	1900-2013	-0.0070	7.2E-17	***	7.1	
Iron	1900-2013	0.0043	9.7E-07	***	6.8	
Cobalt	1900-2013	-0.013	5.3E-05	***	6.8	
PGM	1940-2013	0.0071	3.2E-05	***	6.7	
Manganese	1900-2013	0.012	5.3E-08	***	6.6	
Lead	1900-2013	-0.0030	0.00038	**	6.5	
Lithium	1960-2013	-0.015	1.2E-11	***	6.4	
Arsenic	1925-2013	-0.0059	0.0007	***	5.4	
Gallium	1980-2013	-0.0044	1.5E-07	***	5.3	
REE	1960-2013	0.11	0.0014	**	5.2	
Strontium	1935-2013	0.022	9.0E-05	***	5.2	
Aluminum	1940-2013	-0.0097	2.1E-16	***	4.6	
Tantalum	1964-2013	-0.0033	0.82	**	3.5	
Vanadium	1910-2013	-0.011	6.4E-09	***	3.2	
Magnesium	1950-2013	-0.0021	0.0012	**	3.2	
Germanium	1960-2013	-0.0047	0.093	NS	0.6	

[a] NS = Not Significant (P-value > 0.05), * = Significant (0.01 < P-value < 0.05), ** = Significant (0.001 < P-value < 0.01), *** = Significant (P-value < 0.001)

[b] Scarcity is expressed as Ln (1,000,000 / exhaustion time after 2015 according to Table 3.3, average of scenario 1 and scenario 2)

Scarce
Moderately scarce
Not scarce

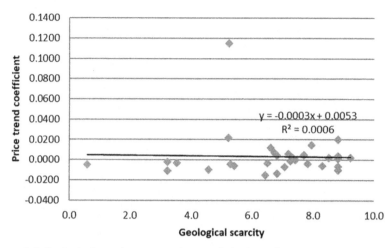

Figure 6.4 Geological scarcity versus price trend. Geological scarcity is expressed as Ln (1,000,000/exhaustion time after 2015).

The significance indicates whether there is a significant correlation between the coefficient of the calculated linear function describing the price development and the observed price trend.

Table 6.2 is graphically presented in Fig. 6.4. A regression analysis of the data shows that there is no significant correlation between geological scarcity and price trend. The *P*-value is .90.

A second observation is that almost all price trend coefficients are near to zero, regardless of geological scarcity. The price trend analysis of 33 minerals for the period 1900 to 2013 demonstrates that none of the minerals considered shows a fast price increase. Although prices of mineral resources show cyclical fluctuations, for a long time, structural price changes (whether an increase or a decrease) have been quite small.

The hypothesis, as explained in Section 6.2, is that the market price of a mineral resource will remain quite stable for a long time. Then, exhaustion of ores will lead to a relatively fast price increase. The greater the geological scarcity of the mineral, the earlier the price increase will start.

Fig. 6.4 demonstrates that there is no significant difference in price trend between geologically scarce minerals and geologically abundant minerals up to now. The conclusion is that geological scarcity of the investigated commodities has not reached a point that the market reacts by a structural price increase.

Table 6.3 Unit values of commodities in the United States, average value between 2000 and 2015, expressed in USD 2019.

Mineral	Unit value per kg[a] (USD 2019)	Scarcity[b]	Mineral	Unit value per kg[a] (USD 2019)	Scarcity[b]
Gold	32,264.13	8.8	Rare-earth elements	18.40	5.2
Platinum group metals	22,342.38	6.7	Antimony	7.26	8.8
Rhenium	3,724.52	7.3	Copper	6.39	9.2
Germanium	1,393.18	0.6	Lithium	4.02	6.4
Gallium	655.12	5.3	Cadmium	3.09	7.1
Indium	614.77	8.3	Aluminum	2.56	4.6
Silver	553.69	8.8	Zinc	2.32	7.8
Tantalum	235.79	3.5	Chromium	2.26	8.0
Selenium	74.19	7.3	Lead	2.26	6.5
Cobalt	42.55	6.8	Manganese	1.38	6.6
Molybdenum	39.05	8.5	Strontium	1.12	5.2
Tungsten	35.26	7.4	Boron (B_2O_3)	1.05	8.8
Vanadium	29.53	3.2	Arsenic	0.92	5.4
Tin	20.98	7.3	Magnesium	0.66	3.2
Nickel	20.07	7.7	Iron (ore)	0.08	6.8
Bismuth	19.55	8.8			

[a] The unit values are the average of the unit values between 2000 and 2015 as provided by the US Geological Survey (2017). These values are expressed in USD (1998). We have transferred the values to USD (2019) on the basis of the inflation of consumer prices in the United States as provided by the US Bureau of Labor Statistics.
[b] Scarcity is expressed as Ln (1,000,000/exhaustion time after 2015 according to Table 3.3, average of scenario 1 and scenario 2).

6.4 Mineral resource prices and scarcity

In Table 6.3, we present the average prices of minerals between 2000 and 2015. It is obvious that the mineral resource prices have no relation with geological scarcity. Some geologically scarce minerals have a relatively low price, for example, boron, zinc, chromium, antimony, and copper. Vice versa, some non-scarce minerals have a relatively high price, for example, platinum group metals, germanium, and gallium.

6.5 Conclusions

The question addressed in this chapter is whether the price mechanism of the free market system can be expected to slow down the extraction of geologically scarce minerals automatically and timely, in order to keep sufficient resources available for future generations. By comparing the real price development of commodities of different geological scarcity over a long period of time, we investigated whether the price trends of mineral resources are related to geological scarcity.

We hypothesized that the prices of geologically scarcer minerals will start rising earlier than the prices of less scarce minerals, but until now the market has not differentiated on the basis of future geological scarcity. Hence, geological scarcity is not yet so critical that the market reacts. Apparently, the market price does not (yet) reflect the large differences in geological scarcity of the minerals considered. This phenomenon might be explained as follows. The time horizon at which geological scarcity will become a reality is at least several decades to centuries away. The time horizon of market prices seems to be some years to about a decade maximum, taking into consideration that the maximum forward time for futures on the London Metal Exchange is 123 months.

Our conclusion is that, so far, viewed over the long term, the prices of all minerals considered have stayed quite stable (shown by the very low slope values in Table 6.2 and the Figures in the Annex, regardless of their scarcity. This conclusion is supported by other researchers (Krautkraemer, 1998; Cuddington, 2010; Fernandez, 2012).

On the basis of the analysis of both price trends and prices, we conclude that geological scarcity is not yet a factor with a discernible influence on the pricing of mineral resources. It remains unclear how near to exhaustion the market will react with price increases linked to geological scarcity.

We conclude that, so far, despite fluctuations in mineral resource prices, there is no significant correlation between the geological scarcity of a mineral resource on the one hand and price trends and prices on the other hand. The price trend of an abundant resource with sufficient geological reserves for thousands of years does not differ significantly from the price trend of a geologically scarce mineral resource whose ores may be exhausted within a century. We therefore argue that it cannot be certain that the price mechanism of the free market system will lead to timely, automatic, and sufficient conservation of geologically scarce non-renewable mineral resources for future generations.

References

Bardi, U., 2009. Mining the oceans: can we extract minerals from seawater? Oil Drum Eur. http://theoildrum.com/node/4558.

Copper Alliance, 2015, 7-12-2015. http://sustainablecopper.org/about-copper/33-more-than-copper.html://.

Cuddington, J.T., 2010. Long term trends in the *Real* real process of primary commodities: inflation bias and the Prebisch-Singer hypothesis. Resour. Policy 35, 72−76.

Dasgupta, P., Heal, G., 1979. Economic Theory and Exhaustible Resources. University Press, Cambridge page 2.

De Bruyn, S., Markowska, A., De Jong, F., Blom, M., December 2009. Resource Productivity, Competitiveness and Environmental Policies. CE Delft. Publication number 09.7951, available from: www.ce.nl.

De Pous, J.W., 11 juli 1962. Nota inzake het aardgas, Slotbeschouwing, 2de kamer zitting 1961-1962.

Farley, J., Costanza, R., 2002. Envisioning shared goals for humanity: a detailed, shared vision of a sustainable and desirable USA in 2100. Ecol. Econ. 43, 245−259.

Fernandez, V., 2012. Trends in real commodity prices: how real is real? Resour. Policy 37, 30−47.

Harmsen, J.H.M., Roes, A.L., Patel, M.K., 2013. The impact of copper scarcity on the efficiency of 2050 global renewable energy scenarios. Energy 50, 62−73.

Krautkraemer, J.A., 1998. Nonrenewable resources scarcity. J. Econ. Lit. 36 (4), 2065−2107.

Norgate, T., Jahanshahi, S., 2010. Smelt, leach or concentrate? Miner. Eng. 23, 65−73.

Seyhan, D., Weikard, H.-P., van Ierland, E., 2012. An economic model of long-term phosphorus extraction and recycling. Resour. Conserv. Recycl. 61, 103−108.

Skinner, B.J., 1976. A second iron age ahead? Am. Sci. 64, 158−169.

Steen, B., Borg, G., 2002. An estimation of the cost of sustainable production of metal concentrates from the earth's crust. Ecol. Econ. 42, 401−413.

Tilton, J.E., Skinner, J.B., 1987. In: McLaren, D.J., Skinner, B.J. (Eds.), The Meaning of Resources, Resources and World Development. John Wiley &Sons, pp. 13−27.

US Geological Survey, 2017. Historical Statistics for Mineral and Material Commodities in the United States.

Worstall, T., 2013. But Why Did Julian Simon Win the Paul Ehrlich Bet? http://www.forbes.com/sites/timworstall/2013/01/13/but-why-did-julian-simon-win-the-paul-ehrlich-bet/.

Yaksic, A., Tilton, J.E., 2009. Using the cumulative availability curve to assess the threat of mineral depletion. Resour. Policy 34, 185−194.

Annex to Chapter 6
Real price development of commodities

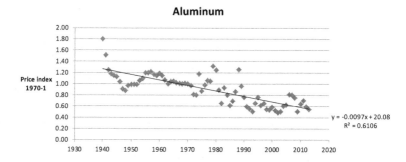

SUMMARY OUTPUT of linear regression analysis

Regression statistics	
Multiple correlation coefficient	0.7814069
R Square	0.6105967
Adjusted R Square	0.6051884
Standard error	0.1677511
Observations	74

ANOVA

	df	SS	MS	F	Significance F
Regression	1	3.177005179	3.177005	112.8983	2.136E-16
Residual	72	2.026110057	0.02814		
Total	73	5.203115236			

	Coëfficiënt	Standard error	T- stat	P-value	Lowest 95%	Highest 95%	Lowest 95,0%	Highest 95,0%
Intercept	20.079859	1.8045551	11.12732	2.66E-17	16.4825444	23.6771741	16.48254441	23.6771741
X	-0.0097	0.000912952	-10.6254	2.14E-16	-0.01152038	-0.0078805	-0.011520382	-0.007880508

Real price development of Aluminum

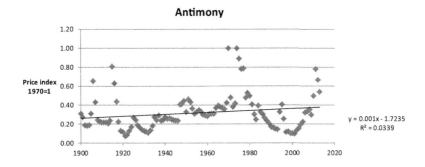

y = 0.001x - 1.7235
R² = 0.0339

SUMMARY OUTPUT of linear regression analysis

Regression statistics	
Multiple correlation coefficient	0.184249
R Square	0.033948
Adjusted R Square	0.025322
Standard error	0.184897
Obeservations	114

ANOVA

	df	SS	MS	F	Significance F
Regression	1	0.13455	0.13455	3.935734	0.049716908
Residual	112	3.828923	0.034187		
Total	113	3.963473			

	Coëfficiënt	standard erro	T- stat	P-value	Lowest 95%	Highest 95%	Lowest 95,0%	Highest 95,0%
Intercept	-1.72346	1.029723	-1.67371	0.096978	-3.76372496	0.31680516	-3.763724963	0.316805164
X	0.001044	0.000526	1.983869	0.049717	1.31386E-06	0.00208665	1.31386E-06	0.002086646

Real price development of Aluminum

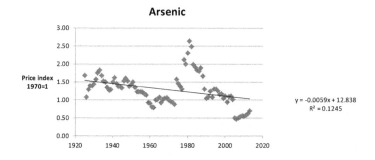

SUMMARY OUTPUT of linear regression analysis

Regression statistics	
Multiple correlation coefficient	0.3528679
R Square	0.1245157
Adjusted R Square	0.1144527
Standard error	0.4044185
Observations	89

ANOVA

	df	SS	MS	F	Significance F
Regression	1	2.023751196	2.023751	12.37357	0.00069464
Residual	87	14.22922485	0.163554		
Total	88	16.25297604			

	Coëfficiënt	Standard error	T- stat	P-value	Lowest 95%	Highest 95%	Lowest 95,0%	Highest 95,0%
Intercept	12.837739	3.2858422	3.906986	0.000184	6.306771599	19.3687059	6.306771599	19.36870594
X	-0.00587	0.001668645	-3.51761	0.000695	-0.00918625	-0.002553	-0.009186255	-0.002553028

Real price development of arsenic

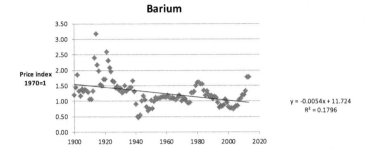

SUMMARY OUTPUT of linear regression analysis

Regression statistics	
Multiple correlation coefficient	0.42377253
R Square	0.17958316
Adjusted R Square	0.17225801
Standard error	0.37965398
Observations	114

ANOVA

	df	SS	MS	F	Significance F
Regression	1	3.533661768	3.533662	24.51597	2.62549E-06
Residual	112	16.14336017	0.144137		
Total	113	19.67702194			

	Coëfficiënt	Standard error	T- stat	P-value	Lowest 95%	Highest 95%	Lowest 95,0%	Highest 95,0%
Intercept	11.7241711	2.114361719	5.545017	1.98E-07	7.534834401	15.9135078	7.534834401	15.91350783
X	-0.0053501	0.001080533	-4.95136	2.63E-06	-0.007491045	-0.0032092	-0.00749105	-0.00320917

Real price development of barium

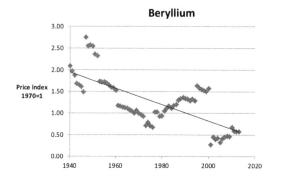

y = -0.0189x + 38.703
R² = 0.5044

SUMMARY OUTPUT of linear regression analysis

Regression statistics	
Multiple correlation coefficient	0.7102406
R Square	0.5044417
Adjusted R Square	0.4975589
Standard error	0.4066188
Observations	74

ANOVA

	df	SS	MS	F	Significance F
Regression	1	12.11779469	12.11779	73.29066	1.37462E-12
Residual	72	11.90439799	0.165339		
Total	73	24.02219267			

	Coëfficiënt	Standard error	T- stat	P-value	Lowest 95%	Highest 95%	Lowest 95,0%	Highest 95,0%
Intercept	38.702585	4.374136686	8.848051	4.01E-13	29.98290248	47.42226672	29.98290248	47.42226672
X	-0.018945	0.002212943	-8.561	1.37E-12	-0.023356416	-0.01453357	-0.02335642	-0.014533574

Real price development of beryllium

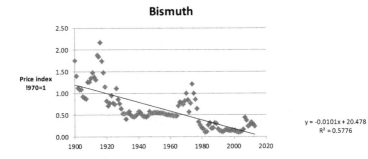

$$y = -0.0101x + 20.478$$
$$R^2 = 0.5776$$

SUMMARY OUTPUT of linear regression analysis

Regression statistics	
Multiple correlation coefficient	0.75997673
R Square	0.57756463
Adjusted R Square	0.57379289
Standard error	0.28808939
Observations	114

ANOVA

	df	SS	MS	F	Significance F
Regression	1	12.70904357	12.70904	153.1293	1.08491E-22
Residual	112	9.295495506	0.082995		
Total	113	22.00453908			

	Coëfficiënt	Standard error	T- stat	P-value	Lowest 95%	Highest 95%	Lowest 95,0%	Highest 95,0%
Intercept	20.4776624	1.604421938	12.76327	1.41E-23	17.29870601	23.6566188	17.298706	23.65661882
X	-0.0101463	0.000819931	-12.3745	1.08E-22	-0.01177086	-0.00852168	-0.01177086	-0.008521684

Real price development of bismuth

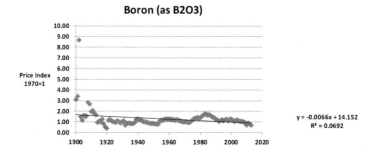

SUMMARY OUTPUT of linear regression analysis

Regression statistics	
Multiple correlation coefficient	0.2630191
R Square	0.069179
Adjusted R Square	0.0608681
Standard error	0.8009369
Observations	114

ANOVA

	df	SS	MS	F	Significance F
Regression	1	5.339774263	5.339774	8.32389	0.004694448
Residual	112	71.84798269	0.6415		
Total	113	77.18775695			

	Coëfficiënt	Standard error	T- stat	P-value	Lowest 95%	Highest 95%	Lowest 95,0%	Highest 95,0%
Intercept	14.151592	4.460562288	3.172603	0.00195	5.313559827	22.98962462	5.313559827	22.98962462
X	-0.006577	0.002279546	-2.88512	0.004694	-0.01109338	-0.00206012	-0.011093381	-0.002060125

Real price development of boron

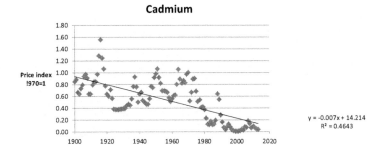

SUMMARY OUTPUT of linear regression analysis

Regression statistics	
Multiple correlation coefficient	0.68138656
R Square	0.46428765
Adjusted R Square	0.4595045
Standard error	0.24932722
Observations	114

ANOVA

	df	SS	MS	F	Significance F
Regression	1	6.034105353	6.034105	97.06742	7.22519E-17
Residual	112	6.962375205	0.062164		
Total	113	12.99648056			

	Coëfficiënt	Standard error	T- stat	P-value	Lowest 95%	Highest 95%	Lowest 95,0%	Highest 95,0%
Intercept	14.2138108	1.388548427	10.23645	9.31E-18	11.46258007	16.9650415	11.46258007	16.9650415
X	-0.00699128	0.00070961	-9.85228	7.23E-17	-0.008397278	-0.0055853	-0.008397278	-0.00558528

Real price development of cadmium

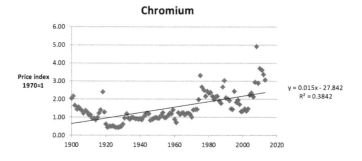

SUMMARY OUTPUT of linear regression analysis

Regression statistics	
Multiple correlation coefficient	0.6198229
R Square	0.3841804
Adjusted R Square	0.378682
Standard error	0.6303891
Observations	114

ANOVA

	df	SS	MS	F	Significance F
Regression	1	27.76623683	27.76624	69.87144	1.93902E-13
Residual	112	44.50772277	0.39739		
Total	113	72.27395961			

	Coëfficiënt	Standard error	T- stat	P-value	Lowest 95%	Highest 95%	Lowest 95,0%	Highest 95,0%
Intercept	-27.84188	3.510750821	-7.93046	1.79E-12	-34.79798274	-20.8857773	-34.7979827	-20.88577729
X	0.0149971	0.00179415	8.358914	1.94E-13	0.011442266	0.01855202	0.011442266	0.018552022

Real price development of chromium

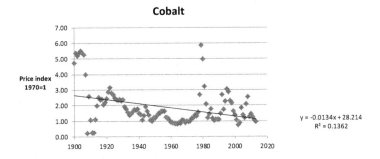

Cobalt

$y = -0.0134x + 28.214$
$R^2 = 0.1362$

SUMMARY OUTPUT of linear regression analysis

Regression statistics	
Multiple correlation coefficient	0.3689963
R Square	0.1361582
Adjusted R Square	0.1284454
Standard error	1.1247233
Observations	114

ANOVA

	df	SS	MS	F	Significance F
Regression	1	22.33156609	22.33157	17.65338	5.34729E-05
Residual	112	141.6802884	1.265003		
Total	113	164.0118545			

	Coëfficiënt	Standard error	T-stat	P-value	Lowest 95%	Highest 95%	Lowest 95,0%	Highest 95,0%
Intercept	28.213902	6.263787777	4.504288	1.64E-05	15.80300992	40.624795	15.80300992	40.62479503
X	-0.0134496	0.003201074	-4.20159	5.35E-05	-0.01979213	-0.00710709	-0.01979213	-0.00710709

Real price development of cobalt

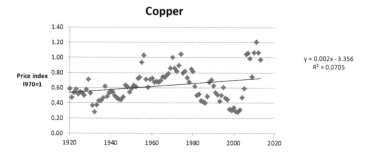

SUMMARY OUTPUT of linear regression analysis

Regression statistics	
Multiple correlation coefficient	0.26552425
R Square	0.07050313
Adjusted R Square	0.0603999
Standard error	0.20188528
Observations	94

ANOVA

	df	SS	MS	F	Significance F
Regression	1	0.284418327	0.284418	6.978278	0.009695218
Residual	92	3.749705203	0.040758		
Total	93	4.03412353			

	Coëfficiënt	Standard error	T-stat	P-value	Lowest 95%	Highest 95%	Lowest 95,0%	Highest 95,0%
Intercept	-3.3559938	1.50925755	-2.22361	0.028623	-6.353509523	-0.35847798	-6.35350952	-0.358477985
X	0.00202723	0.000767411	2.641643	0.009695	0.000503082	0.003551371	0.000503082	0.003551371

Real price development of copper

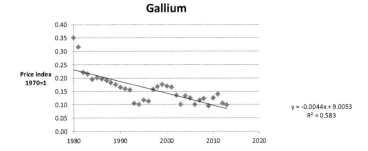

SUMMARY OUTPUT of linear regression analysis

Regression statistics	
Multiple correlation coefficient	0.76352784
R Square	0.58297476
Adjusted R Square	0.56994273
Standard error	0.03789924
Observations	34

ANOVA

	df	SS	MS	F	Significance F
Regression	1	0.064253723	0.064254	44.73397	1.50321E-07
Residual	32	0.045963266	0.001436		
Total	33	0.110216989			

	Coëfficiënt	Standard error	T- stat	P-value	Lowest 95%	Highest 95%	Lowest 95,0%	Highest 95,0%
Intercept	9.00532919	1.322711655	6.808233	1.07E-07	6.311053718	11.6996047	6.311053718	11.69960466
X	-0.0044311	0.000662507	-6.68835	1.5E-07	-0.00578056	-0.00308159	-0.00578056	-0.003081594

Real price development of gallium

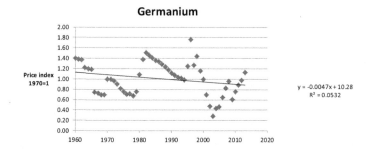

SUMMARY OUTPUT of linear regression analysis

Regression statistics	
Multiple correlation coefficient	0.23056526
R Square	0.05316034
Adjusted R Square	0.03495189
Standard error	0.31282239
Observations	54

ANOVA

	df	SS	MS	F	Significance F
Regression	1	0.285700057	0.2857	2.919542	0.093475935
Residual	52	5.088608211	0.097858		
Total	53	5.374308269			

	Coëfficiënt	Standard error	T- stat	P-value	Lowest 95%	Highest 95%	Lowest 95,0%	Highest 95,0%
Intercept	10.2799076	5.425930604	1.894589	0.063714	-0.608018664	21.16783396	-0.608018664	21.16783396
X	-0.0046669	0.002731318	-1.70867	0.093476	-0.010147703	0.000813879	-0.010147703	0.000813879

Real price development of Germanium

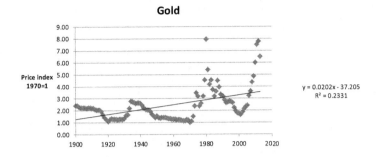

Gold

y = 0.0202x - 37.205
R² = 0.2331

SUMMARY OUTPUT of linear regression analysis

Regression statistics	
Multiple correlation coefficient	0.48278576
R Square	0.23308209
Adjusted R Square	0.22623461
Standard error	1.21901156
Observations	114

ANOVA

	df	SS	MS	F	Significance F
Regression	1	50.58173107	50.58173	34.0391	5.33081E-08
Residual	112	166.4307881	1.485989		
Total	113	217.0125191			

	Coëfficiënt	Standard error	T- stat	P-value	Lowest 95%	Highest 95%	Lowest 95,0%	Highest 95,0%
Intercept	-37.205229	6.788895969	-5.48031	2.65E-07	-50.65655657	-23.7539023	-50.6565566	-23.75390232
X	0.0202417	0.003469428	5.834304	5.33E-08	0.013367469	0.027115923	0.013367469	0.027115923

Real price development of gold

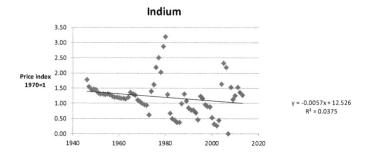

SUMMARY OUTPUT of linear regression analysis

Regression statistics	
Multiple correlation coefficient	0.19354516
R Square	0.03745973
Adjusted R Square	0.02287578
Standard error	0.57751813
Observations	68

ANOVA

	df	SS	MS	F	Significance F
Regression	1	0.856684461	0.856684	2.56856	0.113782523
Residual	66	22.01279438	0.333527		
Total	67	22.86947884			

	Coëfficiënt	Standard error	T- stat	P-value	Lowest 95%	Highest 95%	Lowest 95,0%	Highest 95,0%
Intercept	12.5255409	7.06344707	1.77329	0.080794	-1.577086186	26.628168	-1.577086186	26.628168
X	-0.0057185	0.003568123	-1.60267	0.113783	-0.012842521	0.001405454	-0.012842521	0.001405454

Real price development of indium

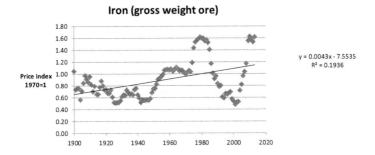

Iron (gross weight ore)

y = 0.0043x - 7.5535
R² = 0.1936

SUMMARY OUTPUT of linear regression analysis

Regression statistics	
Multiple correlation coefficient	0.4399687
R Square	0.1935725
Adjusted R Square	0.1863722
Standard error	0.2926951
Observations	114

ANOVA

	df	SS	MS	F	Significance F
Regression	1	2.303175444	2.303175	26.88415	9.68164E-07
Residual	112	9.595084581	0.08567		
Total	113	11.89826002			

	Coëfficiënt	Standard error	T- stat	P-value	Lowest 95%	Highest 95%	Lowest 95,0%	Highest 95,0%
Intercept	-7.553451	1.630071756	-4.63382	9.76E-06	-10.7832292	-4.32367278	-10.78322925	-4.32367278
X	0.0043193	0.000833039	5.184992	9.68E-07	0.002668742	0.00596986	0.002668742	0.005969862

Real price development of lead

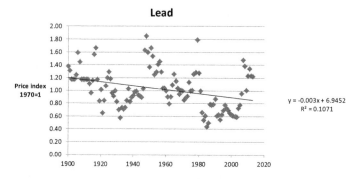

SUMMARY OUTPUT of linear regression analysis

Regression statistics	
Multiple correlation coefficient	0.3273104
R Square	0.1071321
Adjusted R Square	0.09916
Standard error	0.2899422
Observations	114

ANOVA

	df	SS	MS	F	Significance F
Regression	1	1.12972583	1.129726	13.43848	0.000378498
Residual	112	9.415444043	0.084066		
Total	113	10.54516987			

	Coëfficiënt	Standard error	T- stat	P-value	Lowest 95%	Highest 95%	Lowest 95,0%	Highest 95,0%
Intercept	6.9452258	1.614740441	4.301141	3.65E-05	3.745824629	10.14462702	3.745824629	10.14462702
X	-0.0030251	0.000825204	-3.66585	0.000378	-0.00466011	-0.00139004	-0.004660114	-0.001390042

Real price development of lead

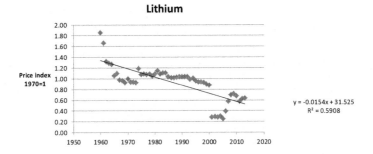

$$y = -0.0154x + 31.525$$
$$R^2 = 0.5908$$

SUMMARY OUTPUT of linear regression analysis

Regression statistics	
Multiple correlation coefficient	0.7686385
R Square	0.5908051
Adjusted R Square	0.5829359
Standard error	0.2035492
Observations	54

ANOVA

	df	SS	MS	F	Significance F
Regression	1	3.11068559	3.110686	75.0788	1.15055E-11
Residual	52	2.154478363	0.041432		
Total	53	5.265163953			

	Coëfficiënt	Standard error	T-stat	P-value	Lowest 95%	Highest 95%	Lowest 95,0%	Highest 95,0%
Intercept	31.525141	3.530577882	8.929173	4.47E-12	24.44051779	38.6097634	24.44051779	38.60976345
X	-0.0153994	0.001777231	-8.6648	1.15E-11	-0.018965629	-0.01183308	-0.018965629	-0.01183308

Real price development of lithium

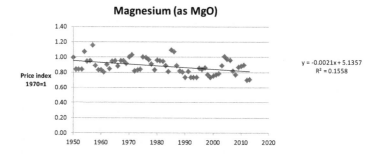

SUMMARY OUTPUT of linear regression analysis

Regression statistics	
Multiple correlation coefficient	0.39469078
R Square	0.15578081
Adjusted R Square	0.14216437
Standard error	0.09379921
Observations	64

ANOVA

	df	SS	MS	F	Significance F
Regression	1	0.100658107	0.100658	11.44064	0.001249187
Residual	62	0.545494047	0.008798		
Total	63	0.646152154			

	Coëfficiënt	Standard error	T- stat	P-value	Lowest 95%	Highest 95%	Lowest 95,0%	Highest 95,0%
Intercept	5.1356836	1.257725293	4.083311	0.000129	2.621526558	7.64984063	2.621526558	7.649840633
X	-0.00214683	0.000634706	-3.3824	0.001249	-0.00341559	-0.00087807	-0.003415592	-0.00087807

Real price development of magnesium

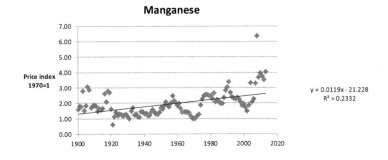

$y = 0.0119x - 21.228$
$R^2 = 0.2332$

SUMMARY OUTPUT of linear regression analysis

Regression statistics	
Multiple correllation coefficient	0.4828851
R Square	0.233178
Adjusted R Square	0.2263314
Standard error	0.7135727
Observations	114

ANOVA

	df	SS	MS	F	Significance F
Regression	1	17.34153749	17.34154	34.05737	5.29257E-08
Residual	112	57.02883256	0.509186		
Total	113	74.37037006			

	Coëfficiënt	Standard error	T- stat	P-value	Lowest 95%	Highest 95%	Lowest 95,0%	Highest 95,0%
Intercept	-21.228007	3.974015526	-5.3417	4.89E-07	-29.102009	-13.3540043	-29.102009	-13.3540043
X	0.0118521	0.002030899	5.835869	5.29E-08	0.007828095	0.015876026	0.007828095	0.015876026

Real price development of manganese

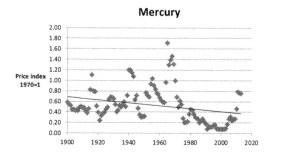

SUMMARY OUTPUT of linear regression analysis

Regression statistics	
Multiple correlation coefficient	0.27611309
R Square	0.07623844
Adjusted R Square	0.06799057
Standard error	0.31989888
Observations	114

ANOVA

	df	SS	MS	F	Significance F
Regression	1	0.945926906	0.945927	9.243408	0.002943046
Residual	112	11.46155311	0.102335		
Total	113	12.40748002			

	Coëfficiënt	Standard error	T- stat	P-value	Lowest 95%	Highest 95%	Lowest 95,0%	Highest 95,0%
Intercept	5.94690646	1.781574773	3.338006	0.001146	2.416944417	9.476868501	2.416944417	9.476868501
X	-0.0027681	0.000910464	-3.0403	0.002943	-0.00457205	-0.000964114	-0.004572049	-0.000964114

Real price development of mercury

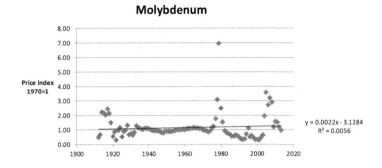

Molybdenum

y = 0.0022x - 3.1284
R² = 0.0056

SUMMARY OUTPUT of linear regression analysis

Regression statistics	
Multiple correlation coefficient	0.074992
R Square	0.005624
Adjusted R Square	-0.00432
Standard error	0.86337
Observations	102

ANOVA

	df	SS	MS	F	Significance F
Regression	1	0.421568066	0.421568	0.565553	0.453797864
Residual	100	74.54084115	0.745408		
Total	101	74.96240921			

	Coëfficiënt	Standard error	T- stat	P-value	Lowest 95%	Highest 95%	Lowest 95,0%	Highest 95,0%
Intercept	-3.1284	5.698583553	-0.54898	0.584244	-14.4342247	8.17743025	-14.4342247	8.177430252
X	0.002183	0.00290341	0.752033	0.453798	-0.00357682	0.00794374	-0.00357682	0.007943742

Real price development of molybdenum

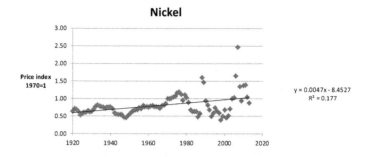

SUMMARY OUTPUT of linear regression analysis

Regression statistics	
Multiple correlation coefficient	0.4206826
R Square	0.1769738
Adjusted R Square	0.1680279
Standard error	0.2783873
Observations	94

ANOVA

	df	SS	MS	F	Significance F
Regression	1	1.533140924	1.533141	19.7826	2.42428E-05
Residual	92	7.129952386	0.077499		
Total	93	8.66309331			

	Coëfficiënt	Standard error	T- stat	P-value	Lowest 95%	Highest 95%	Lowest 95,0%	Highest 95,0%
Intercept	-8.4526511	2.081172647	-4.06148	0.000102	-12.58603958	-4.31926255	-12.58603958	-4.31926255
X	0.0047067	0.001058212	4.447763	2.42E-05	0.002604977	0.006808379	0.002604977	0.006808379

Real price development of nickel

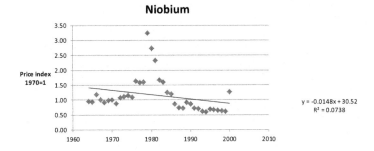

SUMMARY OUTPUT of linear regression analysis

Regression statistics	
Multiple correlation coefficient	0.27157996
R Square	0.07375568
Adjusted R Square	0.04729155
Standard error	0.57648224
Observations	37

ANOVA

	df	SS	MS	F	Significance F
Regression	1	0.92621071	0.926211	2.787006	0.103953556
Residual	35	11.6316121	0.332332		
Total	36	12.55782281			

	Coëfficiënt	Standard error	T- stat	P-value	Lowest 95%	Highest 95%	Lowest 95,0%	Highest 95,0%
Intercept	30.5197054	17.59310915	1.734753	0.091584	-5.196204918	66.23561581	-5.196204918	66.23561581
X	-0.0148184	0.008876314	-1.66943	0.103954	-0.032838285	0.003201465	-0.032838285	0.003201465

Real price development of niobium

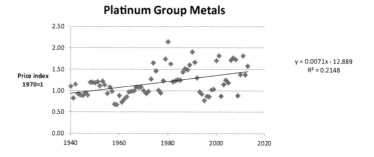

Platinum Group Metals

y = 0.0071x - 12.889
R² = 0.2148

SUMMARY OUTPUT of linear regression analysis

Regression statistics	
Multiple correlation coefficient	0.4634994
R Square	0.2148317
Adjusted R Square	0.20392658
Standard error	0.29491764
Observations	74

ANOVA

	df	SS	MS	F	Significance F
Regression	1	1.713442752	1.713443	19.70008	3.19919E-05
Residual	72	6.26230191	0.086976		
Total	73	7.975744662			

	Coëfficiënt	Standard error	T- stat	P-value	Lowest 95%	Highest 95%	Lowest 95,0%	Highest 95,0%
Intercept	-12.889446	3.172529228	-4.06283	0.000122	-19.21376718	-6.56512431	-19.2137672	-6.565124315
X	0.00712389	0.001605031	4.438478	3.2E-05	0.003924324	0.010323466	0.003924324	0.010323466

Real price development of Platinum Group Metals

REE (as Rare Earth oxide equivalent

$$y = 0.1149x - 223.76$$
$$R^2 = 0.1787$$

SUMMARY OUTPUT of linear regression analysis

Regression statistics	
Multiple correlation coefficient	0.42274696
R Square	0.178715
Adjusted R Square	0.16292105
Standard error	3.9129658
Observations	54

ANOVA

	df	SS	MS	F	Significance F
Regression	1	173.2537137	173.2537	11.31541	0.001449545
Residual	52	796.1876717	15.3113		
Total	53	969.4413854			

	Coëfficiënt	Standard error	T- stat	P-value	Lowest 95%	Highest 95%	Lowest 95,0%	Highest 95,0%
Intercept	-223.76463	67.87071932	-3.29692	0.001766	-359.9571876	-87.572063	-359.957188	-87.57206339
X	0.11492532	0.034164928	3.363839	0.00145	0.04636838	0.18348227	0.04636838	0.183482269

Real price development of Rare Earth Elements

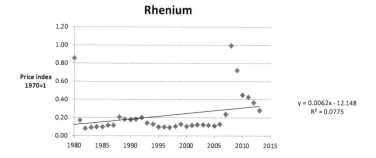

SUMMARY OUTPUT of regression analysis

Regression statistics	
Multiple correlation coefficient	0.2783747
R Square	0.0774925
Adjusted R Square	0.0486641
Standard error	0.2162216
Observations	34

ANOVA

	df	SS	MS	F	Significance F
Regression	1	0.125671814	0.125672	2.688065	0.11089828
Residual	32	1.49605702	0.046752		
Total	33	1.621728834			

	Coëfficiënt	Standard error	T- stat	P-value	Lowest 95%	Highest 95%	Lowest 95,0%	Highest 95,0%
Intercept	-12.14805	7.54629562	-1.6098	0.117263	-27.51934678	3.22325556	-27.5193468	3.223255558
X	0.006197	0.003779717	1.639532	0.110898	-0.001502065	0.013896	-0.00150207	0.013895997

Real price development of rhenium

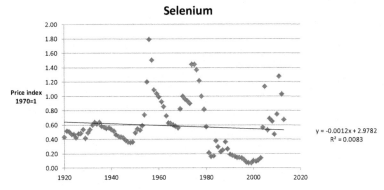

SUMMARY OUTPUT of linear regression analysis

Regression statistics	
Multiple correlation coefficient	0.0910738
R Square	0.0082944
Adjusted R Square	-0.002485
Standard error	0.3650225
Observations	94

ANOVA

	df	SS	MS	F	Significance F
Regression	1	0.10252532	0.102525	0.76947	0.382665924
Residual	92	12.25821359	0.133241		
Total	93	12.36073891			

	Coëfficiënt	Standard error	T- stat	P-value	Lowest 95%	Highest 95%	Lowest 95,0%	Highest 95,0%
Intercept	2.9781625	2.728841974	1.091365	0.277962	-2.44155319	8.397878223	-2.44155319	8.397878223
X	-0.001217	0.001387532	-0.87719	0.382666	-0.00397289	0.001538623	-0.00397289	0.001538623

Real price development of selenium

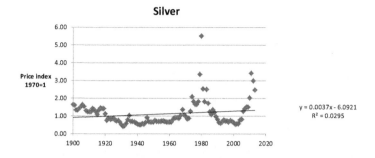

Silver

y = 0.0037x - 6.0921
R² = 0.0295

SUMMARY OUTPUT of linear regression analysis

Regression statistics	
Multiple correlation coefficient	0.17170707
R Square	0.02948332
Adjusted R Square	0.02081799
Standard error	0.70333386
Observations	114

ANOVA

	df	SS	MS	F	Significance F
Regression	1	1.683117356	1.683117	3.402447	0.067742276
Residual	112	55.40399347	0.494679		
Total	113	57.08711083			

	Coëfficiënt	Standard error	T- stat	P-value	Lowest 95%	Highest 95%	Lowest 95,0%	Highest 95,0%
Intercept	-6.0920996	3.916993525	-1.5553	0.122696	-13.85312014	1.668920997	-13.85312014	1.668920997
X	0.00369239	0.002001758	1.844572	0.067742	-0.00027384	0.007658614	-0.00027384	0.007658614

Real price development of silver

Strontium

$y = 0.0217x - 40.465$
$R^2 = 0.1817$

SUMMARY OUTPUT of linear regression analysis

Regression statistics	
Multiple correlation coefficient	0.42625053
R Square	0.18168951
Adjusted R Square	0.17106211
Standard error	1.06282884
Observations	79

ANOVA

	df	SS	MS	F	Significance F
Regression	1	19.31208373	19.31208	17.09631	8.97341E-05
Residual	77	86.97959599	1.129605		
Total	78	106.2916797			

	Coëfficiënt	Standard error	T- stat	P-value	Lowest 95%	Highest 95%	Lowest 95,0%	Highest 95,0%
Intercept	-40.465151	10.35199937	-3.90892	0.000198	-61.07861481	-19.8516863	-61.07861481	-19.8516863
X	0.021682	0.005243824	4.134769	8.97E-05	0.011240213	0.032123788	0.011240213	0.032123788

Real price development of strontium

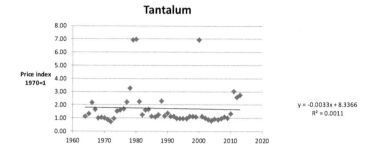

SUMMARY OUTPUT of linear regression analysis

Regression statistics	
Multiple correlation coefficient	0.03330768
R Square	0.0011094
Adjusted R Square	-0.0197008
Standard error	1.46947292
Observations	50

ANOVA

	df	SS	MS	F	Significance F
Regression	1	0.115115864	0.115116	0.05331	0.818381237
Residual	48	103.6488315	2.159351		
Total	49	103.7639474			

	Coëfficiënt	Standard error	T- stat	P-value	Lowest 95%	Highest 95%	Lowest 95,0%	Highest 95,0%
Intercept	8.33662089	28.63657742	0.291118	0.772215	-49.241077	65.91431878	-49.241077	65.91431878
X	-0.003325	0.014400716	-0.23089	0.818381	-0.032279568	0.025629592	-0.032279568	0.025629592

Real price development of tantalum

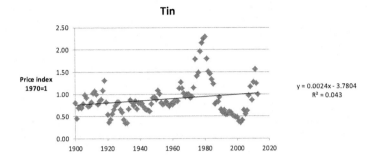

y = 0.0024x - 3.7804
R² = 0.043

SUMMARY OUTPUT of linear regression analysis

Regression statistics	
Multiple correlation coefficient	0.2072576
R Square	0.0429557
Adjusted R Square	0.0344107
Standard error	0.3738131
Observations	114

ANOVA

	df	SS	MS	F	Significance F
Regression	1	0.702450539	0.702451	5.026976	0.02692515
Residual	112	15.65045422	0.139736		
Total	113	16.35290476			

	Coëfficiënt	Standard error	T- stat	P-value	Lowest 95%	Highest 95%	Lowest 95,0%	Highest 95,0%
Intercept	-3.78037	2.081832539	-1.81589	0.072063	-7.9052545	0.34451413	-7.9052545	0.344514127
X	0.0023854	0.001063909	2.242092	0.026925	0.000277382	0.00449338	0.000277382	0.004493381

Real price development of tin

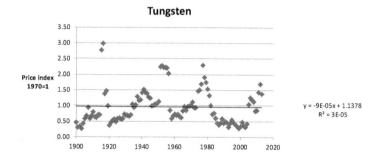

SUMMARY OUTPUT of linear regression analysis

Regression statistics	
Multiple correlation coefficient	0.00537851
R Square	2.8928E-05
Adjusted R Square	-0.00889938
Standard error	0.56568878
Observations	114

ANOVA

	df	SS	MS	F	Significance F
Regression	1	0.001036837	0.001037	0.00324	0.95470902
Residual	112	35.84042558	0.320004		
Total	113	35.84146242			

	Coëfficiënt	Standard error	T- stat	P-value	Lowest 95%	Highest 95%	Lowest 95,0%	Highest 95,0%
Intercept	1.13782224	3.150423207	0.361165	0.718657	-5.10433751	7.37998199	-5.104337505	7.379981989
X	-9.1644E-05	0.001610006	-0.05692	0.954709	-0.00328167	0.00309838	-0.003281666	0.003098377

Real price development of tungsten

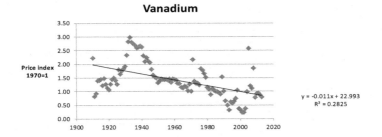

SUMMARY OUTPUT of linear regression analysis

Regression statistics	
Multiple correlation coefficient	0.5314673
R Square	0.2824575
Adjusted R Square	0.2754228
Standard error	0.5316489
Observations	104

ANOVA

	df	SS	MS	F	Significance F
Regression	1	11.34894519	11.34895	40.15186	6.446E-09
Residual	102	28.83035672	0.282651		
Total	103	40.17930191			

	Coëfficiënt	Standard error	T- stat	P-value	Lowest 95%	Highest 95%	Lowest 95,0%	Highest 95,0%
Intercept	22.992945	3.40662858	6.749472	9.22E-10	16.2359129	29.74997621	16.23591294	29.74997621
X	-0.011004	0.001736543	-6.33655	6.45E-09	-0.0144481	-0.00755927	-0.014448118	-0.007559267

Real price development of vanadium

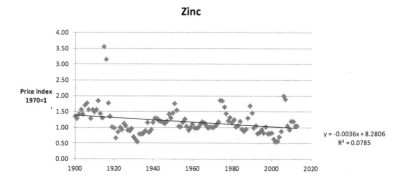

SUMMARY OUTPUT of regression analysis

Regression statistics	
Multiple correlation coefficient	0.2801936
R Square	0.0785084
Adjusted R Square	0.0702808
Standard error	0.412895
Observations	114

ANOVA

	df	SS	MS	F	Significance F
Regression	1	1.626755725	1.626756	9.54208	0.002532731
Residual	112	19.09401683	0.170482		
Total	113	20.72077256			

	Coëfficiënt	Standard error	T- stat	P-value	Lowest 95%	Highest 95%	Lowest 95,0%	Highest 95,0%
Intercept	8.2806358	2.299487057	3.60108	0.000474	3.724496948	12.8367746	3.724496948	12.83677464
X	-0.00363	0.00117514	-3.08903	0.002533	-0.00595843	-0.00130165	-0.00595843	-0.001301649

Real price development of zinc

CHAPTER 7

Thirteen scarce resources analyzed

CHAPTER 7.1

Introduction

In Chapter 3, we investigated which mineral resources are geologically scarce in view of their annual production (and increase thereof) as compared to the available amount of the resource in the Earth's crust. Eight mineral resources were characterized as scarce and five others as moderately scarce. In this chapter, we analyze, for these 13 resources, which technical measures can be taken to get their production rate to a sustainable level, while simultaneously increasing the global service level of the resource to the level of developed countries in 2020. The 13 scarce mineral resources concerned are (in alphabetical order): antimony, bismuth, boron, chromium, copper, gold, indium, molybdenum, nickel, silver, tin, tungsten, and zinc.

In Chapter 5, we stated that *"The extraction of a mineral resource is sustainable, if a world population of 10 billion can be provided with that resource for a period of at least 200/500/1000 years in such a way that during that period every country can enjoy the same service level of that resource as enjoyed by developed countries in 2020 for an affordable price."*

For each of the 13 selected resources, we discuss how to manage their availability for future generations. In this chapter, we focus on technology, and in Chapter 8 we analyze the necessary policy measures to actually realize the technical opportunities.

Governance of the world's mineral resources
ISBN 978-0-12-823886-8
https://doi.org/10.1016/B978-0-12-823886-8.00007-5

Technology can act in two ways to increase the global service level of a resource:

- By increasing the extraction of a raw material, by extracting lower graded ores from deeper in the Earth's crust and more remote places, with a lower impact on the environment and on climate change. This concerns the mining and processing technology. In the past, the development of mining and processing technology greatly increased the amount of mineral resources available for humanity. Technology development has enabled the annual production of practically every mineral resource to increase dramatically over the last century without much of an increase, or even with a decrease, in the real price (see Chapter 6). For the calculation of the ultimately available amount of a resource, we have assumed mining to a depth of 3 km in the Earth's crust, which will be quite challenging. Chemically, in a laboratory environment, a material can be almost completely extracted or isolated from every possible matrix down to a very low concentration. However, this will not happen in mining practice, due to financial and environmental constraints. This subject is discussed in Chapter 3, in which the penultimate sentence of Section 3.5 reads: "The ultimate extractability of a mineral resource is limited to a larger extent by economic, environmental, climatic, and social restrictions than by the mere availability of the resource."
- By increasing the efficient use of mineral resources, so that the same level of services can be maintained with a smaller quantity of primary resources. This is related to increasing the recovery rate of the resource at the mining stage, substitution of the resource with less scarce resources (without diminution of the service level), a higher material efficiency, a lower in-use dissipation rate, less downcycling, and a higher end-of-life recycling rate.

In this introductory section, we discuss the general aspects of each of the types of measures mentioned above.

An important and common issue for each of the 13 resources considered is the size of and increase in their in-use stock. Currently, the size of the global in-use stock of all discussed resources is still growing; however, this growth will probably decrease in the future and the in-use stock will stabilize. As long as a resource is part of the in-use stock, it is not available for recycling.

7.1.1 Recovery efficiency improvement

The recovery efficiency is the ratio of the actual production of a resource by a mine to the total amount of the resource contained in the mined ore. The recovery efficiency differs per resource and per mine. The recovery efficiency depends on whether a resource is the main product or a by-product of the mine, where the recovery efficiency of a by-product is usually lower than the recovery efficiency of the main product. The recovery efficiency also depends on the ore grade: the lower the ore grade, the more difficult it is to achieve a high recovery efficiency. Another factor determining the recovery efficiency of a resource is the matrix in which the resource is included (oxide, sulfide, etc.). Furthermore, the market price and technological developments also impact on the recovery efficiency. Technically, it is always possible to improve the recovery efficiency, but at a price. In practice, the recovery efficiency is a balance between the extra energy costs and the additional profit. For each of the 13 elements discussed in this chapter, we explore whether the current recovery efficiency of the element can be improved.

7.1.2 Substitution

Four main factors determine the potential for substitution of a material:
1. *The performance of the substitute compared with the performance of the original.* An important condition for the adequate applicability of a substitute is that the services provided by the original product are maintained. For some uses, the performance of the substitute may matter less than for others. A 100% equal performance compared with the original is not always necessary, as each specific application has its own requirements.
2. *The environment, health, and safety (EHS) properties of the substitute compared with the original.* The EHS properties of the substitute and the original should encompass all aspects, from cradle to grave, and in all stages, from extraction to end-of-life.
3. *The costs of the substitute compared with the original.* The costs of a substitute will depend on its availability, accessibility, and technology. While the effect of prices may be a relative factor, it can be a decisive element for substitutability in practice.
4. *The geological availability of the substitute compared with the geological availability of the original.* The aim of our investigation of the possible extraction

Table 7.1.1 Types of substitution (Ziemann and Schebek, 2010).

Substitution type	Explanation	Example
Material substitution	Material A is replaced by material B	Zinc in roof gutters is replaced by plastic
Technological substitution	Reduction of material consumption by technological progress	Reduction of nickel content in certain types of stainless steel
Functional substitution	Product A is replaced by product B or service C with the same function	Silver as a photochemical in photography is replaced by digital photography
Quality substitution	Product A is replaced by product A′ with a lower, but still sufficient quality	Copper is replaced by aluminum in high-voltage transport of electricity
Nonmaterial substitution	A product is replaced by a service with the same function	Travel (by car or airplane) is replaced by videoconferencing

reduction of a material is to conserve scarce materials for future generations. Therefore, substitutes should not be scarcer than the original. Five types of substitution are presented in Table 7.1.1.

Note that an application can be so specific that it is very difficult or impossible to substitute the material, for example the application of boron as a micronutrient in fruit and seed production. In such an application, efficient use is the only option to reduce primary boron use. Substitution is not applicable in such a case and recycling can only take place to a limited extent.

7.1.3 Material efficiency improvement

Material efficiency (or resource productivity) reflects the quantity of services that can be provided by a given amount of a material. For example, lightweight packaging may result in reduced material use to package the same product. Table 7.1.2 provides a general overview of the possibilities for material efficiency increase.

Recycling from end-of-life products is addressed separately in Section 7.1.6.

Product designers do not usually focus primarily on resource conservation through design for recycling, reuse, maintenance, repair, and waste minimization in general (Ordoñez and Rahe, 2013). Hence, potential for material efficiency may exist in many products and applications. In general, lightweight design, product lifetime extension, and more intensive product

Table 7.1.2 Overview of possibilities for material efficiency (ME).

ME in production process	Prevention of material loss
	Process optimization
	ME in resources purchase
	Recycling of production waste
ME in products	Lightweight products
	Design for recycling
	Design for reuse and multipurpose use
	Design for longer use, maintenance, repair, remanufacturing
ME during consumption	Reduction of in-use dissipation
	Longer use
	Reuse
	Shared use

use are the most effective means to increase material efficiency (Allwood, 2013). Product service systems (PSS), such as product lease instead of product ownership, may have some environmental gains, but generally do not drastically improve material efficiency (Tukker, 2004). The most that can be expected from PSS is the promise of a functional result; for example, international travel can be replaced by videoconferencing. In this case, the functional result is an adequate meeting with effective communication.

How can the potential effect of material efficiency be quantified? The current literature on material efficiency improvement mainly provides examples for specific materials or products, but no meta-studies are known to us that provide a general overview of potentials. Based on the variety found in the literature, we provide our own expert estimate of the order of magnitude of the improvement potential of various types of measures in Table 7.1.3. The material efficiency potential indicates the reduction percentage of material use for providing the same quantity of services compared with the original material use. Considering a specific material, a

Table 7.1.3 Estimated material efficiency improvement potential range (own expert judgment based on literature).

	Estimated material efficiency improvement potential range
Material efficiency potential in production processes	1%—10%
Material efficiency potential in products	10%—50%
Material efficiency potential during consumption (excluding recycling of end-of-life products)	10%—50%

material efficiency improvement potential of 25% means that only 75% of the original quantity of that material would be needed to provide the same services.

Although it differs for particular materials and products, material efficiency has a large potential. Table 7.1.3 shows the wide spread found for a variety of applications in the literature. However, this potential will not always be easily realized in practice. Therefore, if we have no data for specific applications of the materials we study, we assume a conservative default material efficiency potential of 10%, excluding the impact of dissipation reduction and increased recycling of materials from end-of-life products. Material efficiency is an important option if substitution and recycling are not sufficient to reduce the use of a material to a sustainable level. More research may then be necessary for specific materials and applications.

7.1.4 Reduction of in-use dissipation

In-use dissipation concerns a variety of applications; for example, the use of a raw material in food, fertilizers, detergents, paints, lubricating oil, or for anticorrosion. Some mineral resources have substantial in-use dissipation, others very little or none. In some cases, it is possible to substitute the material for another material—or to use less of it, for example in paint, lubricating oil, or for anticorrosion. Sometimes this is not possible, such as if an element is applied in food or as a micronutrient.

In the following sections, we explore for each element how much of it is used in a dissipative way and consider the possibilities to reduce this.

7.1.5 Decrease in downcycling

Downcycling is the phenomenon that an element is recycled together with another element and loses its specific functions, while being permanently included in the flow of the other element. In reality, therefore, downcycling is a form of disposal (other than in landfills or incinerators), as it will be very difficult to reuse the element once it has been downcycled. Take as an example stainless steel, which contains a variety of metals, the relative amounts of which depend on the type of stainless steel. The extent by which the different types of steel are separated from each other in the end-of-life stage depends on whether or not this is economic. Hence, it depends on the market price of metal A, and of its concentration in steel B, whether and to what extent it is financially attractive to keep that steel B separated from other types of steel in order to recycle steel B in its original application together with metal A, rather than recycling steel B together with a number of other types of steel, in which metal A has no function.

7.1.6 Increase in recycling

The recycling potential of a specific material from a specific end-of-life product depends on the following factors (Graedel and Erdmann, 2012; Worrell and Reuter, 2014):

- *Concentration.* The higher the concentration, the higher the recycling potential. As a general rule, the concentration should be at least as high as the minimally profitable concentration in virgin ore.
- *Material composition.* Alloys, composites, and laminates of various materials make it difficult to isolate the mono-materials, which may limit (or even inhibit) recycling or result in downcycling.
- *Product composition.* The more complex the composition or assemblage of the product, the lower the recycling potential (e.g., indium in a mobile phone).
- *Dispersed use.* Dispersed use of materials inhibits the (economic) recoverability of materials (e.g., in paints).
- *Contamination.* The more a product (or waste product) is contaminated, the lower the recycling potential.

The UNEP International Resource Panel (2011) made an order of magnitude expert estimate of the end-of-life recycling rate of 60 metals on a global scale between 2000 and 2005. The UNEP experts chose five ranges of recycling rates (see Table 7.1.4).

Table 7.1.4 shows large differences between the considered elements.

In practice, recycling rates higher than 85%–90% are not feasible. One reason for this is that there will be losses at every step of the recycling chain (collection, sorting, processing). Moreover, it can be costly to recycle a material from a product because of its low concentration, or a complex product composition. To enable a high recycling rate of a specific material, an infrastructure needs to be built that includes many stakeholders (e.g., manufacturers, households, waste management companies, recycling traders and processors, local governments), and that encompasses the generation of new (international) markets for recycled materials. Building such an infrastructure is capital- and time-intensive, and is affected by changes in product and material compositions. However, recycling is a core factor for

Table 7.1.4 Order of magnitude estimate of the end-of-life recycling rates of 60 metals in the period 2000—05 (UNEP, 2011).

<1%		1%—10%	10%—25%	25%—50%	>50%	
As	Sc	Hg	Cd	Ir	Ag	Ni
B	Se	Sb	Ru	Mg	Al	Pb
Ba	Sr	W	W	Mo	Au	Pd
Be	Ta				Co	Pt
Bi	Te				Cr	Re
Ga	Tl				Cu	Rh
Ge	V				Fe	Sn
Hf	Y				Mn	Ti
In	Zr				Nb	Zn
Li	Lanthanides					
Os						

primary resource use reduction and material efficiency, and will always be a part of the portfolio of measures to sufficiently reduce primary resource use.

We analyze the current end-of-life recycling rates of 13 selected scarce raw materials in detail, and explore to what extent these would need to be increased to achieve a sustainable use of the virgin raw material, and what the limitations are.

7.1.7 In-use stock

The use of a product in a country often follows an S-curve, for instance for public roads. It may also follow a bell-shaped curve, such as for canals and monochrome TVs (see Fig. 7.1.1A and D).

The single curves have an S-form because, from a certain GDP/capita, nearly every household or adult person in developed countries owns certain applications, such as a house, a washing machine, a freezer, a mobile phone, and a car. Richer persons and households own more expensive houses, cars, and household appliances, but not necessarily a larger number. This is why an increasing GDP/capita does not lead to ever-growing resource consumption, but why, from a certain GDP/capita level, we observe a decoupling of GDP growth and resource use. This is not yet the case in developing countries, where material consumption has not reached the saturation level. The S-curves are therefore closest to the starting point in the poorest countries. Globally, material consumption is at a level of about 25% of the saturation level in developed countries, but in the poorest countries it may be below 10% of the level in developed countries (see also Section 5.4.3).

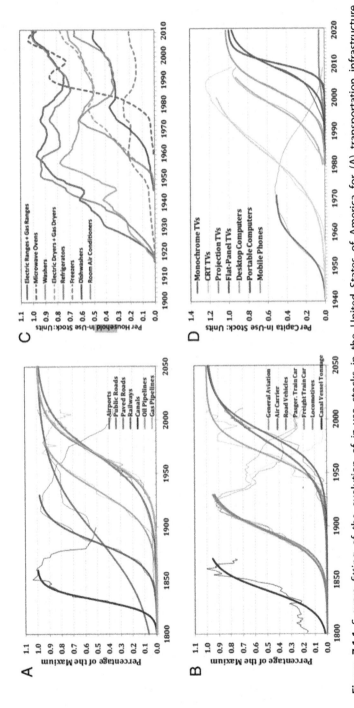

Figure 7.1.1 S-curve fitting of the evolution of in-use stocks in the United States of America for (A) transportation infrastructure, (B) transportation facilities, (C) home appliances, and (D) electronic products. The *thin lines* in (A), (B), and (D), as well as *all lines* in (C) show empirical data (Chen and Graedel, 2015).

Certain resource applications disappear from the market, such as the application of silver in photography, mercury in thermometers, and lead in the water supply infrastructure. Meanwhile, new products continue to appear on the market, such as wind turbines, PV cells, and electric cars. That means that both the consumption and in-use stock of a mineral resource in a country consist of a superposition of S-curves and bell-shaped curves. Note that this superposition will not necessarily also have the form of an S-curve. See also Fig. 5.2, in which the per capita consumption of a number of commodities is presented dependent on the GDP per capita. Fig. 5.2 clarifies that, at low GDPs per capita, mineral consumption per capita increases with the increase in GDP per capita even not in the richest countries. From a certain level of GDP/capita, decoupling may take place, the extent of which differs per element. The per capita consumption of some materials, for example platinum, rare earth elements, gallium, cobalt, lithium, and indium have not yet decoupled from GDP increase per capita. Decoupling depends on whether and to what extent an element obtains new technology applications over time.

The in-use stock of a mineral resource, both in a particular country and globally, depends on the annual consumption of that mineral resource and the average lifetime of the mix of applications of that resource. Note that the lifetimes of different applications of the same mineral resource differ (see Fig. 7.1.2).

Average lifetime estimates of 15 resources are presented in Table 7.1.5. Comparable figures can be found in other publications.

The total in-use stock can be calculated by summing the annual difference of inflow minus outflow minus dissipation according to the following formula:

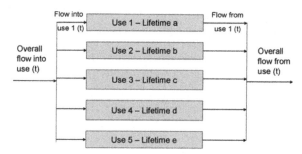

Figure 7.1.2 Different applications of the same mineral resource have different lifetimes. *(Derived from UNEP International Resource Panel, 2010. Metal Stocks in Society, Scientific Synthesis, 2010.)*

Table 7.1.5 Average in-use lifetimes of mineral resources.

Average in-use lifetime of mineral resources (rounded to 5 years)			
Aluminum	25	Lead	45
Antimony	10	Molybdenum	20
Cadmium	5	Nickel	25
Chromium	25	Platinum	25
Cobalt	15	Silver	25
Copper	30	Tin	15
Gold	Very long	Zinc	20
Iron	30		

Derived from Gerst, M.D., Graedel, T.E., 2008. In-use stocks of metals: status and implications. Environ. Sci. Technol. 42 (19), 7038–7045.

$$S_t = \sum\nolimits_{t_0}^{t} \left(\text{Inflow}_t - \text{Outflow}_t - \text{Dissipation}_t \right) + S_0$$

where S_t = the in-use-stock in year t, and S_0 = the in-use-stock in year 0.

As long as the global annual consumption of a resource is increasing, the global in-use stock of that resource will also increase. The size of the in-use stock of a resource per capita in a country depends on the consumption per capita of that resource. A four times higher consumption leads to a four times higher in-use stock, assuming that the mix of applications of the resource does not lead to a change in average lifetime. Higher lifetimes result in a larger in-use stock. If we assume additionally that, ultimately, the global consumption of a resource will evolve according to an S-curve in time, then the in-use stock of that resource will also evolve as an S-curve (see Fig. 7.1.3).

The global in-use stock of a resource will continue to increase until the consumption of that resource has reached the same level everywhere. We call this the resource consumption saturation point. Then, an average lifetime period after the resource consumption saturation point has been reached, the annual output of end-of-life products will also stabilize. Table 7.1.6 gives the proportion of the annual consumption representing the growth of the in-use stock. For example, in 2000, for the given assumptions, the net growth of the in-use stock of a resource with an average lifetime of 15 years will be 51% of the consumption of that resource. This means that the relative amount of the resource in end-of-life products in that year is 49% of the consumption minus the amount lost through in-use dissipation.

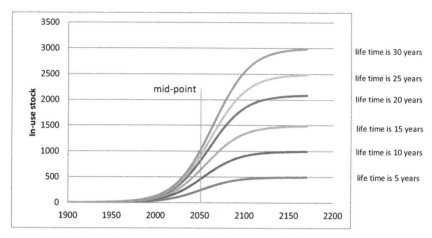

Figure 7.1.3 Schematic presentation of the in-use stock of a resource over time for different lifetimes. Annual consumption is 100. In all cases, we have assumed that the consumption level develops as an S-curve, being near zero in 1900 and reaching saturation in 2200. The assumed midpoint is in 2050 and the steepness coefficient of the consumption S-curve is 0.05. There is no in-use dissipation.

Table 7.1.6 Proportion of consumption to in-use stock at different product lifetimes at different points in time.

Average in-use lifetime	Build-up of in-use stock as percentage of annual consumption					
	1900	1950	2000	2050	2100	2150
5 years	100%	22%	21%	12%	2%	0%
10 years	100%	39%	37%	24%	5%	0%
15 years	100%	53%	51%	36%	8%	1%
20 years	100%	65%	63%	48%	12%	1%
25 years	100%	71%	70%	55%	16%	2%
30 years	100%	78%	76%	64%	21%	2%

Assumptions are: start of consumption around 1900, in-use stock development according to sigmoid function with midpoint in 2050, and steepness coefficient of consumption S-curve is 0.05.

Once the consumption of a resource has reached its final level, the ultimate in-use stock will be the product of annual consumption and average lifetime, in formula:

$$S = C \times L,$$

where S = in-use-stock, C = annual consumption, L = average lifetime.

Table 7.1.7 Flows from in-use stocks of seven metals.

Metal	Year	Outflow from in-use stock as % of inflow	Source
Chromium	2000	22%	Johnson et al. (2006)
Iron	2000	36%	Wang et al. (2007)
Nickel	2000	41%	Reck et al. (2008)
Silver	1997	52%	Johnson et al. (2005)
Lead	2000	89%	Mao et al. (2008)
Copper	1994	33%	Graedel et al. (2004)
Zinc	2010	65%	Meylan and Reck (2016)

Data on the build-up of in-use stocks of seven metals are presented in Table 7.1.7.

In Chapters 7.2–7.14, we analyze the build-up of in-use stock for each of the 13 mineral resources considered.

7.1.8 Discussion

Every resource requires a specific mix of measures to enable a sustainable production rate. Some elements are not or hardly substitutable in their applications according to current knowledge and will require a focus on maximizing material efficiency and recycling. Other elements are relatively easily substitutable for other elements or other products. In that case, recycling may still be important, but is less crucial. Every element requires a specific approach, and there is no panacea for all elements.

Generally, the costs of any measure to reduce the use of primary materials (e.g., recycling) must be lower than the costs of primary production of the raw material to ensure that the reduction measure is financially viable. Governments may promote substitution, material efficiency improvement, and recycling by an array of policy measures, such as subsidizing secondary (recycled) resources and/or taxing geologically scarce primary resources. Such supporting policy measures will be very important in promoting the sustainable use of primary resources. Without adequate policy measures, technological possibilities will not be used, or at least not fully used, because they are not financially viable.

In the following sections of this chapter, we discuss the specific approaches for 13 geologically scarce elements. In that framework, we systematically discuss the future potential of the technical approaches (increase in the recovery efficiency at the mining stage, substitution, material efficiency improvement, dissipation reduction, and increased recycling). The objective is to investigate whether it is possible to achieve a sustainable production level, while simultaneously increasing the service levels of the considered elements globally to the service level of developed countries in 2020.

For each selected resource, our approach consists of five main consecutive steps:

1. Investigate the possibilities of increasing the recovery efficiency at the mining and processing stage
2. Investigate the possibilities of substitution
3. Find out *whether* and to what extent the remaining quantity of the material requires further reduction, first through material efficiency
4. Investigate the possibilities of dissipation reduction
5. Investigate further recycling measures on top of what is already being done

The intention is not to propose a blueprint of measures to be taken for a certain material, but to investigate whether and how a sustainable production rate of that material can be combined with a global service level that is equal to the service level of that material in developed countries in 2020.

References

Allwood, J.M., 2013. Transitions to material efficiency in the UK steel economy. Phil. Trans. R. Soc. A 371, 20110577.

Chen, W., Graedel, T.E., 2015. In-use product stocks link manufactured capital to natural capital. Proc. Natl. Acad. Sci. U.S.A. 112 (20), 6265–6270.

Gerst, M.D., Graedel, T.E., 2008. In-use stocks of metals: status and implications. Environ. Sci. Technol. 42 (19), 7038–7045.

Graedel, T.E., Van Beers, D., Bertram, M., Fuse, K., Gordon, R.B., Gritsinin, A., Kapur, A., Klee, R.J., Lifset, R.J., Memon, L., Rechberger, H., Spatari, S., Vexler, D., 2004. Multilevel cycle of anthropogenic copper. Environ. Sci. Technol. 38, 1242–1252.

Graedel, T.E., Erdmann, L., April 2012. Will metal scarcity impede routine industrial use? Mater. Res. Soc. Bull. 37, 325–331.

Johnson, J., Jirikowic, J., Bertram, M., Van Beers, D., Gordon, R.B., Henderson, K., Klee, R.J., Lanzano, T., Lifset, R., Oetjen, L., Graedel, T.E., 2005. Contemporary in-use silver cycle: a multilevel analysis. Environ. Sci. Technol. 39, 4655–4665.

Johnson, J., Schewel, L., Graedel, T.E., 2006. The contemporary in-use chromium cycle, Environ. Sci. Technol. 40, 7060–7069.

Mao, J.S., Dong, J., Graedel, T.E., 2008. The multilevel cycle of in-use lead, results and discussion. Resour. Conserv. Recycl. 52, 1050–1057.

Meylan, G., Reck, B.K., 2016. The anthropogenic cycle of zinc: status quo and perspectives. Resour. Conserv. Recycl. https://doi.org/10.1016/j.resconrec.2016.01.006.

Ordoñez, I., Rahe, U., 2013. Collaboration between design and waste management: can it help close the material loop? Resour. Conserv. Recycl. 72, 108–117.

Reck, B.K., Müller, D.B., Rostowski, K., Graedel, T.E., 2008. In-use nickel cycle: insights into use trade and recycling. Environ. Sci. Technol. 42, 3394–3400.

Tukker, A., 2004. Eight types of product service system: eight ways to sustainability? Experiences from SusProNet. Bus. Strategy Environ. 13, 246–260.

UNEP International Resource Panel, 2010. Metal Stocks in Society, Scientific Synthesis.

UNEP International Resource Panel, 2011. Recycling Rates of Metals, a Status Report.

Wang, T., Müller, D.B., Graedel, T.E., 2007. Forging the in-use iron cycle. Environ. Sci. Technol. 41, 5120–5129.

Worrell, E., Reuter, M. (Eds.), 2014. Handbook of Recycling – State-of-The-Art for Practitioners, Analysts and Scientists, first ed. Elsevier, Amsterdam.

Ziemann, S., Schebek, S., 2010. Substitution knapper Metalle – ein Ausweg aus der Rohstoffknapheit? Chem. Ing. Tech. 82 (11), 1965–1975.

CHAPTER 7.2

Antimony[1]

7.2.1 Introduction

Antimony is an element that is used in many applications, for example as a component in flame-retardants, as a catalyst to produce polyester, in lead-acid batteries, and in lead alloys. Comparing the ultimately available resources of antimony with its current pace of extraction, antimony is one of the scarcest mineral resources, and the ultimately available global resources may be exhausted within a century (see Table 3.2). This does not mean that antimony will have disappeared from the Earth's crust within a century, but that the relatively easily extractable antimony-containing ores will be exhausted.

Antimony is also part of the EU's list of critical raw materials, because the EU considers antimony to be essential for the European economy, and because more than 70% is produced in a single country (China) and more than 90% in just three countries (China, Russia, and Tajikistan).

In this chapter, we investigate whether and how the sustainable extraction of antimony can be achieved, while simultaneously increasing and maintaining antimony's service level globally at a level that is equal to its service level in developed countries in 2020.

7.2.2 Properties and applications

Antimony has a density of 6.68 g/cm^3, a melting point of 630.5°C, and a boiling point of 1587°C. It is a poor conductor of heat and electricity, and is used to increase the strength, hardness, and corrosion resistance of alloys (US Geological Survey, 2017b).

The global end uses of antimony in 2010 are presented in Table 7.2.1. Note that this concerns the consumption of antimony, including primary and secondary production. There are two main types of application of antimony: nonmetallurgical applications and metallurgical applications.

---

[1] This section is based on the publication of Henckens, M.L.C.M., Driessen, P.P.J., Worrell, E., 2016. How can we adapt to geological scarcity of antimony? Investigation of antimony's substitutability and of other measures to achieve a sustainable use. Resour. Conserv. Recycling 108, 54–62.

Table 7.2.1 Estimated global consumption of antimony by end use in 2010 (tons Sb) (Roskill Consulting Group, 2011).

	Tons Sb	%	Main use
Nonmetallurgical applications			
Flame-retardants	103,500	51.9%	Plastics
PET catalyst	11,400	5.7%	PET
Heat stabilizer	2600	1.3%	PVC
Glass	1700	0.9%	Cathode ray tubes and solar glass
Ceramics	2500	1.3%	Construction
Other	1840	0.9%	Various
Subtotal	123,540	61.9%	
Metallurgical applications			
Lead–acid batteries	53,000	26.6%	Automotive
Lead alloys	23,000	11.5%	Construction
Subtotal	76,000	38.1%	
Total	199,540	100%	

7.2.2.1 Nonmetallurgical applications

7.2.2.1.1 Antimony in flame-retardants

About 50% of worldwide antimony use is in flame-retardants. Antimony-containing flame-retardants are mainly applied in plastics. In its application in flame-retardants, antimony trioxide (Sb_2O_3) is almost always used in combination with halogen-bearing compounds, usually in a proportion of about 25%—35% (Butterman and Carlin, 2004). According to Alaee et al. (2003), the brominated flame-retardant content (of which in the order of 30% is antimony trioxide) is about 10%—15% depending on the application, which means that these plastics contain 3%—5% antimony trioxide.

Sb_2O_3 does not have flame-retardant properties in itself; however, it improves the flame-retardant properties of the used halogen hydrocarbons. It can therefore be described as a synergist in this application.

The plastics with antimony trioxide-containing flame-retardants are mainly commodity plastics such as flexible PVC, polyethylene, polypropylene, polybutylene, polystyrene, PET, ABS, and polyurethanes. In PVC, antimony trioxide can be used as such because PVC is a halogenated compound itself.

Table 7.2.2 Flame-retardant applications.

Application	Proportional use of flame-retardants		
	Worldwide (Keyser, 2009)	EU-2006 (Cusack, 2007)	Denmark (Lassen et al. (1999)
Electric and electronic equipment	39%	56%	70%
Building and construction	34%	31%	15%
Textiles, adhesives, and coatings	15%	7%	1%
Transportation	12%	6%	12%
Total	100%	100%	98%

There are four main application areas for flame-retardants: (1) electric and electronic equipment (e.g., televisions, computers, refrigerators), (2) building and construction (e.g., wiring, insulation materials, paints), (3) textile and coatings (e.g., foam upholstery carpets, foam mattresses, curtains), and (4) transportation (e.g., seat covers and fillings, insulation panels, carpets, cable wiring). Table 7.2.2 clearly illustrates that the majority of flame-retardants (with or without antimony trioxide) are applied in electric and electronic equipment. More than 50% of the brominated flame-retardants used in Europe, including antimony trioxide, are used in electric and electronic equipment (EFRA, 2013).

According to Camino (2008), 30% of the plastics in electric and electronic equipment in the EU in 2008 contained flame-retardants. Of these flame-retarded plastics, 41% contained halogenated flame-retardants (with antimony). According to Cusack (2007), of the halogenated flame-retardants used in electric and electronic equipment, 59% are used in casings, 30% in printed circuit boards, 9% in connectors and relays, and 2% in wires and cables.

According to Flame-retardants On Line (2013), total flame-retardant consumption was about two million tons in 2012 with a global annual growth of about 4%–5%.

7.2.2.1.2 Antimony as PET catalyst
Antimony trioxide and antimony triacetate are used as catalysts for the polycondensation of polyethylene terephthalate (PET). PET is mainly used for producing synthetic fibers (>60%) used in textiles (described as polyester

in that application), and for producing plastic bottles (about 30%) that are mainly used as containers for beverages, food, and other liquids (Ji, 2013). The third major application of PET is in plastic films, which are mainly used for packaging purposes. PET is one of the polymers with the highest production volume in the world, along with polyethylene (PE) and polypropylene (PP). It is produced by the polymerization reaction between terephthalic acid and ethylene glycol. Since the start of PET production, antimony compounds (such as antimony trioxide, antimony acetate, or antimony glycolate) have been used as the polymerization catalyst and they are still used for > 90% of global PET manufacturing (Thiele, 2004), despite certain disadvantages. The resulting antimony content in PET is between 150 and 300 ppm (Thiele, 2004).

7.2.2.1.3 Antimony as heat stabilizer

Many common plastics are susceptible to the degrading effects of heat and ultraviolet light (UV). Heat is relevant both during the production and processing (for instance, extrusion) of the plastic and during use. The effects of heat and UV may include oxidation, chain scission, uncontrolled recombination, and cross-linking reactions. Plastics can be protected from these effects by adding stabilizers. The majority of heat and UV stabilizers are used in polyvinylchloride (PVC); according to Butterman and Carlin (2004), in the United States this may be 85%−90%, and Markarian (2007) provides comparable figures. PVC is mainly used for construction materials used in residential housing and industrial buildings, such as window profiles, roofing membranes, wall and floor coverings, sewer and clean water pipes, cable insulation, and conduit ducts.

Antimony, in the form of antimony mercaptide, is used as a heat stabilizer (mostly) in PVC, although it competes with some other stabilizer families. Antimony mercaptides are sensitive to photodecomposition and are therefore mainly applied in underground applications of PVC.

7.2.2.1.4 Antimony in ceramics

In ceramics, antimony is applied as pigment:
- Antimony trioxide + tin oxide for gray to blue color
- Antimony trioxide + titanium dioxide for yellow color
- Antimony trioxide or sodium antimonite together with titanium dioxide as an opacifier

(Butterman and Carlin, 2004).

7.2.2.1.5 Antimony in glass

Antimony's most important use in glass (as antimony oxide and sodium antimonite) is that of a fining agent/decolorant/antisolorant in high-quality transparent glass, for instance in cathode ray tubes and solar glass. As a fining agent, the used antimony compound is added to the high-temperature glass melt. This causes large gas bubbles, effectively stirring and cleaning the viscous glass melt to eliminate gas seeds in the final product. The fining agent releases gases that encapsulate smaller seeds in the glass melt. There are several fining agents in use, with the type of fining agent depending on the specific glass, the melt temperature, and the viscosity of the melt. At present, antimony (III) compounds in combination with a strong oxidizing agent, such as sodium or potassium nitrate, are used as fining agents in glass melts for PV modules, in combination with their function as a decolorant.

An antisolorant prevents glass from colorizing under the influence of UV light.

Antimony sulfide is used as a glass colorant for amber, green, or red glass (Butterman and Carlin, 2004).

7.2.2.1.6 Antimony in other nonmetallurgical uses

Antimony compounds are used in a large variety of nonmetallurgical applications, for example as an antisolorant (color stabilizer) in pigments (for instance, in striping applied to road pavements), for coloring rubber black, in antistatic plastic coatings of electronic equipment, as an additive to some lubricants to increase their chemical stability, in fluorescent lamps, as a vulcanization agent for the production of red rubber, as a primer for ammunition, as a lubricant of friction material in automotive brake and clutch linings, in brake pads of cars, and in fireworks and matches (Butterman and Carlin, 2004).

7.2.2.2 Metallurgical applications

7.2.2.2.1 Antimony in lead-acid batteries

Antimony is used to strengthen the lead plates in lead-acid batteries, which would otherwise be too soft. However, antimony increases the water electrolysis and self-discharge of the battery and, hence, water loss. This phenomenon necessitates venting and the periodic addition of water to the electrolyte. Therefore, battery producers have focused on decreasing the antimony content of lead plates as much as possible in combination with higher liquid quantities and a closed system to produce "low-maintenance" and "maintenance-free" batteries. Maintenance-free batteries are antimony-free and do not need periodic water refilling. Over the last few decades, antimony has been replaced by calcium in lead-acid batteries. Calcium-calcium lead-acid batteries have a much lower self-discharge. As a result of these efforts, the antimony content of automotive batteries is being reduced. However, the required technology for calcium-calcium lead-acid batteries is more complicated and therefore these batteries are still more expensive than the traditional antimony-containing lead-acid batteries. Nevertheless, experts expect that antimony use in car batteries will eventually be completely eliminated (Roskill Consulting Group, 2011). Despite this development toward antimony-free batteries, global antimony use in car batteries is currently still increasing because of the strong growth in the number of cars in countries such as China, India, and Brazil. According to Roskill Consulting Group Limited (2011), global antimony use in lead-acid batteries increased from 40,000 tons in 2000 to 53,000 tons in 2010. Antimony-containing lead-acid batteries remain in use for other applications because of their longer lifetime compared to calcium-calcium lead-acid batteries. Such applications include load-leveling batteries, emergency power supply, traction batteries for forklifts, and baggage carts. However, because of the inherent maintenance requirements of antimony-containing batteries, they are also being gradually replaced in these applications by antimony-free batteries. Car batteries previously contained up to 7% antimony, but conventional batteries with antimonial lead grids contain only 1.6% (Butterman and Carlin, 2004). According to the same source, nonautomotive batteries may still contain a substantial amount (up to 11%) of antimony.

7.2.2.2.2 Antimony in alloys

Antimony itself is hard and brittle, but as an alloying agent it hardens and strengthens metals.

The use of antimony-containing lead alloys is increasing because of lead use in construction in fast-growing economies in Asia, Latin America, and Africa (Roskill Consulting Group (2011). Lead sheets used on roofs and for gutters contain up to 6% antimony in order to harden and strengthen the lead. The weight percent of antimony in a number of alloys is presented in Table 7.2.3.

7.2.2.3 Future developments

A growth area is the use of antimony in glass panels for photovoltaic solar cells (Roskill Consulting Group, 2011). However, in the longer term, the

Table 7.2.3 Principal antimony alloys (Butterman and Carlin, 2004).

Antimony-containing alloys	Antimony (weight%)
Battery grids	1.6
Bearing metal	6–16
Britannia metal	2–10
Sheets, pipes, pumps, valves for the chemical industry	4–15
Leaded roofing and gutters	6
Sheet and pipe	2–6
Bullets	0.5–1.5
Fragmentation ammunition	12–15
Collapsible tubes	1–4
Electrical cable covering	0.5–1.0
Pewter	1–8
Solder (filler)	2–5
Solder (plumber)	0–2
Specialty castings	11
Type metal	4–23

global use of antimony is expected to decline due to its frequent use together with halogenated hydrocarbons or lead. Worldwide, the use of both halogenated hydrocarbons and lead is under scrutiny for environmental and health reasons. The problem is that toxic gases may be released by halogenated flame-retardants in the case of fire, and also the ecotoxic properties of these substances as such are a problem. Mainly for these reasons, the application of antimony in flame-retardants is lower in Europe and the United States than in other parts of the world. This illustrates that environmental regulations are important determinants for the use of specific flame-retardants. Polybrominated biphenyls and polybrominated diphenyl ethers have been banned from use in electric and electronic equipment by a European Union Directive (June 2011), and the application of halogenated hydrocarbons as flame-retardants in building cables may be further affected by the European Construction Product Directive (March 2011) requesting testing of acidity, toxicity, and smoke properties. The 2012 EU Directive on waste electrical and electronic equipment (WEEE) obliges Member States to adopt appropriate measures to minimize the disposal of WEEE in the form of unsorted municipal waste, with minimum recycling targets for various types of WEEE varying between 70% and 85%. Within this framework, plastics that contain brominated flame-retardants have to be removed from the separately collected WEEE and disposed of or recovered in compliance with the EU Waste Directive (2008). Annex VII of the WEEE Directive prescribes selective treatment for plastic containing brominated flame-retardants. This means that these substances are to be removed from collected WEEE and treated separately.

Apart from the environmental concerns with regard to the use of halogenated hydrocarbons in flame-retardants, the price also plays a role. According to US Geological Survey (2015, p. 19), the flame-retardant industry "began substituting for antimony trioxide in 2011 following a significant increase in price of antimony trioxide."

The use of antimony compounds as a catalyst for the polycondensation of PET is also under discussion because of the migration of small quantities of antimony to food and beverages in PET bottles and PET containers. Furthermore, the use of antimony in lead-acid batteries will probably continue to decline as explained in Section 7.2.2.2.1.

7.2.3 Production and resources

7.2.3.1 Production

Fig. 7.2.1 shows the increase in global antimony production since 1900. In recent years, the annual amount of extracted antimony has shown relatively large variations (Fig. 7.2.1).

The production increase of antimony between 1980 and 2015 was 2.16% annually (see Table 3.2.)

China is the main antimony supplier, although its contribution is declining (see Fig. 7.2.2 and Table 7.2.4).

7.2.3.2 Resources

The average concentration of antimony in the Earth's continental crust is 0.2 ppm (see Table 2.2). It is typically found in sulfides, oxides, antimonites, and antimonates (US Geological Survey, 2017a, b, c). According to the US Geological Survey, the resource base of antimony is 4.3 Mt (US Geological Survey, 2009), but it does not provide a figure for the identified resources of antimony.

According to the generic approach that the extractable global resources of a resource are a maximum of 0.01% in the upper 3 km of the Earth's continental crust, antimony resources are rounded to 20 Mt (see Table 2.2).

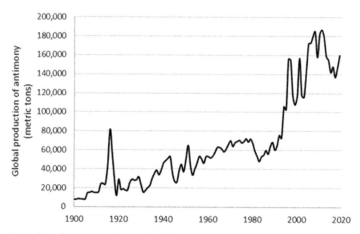

Figure 7.2.1 Development of annual global antimony production between 1900 and 2019 (metric tons). Antimony production in 2020 was 160,000 metric tons. *(Derived from US Geological Survey (2017a, c, 2018, 2019, 2020).)*

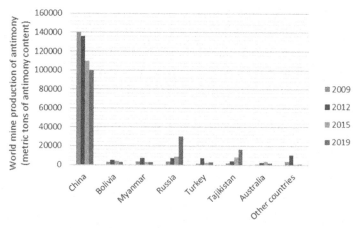

Figure 7.2.2 Antimony-producing countries in 2009, 2012, 2015, and 2018 (US Geological Survey, 2011, 2014, 2017, 2020).

On the basis of an extrapolation of the results of an assessment by the US Geological Survey (2000) of deposits of gold, silver, copper, lead, and zinc in the United States in 1998, the global amount of antimony resources can be estimated at about 100 Mt (see Tables 2.4 and 2.8). We will consider both the lower and higher estimates (20 and 100 Mt) in the analysis of measures necessary to achieve a sustainable production level of antimony.

Table 7.2.4 Main antimony-producing countries (US Geological Survey, 2011, 2014, 2017a, 2020).

	2009	2012	2015	2019
China	88%	75%	78%	63%
Myanmar	2%	4%	2%	2%
Russia	2%	4%	6%	19%
Bolivia	2%	3%	3%	2%
Tajikistan	1%	2%	6%	10%
Turkey	1%	4%	2%	2%
Australia	1%	1%	3%	1%
Other countries	2%	6%	1%	0%
Total	100%	100%	100%	99%

7.2.4 Current stocks and flows

7.2.4.1 Recovery efficiency

The antimony recovery efficiency from a typical stibnite ore mine is about 94% (Anderson, 2012).

7.2.4.2 Dissipative use

Only a small part of antimony use is dissipative. This concerns the *"other"* nonmetallurgical uses of antimony applications such as color stabilizer in pigments applied to road pavements, additive to lubricants, primer for ammunition, fireworks, matches, and friction material lubricant in automotive brakes. This includes about 0.9% of the antimony applications.

7.2.4.3 In-use stock

We estimate the average lifetime of antimony-containing products to be about 10 years (see Table 7.2.5).

We estimate that the current antimony in-use stock is about two million tons. For this estimation, we have used the formula in Chapter 7.1 for the in-use stock calculation.

With a recycled content of consumed antimony of 16% (UNEP, 2011a), and taking into account the annual changes in primary antimony production, this leads to the conclusion that in the 20 years between 1999 and 2018, the annual build-up of in-use stock of antimony was an average of about 20% of the annual antimony consumption.

The further increase in the antimony in-use stock will gradually slow down until about ten years after the moment that global antimony consumption has stopped growing. That is probably when the global antimony

Table 7.2.5 Estimation of average lifetime of antimony.

Application	End use (Roskill Consulting Group, 2011)	Lifetime (own estimate, years)	Contribution to average lifetime (years)
Flame-retardants	52%	10	5.2
PET catalyst	6%	2	0.1
Heat stabilizer	1%	10	0.1
Glass	1%	10	0.1
Ceramics	1%	10	0.1
Other nonmetallurgical applications	1%	0	0.0
Lead–acid batteries	27%	5	1.3
Lead alloys	12%	20	2.3
Total/average	100%	9.3	Rounded: 10 years

consumption level has caught up with the present level of antimony consumption in developed countries, unless new antimony applications would be developed and adopted.

7.2.4.4 Recycling

A substantial proportion of PET bottles is recycled and scrapped. According to Thiele (2009), in 2007 about 24% of PET in bottles (30% of the PET applications) was recycled. Neither PET in polyester-containing textiles (about 60% of the total PET use) nor PET in other applications (about 10%) is recycled. This means that, overall, PET recycling is 0.24 × 30 = 7%. In 2007, 72% of the recycled PET flakes was used in polyester fiber in various textiles (Noone, 2008), and 10% was used again in bottles (Thiele, 2009). According to the same author, recycled PET fiber accounted for about 8% of the world PET fiber production in 2007.

Antimony recycling from other nonmetallurgical uses, such as from flame-retardants, heat stabilizers, glass, ceramics, and chemicals, remains nonexistent. Due to the type of use in these nonmetallurgical applications, it is currently not economical, although it is technically feasible, to recycle antimony from end-of-life products. This means that antimony contained in end-of-life products is eventually disposed of in landfills or incinerators.

The recycling rate of lead–acid batteries is high, especially in developed countries. This is an important source of secondary antimony. According to

Table 7.2.6 Current recycling rates of antimony in developed countries. Antimony end use is based on the data of Roskill Consulting Group (2011).

	Current distribution of antimony end use (2010) (Roskill Consulting Group, 2011) (%)	Antimony end-of-life recycling rate (%)	Reference	Contribution to antimony recycling (%)
Nonmetallurgical applications				
Flame-retardants	51.9	0		0.0
PET catalyst	5.7	7	Thiele (2009)	0.4
Heat stabilizer	1.3	0		0.0
Glass	0.9	0		0.0
Ceramics	1.3	0		0.0
Other	0.9	0		0.0
Metallurgical applications				
Lead–acid batteries	26.6	90	Carlin (2006)	23.9
Lead alloys	11.5	10	Calculated	1.2
Subtotal				25.1
Total	100.0	26		25.5

Carlin (2006), about 95% of secondary antimony in the United States originated from lead-acid batteries by 2000. Metallurgical antimony applications are currently the only source of secondary antimony. This implies that 5% of secondary antimony in the United States results from other antimony-containing scrap of metallurgical applications (such as lead sheets, pipes, tubes, gutters, etc.). On the basis of these figures, it can be calculated that the end-of-life recycling rate of antimony from antimony-containing alloys was about 10% in 2000 in the United States. A large proportion of the lead grids from batteries and collected end-of-life antimony-containing alloys are scrapped and recycled in lead smelters. The resulting secondary antimony is mostly used again in lead-acid batteries, although this may change in the future with the growing use of low-maintenance and maintenance-free batteries. Globally, the recycling of antimony from batteries and alloys is lower.

The above recycling data are summarized in Table 7.2.6. The result is an end-of-life recycling rate of about 26% in developed countries. In

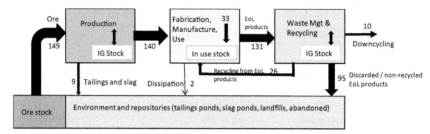

Figure 7.2.3 Global antimony flows in 2018. The amounts are in thousands of tons. The starting point is the global antimony production of 140,000 tons in 2018. Further assumptions, explained in the text, are a build-up of the in-use stock of 20% of the total consumption, end-of-life recycling rate of 20%, downcycling of 8% of end-of-life products, dissipation of 1% of the consumption, and a recovery efficiency of 94%. IG stock is industry and government stocks. The arrow widths are an indicative representation of the relative volumes of the flows.

developing countries, the recycling rate of antimony from metallurgical applications is lower. Based on a recycled content of the consumed antimony of 16% globally (UNEP, 2011), it can be calculated that the global end-of-life recycling rate is about 20%. We use an end-of-life recycling rate of 20% for the current antimony flow scheme, which is represented in Fig. 7.2.3. This could, for example, be the result of 70% end-of-life recycling of lead-acid batteries and 5% end-of-life recycling of lead alloys.

7.2.4.5 Downcycling

According to Table 7.2.6, about 10% of antimony in lead alloys is recycled in the antimony cycle. It is estimated that, of the remainder of antimony in lead alloys, 70% is downcycled to the lead cycle, and 20% is disposed of. Hence, antimony downcycling is about 70% of 11.5%, which is about 8% of antimony end-of-life products.

7.2.4.6 Current antimony flows summarized

The above figures regarding the current recovery efficiency at the mining stage, the in-use dissipation rate, the increase in the in-use stock, and the end-of-life recycling rate are the point of the departure for the antimony flow scheme presented in Fig. 7.2.3.

7.2.5 Sustainable antimony flows

7.2.5.1 Introduction

In this section, we investigate whether and under which conditions the same service level of antimony per capita can be reached globally as in developed countries in 2020, while simultaneously reducing the use of primary antimony to a sustainable level. Antimony's consumption level in developed countries in 2020 was about 90 g/capita/year (see Table 5.3).

A reduction in the use of primary raw materials is realized by a mixture of (1) improvements in recovery efficiency, (2) substitution, (3) material efficiency, (4) dissipation, and (5) recycling. The approach will be to investigate these five types of economizing measures in the indicated order. For each of these measures, we investigate the maximum, but realistic, reduction of the use of primary antimony that can be achieved. What can be adequately substituted does not need further attention in the framework of improving material efficiency, dissipation reduction, and recycling. The resulting mix of measures should not be considered as a blueprint, but for the purpose of estimating the theoretically achievable extraction reduction. In practice, other mixes of measures may be more economically viable.

7.2.5.2 Increase in recovery efficiency

The recovery efficiency of antimony is already quite high, at 94%. In the future, ore grades can be expected to decline further. Therefore, we assume that the recovery efficiency of antimony will be barely improvable economically.

7.2.5.3 Substitutability of antimony-containing products

Antimony is unique in the sense that it is relatively easily substitutable in many of its applications. According to the research of Henckens et al. (2016), antimony can be substituted in more than 90% of its uses (see Table 7.2.7).

7.2.5.4 Material efficiency improvement

In a situation with a great variety of applications, such as in the case of antimony, it is difficult to provide a reliable estimate of the possible improvement in material efficiency. We follow the general conclusions of Section 7.1.3 and assume that the material efficiency improvement potential of the antimony applications remaining after maximum substitution is 10%. The result is presented in Table 7.2.8.

Table 7.2.7 Antimony substitutability in various applications.

Column	1	2	3	4
	Current distribution of antimony consumption (2010)	Substitutability	References	Remaining antimony consumption after substitution
	% of total consumption	(%)		% of total consumption
Nonmetallurgical applications				
Flame-retardants	51.9	95%	Henckens et al., 2016	2.6
PET catalyst	5.7	100%	Thiele, 2001, 2004, 2009; Yang et al., 2012, 2013; Butterman and Carlin, 2004; Gross et al., 2010; Furlong, 2014	0.0
Heat stabilizer	1.3	100%	Henckens et al., 2016	0.0
Glass	0.9	80%	Henckens et al., 2016	0.2
Ceramics	1.3	100%	Henckens et al., 2016	0.0
Other	0.9	100%	(a)	0.0
Metallurgical applications				
Lead-acid batteries	26.6	100%	May, 1992; Toniazzo, 2006, Misra, 2007	0.0
Lead alloys	11.5	50%	Henckens et al., 2016	5.8
Total	100.0	91%		8.5

The substitution figures are the maximum achievable within a period of 10 years. (a) It is assumed that the "*other nonmetallurgical applications*" can be 100% substituted or abandoned. The "other" nonmetallurgical applications are specified in Section 7.2.2.

Table 7.2.8 The reduction effect of material efficiency measures on the remaining antimony end use after maximum substitution.

Column	1	2	3
	Remaining antimony consumption after substitution (column 4 of Table 7.2.7)	Reduction through material efficiency	Remaining antimony consumption after substitution and material efficiency measures
	% of original consumption	(%)	% of original consumption
Nonmetallurgical applications			
Flame-retardants	2.6	10%	2.3
PET catalyst	0.0		
Heat stabilizer	0.0		
Glass	0.2	10%	0.2
Ceramics	0.0		
Other	0.0		
Metallurgical applications			
Lead–acid batteries	0.0		
Lead alloys	5.8	10%	5.2
Total	8.5	10%	7.7

The conclusion is that material efficiency might further reduce the antimony end use from 8.5% of the original antimony consumption to 7.7% of the original antimony consumption.

7.2.5.5 Dissipation reduction

Dissipation reduction is only relevant for the "other nonmetallurgical applications," which are less than 1% of total antimony use. However, these applications are assumed to have been substituted or abandoned (see Table 7.2.7).

7.2.5.6 In-use stock

As a result of all of the antimony-economizing measures, the in-use stock of antimony will stop increasing and will decline to a new stable level which, at a sustainable antimony consumption rate, will be considerably lower than it is currently. This means that any further accumulation of antimony in the in-use stock does not need to be taken into account.

7.2.5.7 Increase in recycling

Recycling will focus on the recycling of antimony from the remaining applications of antimony after substitution, that is, in flame-retardants, glass, and lead alloys. It is no longer necessary to consider the increased recycling of antimony from other applications of antimony because antimony can be 100% substituted by other products in all of these applications.

The EU WEEE Directive (2012) facilitates the opportunity to concentrate antimony from WEEE. In principle, various recycling routes are possible. Plastics that contain brominated flame-retardants may be separated per type of material and reused in the original products. Another possible route is incineration, whereby antimony will mainly be concentrated in the incineration fly ash where it can be recovered, from a technical point of view. If we assume that 50% of antimony in non-substituted flame-retardants can be recovered in this way and, additionally, that downcycling of antimony in lead alloys to the lead cycle is reduced from 70% to 40%, it is possible to recycle about 50% of the remaining antimony. This would lead to a maximum amount of about 4% of the original antimony consumption finally being downcycled or disposed of after all of the measures have been taken.

7.2.6 Conclusions

The relatively easy substitutability of antimony in its most important applications (up to about 90% of its current applications) makes decreasing the antimony production rate to a sustainable level achievable, while simultaneously increasing the antimony service level globally to the same level as in developed countries in 2020 (90 g/capita/year). Table 7.2.9 presents some combinations of substitution, material efficiency improvement, and recycling.

The conclusions are as follows:

- A sustainability ambition of 1000 years is only achievable with a minimum of 41 Mt of ultimately available antimony resources, in combination with 90% substitution and 50% end-of-life recycling of remaining antimony. With 100 Mt of ultimately available resources the requirements are less far reaching.
- Sustainability ambitions of 500 and 200 years are achievable with measures, of which the magnitude depends on the ultimately available antimony resources. The combination of the lowest sustainability ambition (200 years) with the highest amount of ultimately available resources (100 Mt) still requires measures to be taken.

Table 7.2.9 Different mixes of technical measures to achieve a sustainable antimony production rate, while realizing a global antimony service level that is the same as the antimony service level in developed countries in 2020. The boldfaced rows provide the necessary measures at different sustainability ambitions and assumed ultimately available resources.

		Sustainability ambition (years of sufficient availability)								
		1000		**500**			**200**			
Ultimately available antimony resources[a,b]	Mt	41	100	20	100	100	20	20	100	100
Sustainable antimony production[c]	kt/year	41	100	41	200	200	100	100	500	500
Aspired future antimony service level	g/capita/year	90	90	90	90	90	90	90	90	90
Assumed future substitution	%	**90%**	75%	90%	51%	69%	85%	75%	23%	**0%**
Assumed increase in material efficiency	%	10%	10%	10%	10%	10%	10%	10%	10%	10%
Aspired future antimony consumption level	g/capita/year	8.1	20	8.1	40	25	12	20	62	81

Continued

Table 7.2.9 Different mixes of technical measures to achieve a sustainable antimony production rate, while realizing a global antimony service level that is the same as the antimony service level in developed countries in 2020. The boldfaced rows provide the necessary measures at different sustainability ambitions and assumed ultimately available resources.—cont'd

		Sustainability ambition (years of sufficient availability)									
		1000			**500**				**200**		
Corresponding annual antimony consumption	kt/year	81	199	124	81	397	248	124	199	624	810
Assumed in-use dissipation[d]	%	1%	1%	1%	1%	1%	1%	1%	1%	1%	1%
Antimony in end-of-life products	kt/year	80	197	123	80	393	245	123	197	617	802
Required antimony recycling	kt/year	40	99	24	40	197	48	24	99	124	310
Required antimony recycling rate[e]	**%**	**50%**	**50%**	**20%**	**50%**	**50%**	**20%**	**20%**	**50%**	**20%**	**39%**

It is assumed that the antimony recovery efficiency at the mining stage will remain at 94%. The assumed maximum substitutability in the future is 90%, the assumed maximum end-of-life recycling rate of antimony from remaining end-of-life products is 50%, and the assumed future in-use dissipation of antimony is a minimum of 1%. The assumed future material efficiency increase is 10%.
[a]Low estimate: 20 Mt; high estimate: 100 Mt (see Section 7.2.3.2).
[b]Current recovery efficiency: 94% (see Section 7.2.4).
[c]It is assumed that the world population stabilizes at 10 billion people.
[d]Current in-use dissipation: 1% (see Section 7.2.4).
[e]Current antimony end-of-life recycling rate: 20% (see Section 7.2.4).

References

Alaee, M., Arias, P., Sjödin, A., Bergman, A., 2003. An overview of commercially used brominated fire retardants, their applications, their use patterns in different countries/regions and possible modes of release. Environ. Int. 29, 683–689.

Anderson, C.G., 2012. The metallurgy of antimony. Chem. Erde 72, 3–8.

Butterman, W.C., Carlin, J.F., 2004. Mineral Commodity Profiles, Antimony, US Geological Survey.

Camino, G., 2008. Introduction to Flame Retardants in Polymers. Elawijt Center.

Carlin, J.F., 2006. Antimony Recycling in the United States in 2000, US GEOLOGICAL SURVEY, 1196-Q.

Cusack, P., 2007. EU brominated fire retardant consumption (by industry sector). In: 5th International Conference on Design and Manufacture for Sustainable Development. Loughborough University.

EFRA, 2013. Website. Available from: www.flameretardants.eu.

European Union, 2011. Regulation No 305/2011 of the European Parliament and of the Council of 9 March 2011 Laying Down Harmonized Conditions for the Marketing of Construction Products and Repealing Council Directive 89/106/EEC.

Flame retardants on line. Available from: http://www.flameretardants-online.com/web/en/106/.

Furlong, K., 2014. Titanium — a Real Alternative to Antimony Catalysts in PET Bottles. http://www.jmcatalysts.com/pct/news2.asp?newsid=76.

Gross, R.A., Ganesh, M., Lu, W., 2010. Enzyme-catalysis breathes new life into polyester condensation polymerizations. Trends Biotechnol. 28, 435–443.

Henckens, M.L.C.M., Driessen, P.P.J., Worrell, E., 2016. How can we adapt to geological scarcity of antimony? Investigation of antimony's substitutability and other measures to achieve a sustainable use. Resour. Conserv. Recycl. 108, 54–62.

Ji, L.N., 2013. Study on preparation process and properties of polyethylene terephthalate (PET). Appl. Mech. Mater. 312, 406–410.

Lassen, C., Lokke, S., Andersen, L.I., 1999. Brominated Flame Retardants — Substance Flow Analysis and Assessment of Alternatives. Danish Environmental Protection Agency. Environmental project no 494.

Markarian, J., 2007. PVC Additives, What Lies Ahead? Plastic Additives and Compounding November/December 2007.

May, G.J., 1992. Gelled-electrolyte lead/acid batteries for stationary and traction applications. J. Power Sources 40, 187–193.

Misra, S.S., 2007. Advances in VRLA battery technology for telecommunications. J. Power Sources 168, 40–48.

Keyser, C., 2009. Fire Retardant Global Industry Overview. Personal email communication with Mike Kiser, (Placed on internet).

Noone, A., 2008. Collected PET bottles. In: Proceedings of the 13thInternational Polyester Recycling Forum 2008.

Roskill Consulting Group Limited, 2011. Study of the Antimony Market.

Thiele, U.K., 2001. The current status of catalysis and catalyst development for the industrial process of PET polycondensation. Int. J. Polym. Mater. 50, 387–394.

Thiele, U.K., 2004. Quo vadis polyester catalyst? Chem. Fibers Int. 54, 162–163.

Thiele, U.K., 2009. In: 13th International Polyester Recycling Forum, 2009, pp. 22–23.

Toniazzo, V., 2006. The key to success: gelled-electrolyte and optimized separators for stationary lead-acid batteries. J. Power Sources 158, 1124–1132.

UNEP, 2011. Recycling Rates of Metals, a Status Report.

US Geological Survey, 2000. 1998 Assessment of undiscovered deposits of gold, silver, copper, lead and zinc in the United States. Circular 1178, 2000.

US Geological Survey, 2009. Mineral Commodity Summaries, Antimony.

US Geological Survey, 2011. Mineral Commodity Summaries, Antimony.

US Geological Survey, 2014. Mineral Commodity Summaries, Antimony.

US Geological Survey, 2015. Mineral Commodity Summaries, Antimony.

US Geological Survey, 2017a. Mineral Commodity Summaries, Antimony.

US Geological Survey, 2017b. Antimony, Section C of Critical Mineral Resources of the United States — Economic and Environmental Geology and Prospects for Future Supply. Professional Paper 1802-C.

US Geological Survey, 2017c. Antimony Statistics.

US Geological Survey, 2018. Mineral Commodity Summaries, Antimony.

US Geological Survey, 2019. Mineral Commodity Summaries, Antimony.

US Geological Survey, 2020. Mineral Commodity Summaries, Antimony.

Yang, Y.K., Bae, S.B., Hwang, Y.T., 2013. Novel catalysts based on zirconium(IV) for the synthesis of poly(ethylene terephthalate-co-isophthalate)polyesters. Tetrahedron Lett. 54, 1239—1242.

Yang, Y.K., Yoon, S.W., Hwang, Y.T., Song, B.G., 2012. New titanium-based catalysts for the synthesis of polyethylene terephthalate. Bull. Korean Chem. Soc. 33 (10), 3445—3447.

CHAPTER 7.3

Bismuth

7.3.1 Introduction

We investigate whether and under which conditions a sustainable extraction rate of bismuth is feasible while simultaneously increasing the global service level of bismuth to the service level in developed countries in 2020.

7.3.2 Properties and applications

Bismuth is a brittle metal with a silvery white color. Its density is 9.7 g/cm^3, its melting point is 271°C, and the boiling point is 1564°C. Bismuth has a number of properties that are decisive for its various applications. It has a low toxicity for humans, and therefore is also called "green bismuth". In comparison with lead, antimony, and polonium, which are bismuth's neighbors in the periodic table of elements, bismuth has surprisingly low toxicity. This is an important reason for bismuth being used in a number of pharmaceutical and cosmetic products, and also, for example, for replacing lead-tin solder with bismuth-tin solder and lead with bismuth in shotgun bullets and fishing weights. Bismuth is diamagnetic; that is to say, bismuth is repelled by a magnetic field. This property is used in magnetic levitation trains (Maglev). Bismuth expands at a lower temperature and on solidification (as does water), which explains its applications in smelt fuses and temperature-sensitive electronic circuits. Bismuth also has a low thermal conductivity (Fig. 7.3.1).

The applications of bismuth can be subdivided as follows (Umicore, 2018; Mohan, 2010; Metalpedia, 2018):

Pharmaceutical, cosmetic, and chemical products

- Medicines, especially for treating infection problems in the stomach (peptobismol) and eyes (eye drops). The active component in peptobismol is bismuth oxide salicylate
- Cosmetics, as bismuth oxychloride to give a shine to lipsticks, nail varnishes, and hair lacquers

Figure 7.3.1 The chemical element bismuth in synthetically made bismuth crystals. The surface is oxide free. Additionally, a high-purity (99.99%) 1 cm^3 bismuth cube for comparison (Wikipedia, 2018).

Metallurgical additives

- Alloys to improve the machinability of cutting steels
- Replacement of lead in all kinds of applications due to the relatively low toxicity of bismuth
- As a mold for precision casting in combination with lead (lead shrinks and bismuth expands when solidified), as lead and bismuth compensate each other

Fusible alloys, solder, and cartridges

- Smelt fuses, such as in sprinkler systems, electrical fuses, and in the car and aviation industries
- Temperature-sensitive electronic circuits
- Peltier thermoelectric devices for refrigeration purposes

Other applications

- Pigments, for example for bright yellow colors in the automotive industry
- Fireworks ("dragon's eggs")
- Magnetic levitation trains (Maglev)
- Lewis acid catalysts in organic synthesis
- Bismuth telluride is a semiconductor and is used in refrigerators, and as a detector in infrared spectrophotometers

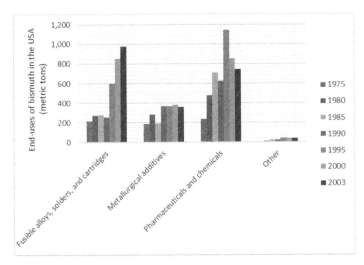

Figure 7.3.2 End uses of bismuth in the United States (US Geological Survey, 2005).

An important potential new application is for solar generation of hydrogen fuel by using bismuth vanadate to split water into hydrogen and oxygen using sunlight (US Geological Survey, 2017c).

The distribution of bismuth within these applications in the United States is shown in Fig. 7.3.2.

The distribution in 2003 (the last year for which data are available) is presented in Table 7.3.1.

Table 7.3.1 Bismuth applications in 2003 in the United States (US Geological Survey, 2005).

Fusible alloys, solders, and cartridges	46%
Metallurgical additives	17%
Pharmaceuticals and chemicals	35%
Other applications	2%
Total	100%

7.3.3 Production and resources

7.3.3.1 Production

Global bismuth production development is presented in Fig. 7.3.3. Since 1937, bismuth production has increased by 3.9% annually on an average basis (US Geological Survey, 2017a). The most important bismuth-producing countries are, in order of importance: China (72% of world

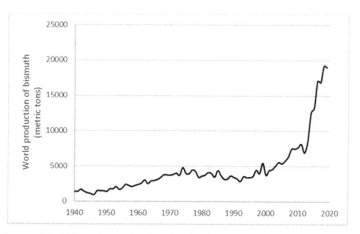

Figure 7.3.3 Global bismuth production (US Geological Survey, 2017a, b, 2018, 2019). Global bismuth production in 2019 was 19,000 metric tons (US Geological Survey, 2019).

production in 2016), Vietnam (20% of world production in 2016), and Mexico (7% of world production in 2016) (US Geological Survey, 2017b; US Geological Survey, 1996–2017). The total mined production of bismuth was 10,200 metric tons in 2016 (see Fig. 7.3.4).

7.3.3.2 Resources

The average bismuth concentration in the Earth's crust is about 50 ppb by weight.[2] More than 50% of bismuth is produced as a by-product of lead production (Mudd et al., 2014). Furthermore, bismuth is produced to a lesser extent as a by-product of the extraction of copper, tin, and tungsten. In 2018, there was one dedicated bismuth mine (in China). The bismuth concentration in bismuth-containing ores is 0.003%–0.06%, but can exceed 2% in ores of proper bismuth deposits (Avdonen et al., 2005).

The most recent estimate of the global bismuth reserves is 370,000 tons (US Geological Survey, 2017b). In 2009, the reserve base of bismuth was estimated at 680,000 metric tons (US Geological Survey, 2009). According to the 0.01% approach (see Table 2.2), bismuth resources in the upper 3 km of the Earth's crust are about 6 Mt. On the basis of an extrapolation of the results of an assessment by the US Geological Survey (2000) of deposits of

---

[2] The scientific literature provides different values for the crustal abundance. We have taken the average value of seven sources. Refer to Table 2.2 in this book.

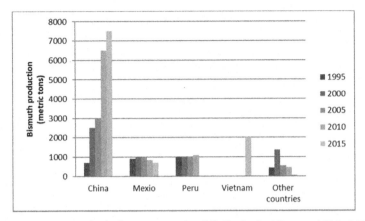

Figure 7.3.4 Bismuth-producing countries (US Geological Survey, 1996–2017).

gold, silver, copper, lead, and zinc in the United States in 1998, the global amount of boron resources can be estimated at about 20 Mt (see Table 2.3).

We consider both the lower estimate of 6 Mt and the higher estimate of 20 Mt in the analysis of the necessary measures to achieve the sustainable production of bismuth.

7.3.4 Current stocks and flows

There are virtually no quantitative data on bismuth flows, so we have to make estimates.

7.3.4.1 Recovery efficiency

Almost all bismuth is produced as a by-product of lead, copper, tin, and tungsten production. Ore beneficiation is therefore not primarily aimed at the production of bismuth. Bismuth is removed from lead bullion by electrorefining or pyrorefining. In the case of electrorefining, bismuth remains as slime at the anode with up to 20% bismuth. In the case of pyrorefining, bismuth is selectively removed from the lead bullion by the addition of calcium and magnesium. A precipitate of a calcium–magnesium–bismuth compound forms, which can then be separated from the lead bullion.

During the production of copper, 50%–90% of bismuth is concentrated in the flue dust and the remainder in the waste slag (Krenev et al., 2015;

Vitkova et al., 2011; Yu and Pan, 1997; Yao, 2003). The copper smelter converter dust contains bismuth as well as much lead and silver. The typical bismuth content of copper smelter converter dust is 2%–3% (Ha et al., 2015; Chen et al., 2012). Because of the bismuth content, the copper smelter converter flue dust cannot be fed to the lead smelters and needs separate treatment for bismuth removal.

Because bismuth is a by-product, and because the market price of bismuth is low, the recovery efficiency is relatively low at 50%–70% (Krenev et al., 2015; Chen et al., 2012; Ha et al., 2015). The bismuth recovery efficiency will remain low as long as there is no conflict between supply and demand. The market price of bismuth has declined significantly over time, but has remained relatively stable during the last 40 years at between 10 and 20 USD per kg (expressed in USD of 1998) (see Fig. 7.3.5). As long as the price of primary bismuth remains relatively low, there is insufficient driving force for the involved mining companies to increase the recovery efficiency of bismuth. On the other hand, bismuth is a disturbing element in both lead and copper refining, and has negative effects on the end product. Hence, it is important to keep bismuth out of the lead and copper cycles.

For the flow scheme of Fig. 7.3.6 we have assumed a bismuth recovery efficiency of 60%.

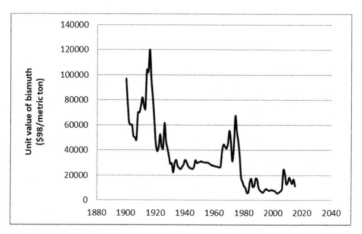

Figure 7.3.5 Historic development of the unit value of bismuth (US $98/metric ton) (US Geological Survey, 2017a).

7.3.4.2 Dissipative use

In its pharmaceutical and chemical applications, including cosmetic applications, bismuth is dissipated through its use. Based on bismuth's use in the United States, this affects about 35% of the applications (see Table 7.3.1).

7.3.4.3 In-use stock

Based on the end uses, we estimate the average lifetime of bismuth-containing products to be about six years (rounded) (see Table 7.3.2).

Based on an average lifetime of 6 years for bismuth-containing products and the annual bismuth production figures, it can be calculated using the in-use stock formula from Chapter 7.1 that the build-up of bismuth in the in-use stock is currently about 30% of the inflow.

7.3.4.4 Recycling

Bismuth can be recycled from bismuth-containing new and old alloy scrap. However, less than 5% of bismuth's apparent consumption consists of recycled bismuth in the United States (US Geological Survey, 2019). The amount of bismuth contained in products in the form of metallurgical additives and fusible alloys for instance is so low that recycling is barely worthwhile. We assume that—at a global scale—bismuth recycling is even lower than in the United States. In Fig. 7.3.6, which represents current bismuth flows, we have assumed that the global end-of-life recycling rate of

Table 7.3.2 Estimation of the average lifetime of bismuth.

Application	End use (US Geological Survey, 2005)	Lifetime (own estimate, years)	Contribution to average lifetime (years)
Fusible alloys, solders, and cartridges	46%	10	4.6
Metallurgical additives	17%	10	1.7
Pharmaceuticals and chemicals	35%	0	0
Other applications	2%	5	0.1
Total	100%	6.4	Rounded: 6 years

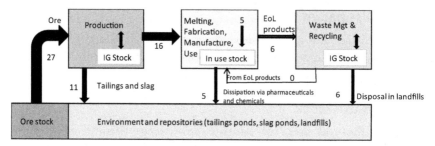

Figure 7.3.6 Simplified global anthropogenic bismuth cycle for 2018. Flows are in kt. Points of departure are: bismuth EoL (end-of-life) recycling rate: 0%; bismuth recovery efficiency: 60%; build-up of bismuth in anthropogenic stock: 30%; bismuth dissipation through use: 35%. The global mine production of bismuth was 16,000 metric tons in 2018. The arrow widths are a rough indication of flow magnitudes. IG stock is industrial, commercial, and government stocks.

bismuth is close to 0%. This is supported by the International Resources Panel of UNEP (2011), see Table 7.1.4).

7.3.4.5 Current bismuth flows summarized

The current recovery efficiency, in-use dissipation rate, increase in in-use stock, and end-of-life recycling rate of bismuth have been used as the basis for Fig. 7.3.6.

7.3.5 Sustainable bismuth flows
7.3.5.1 Introduction

In this section, we investigate whether and under which conditions, and for different sustainability ambitions, a service level of bismuth per capita can be reached globally that is the same as the bismuth consumption in developed countries in 2020, while simultaneously reducing the use of primary bismuth to a sustainable level. The consumption level of bismuth in developed countries in 2020 is estimated at 9 g/capita/year (see Table 5.3).

A reduction in the use of primary raw materials can be realized by a mixture of (1) improvements in recovery efficiency, (2) substitution, (3) material efficiency improvement, (4) dissipation reduction, and (5) recycling. The approach will be to investigate these five types of economizing measures in the indicated order. For each of these measures, we investigate the maximum, but realistic, reduction of the use of primary bismuth that can be achieved. Bismuth applications that can be adequately substituted do not need further attention in the framework of improving material efficiency, dissipation reduction, and recycling. The resulting mix of measures should not be considered as a blueprint of measures, but only for the purpose of estimating the theoretically achievable bismuth extraction reduction. In practice, other mixes of measures may be more economically viable.

7.3.5.2 Increase in recovery efficiency

The recovery efficiency of bismuth is relatively low, at between 50% and 70%. We assume that it will be possible to increase bismuth's recovery efficiency to a maximum of 70% in the future.

7.3.5.3 Substitutability of bismuth-containing products

According to US Geological Survey (2019), bismuth can be replaced in pharmaceutical applications by alumina, antibiotics, calcium carbonate, and magnesia. For pigment uses, bismuth can be substituted with titanium dioxide–coated mica flakes and fish-scale extracts. Cadmium, indium, lead, and tin can partially replace bismuth in low-temperature solders, although indium is also a scarce raw material. Bismuth alloys for holding metal shapes during machining can be substituted with resins, and bismuth alloys in triggering devices for fire sprinklers can be replaced with glycerin-filled glass bulbs. In free-machining alloys, lead, selenium, or tellurium can replace bismuth. All of these substitution possibilities are derived from US Geological Survey (2019). Summarizing, we conclude that the substitutability of bismuth in pharmaceuticals and chemicals is 100%. However, we estimate that it will be difficult to replace bismuth in metallurgical applications and other applications in which the temperature sensitivity of bismuth is important. In other applications, we therefore assume that the substitutability of bismuth is no more than 10%. Based on this, we estimate that the overall substitutability of bismuth in its current applications is a maximum of about 35% (see Table 7.3.3).

7.3.5.4 Material efficiency improvement

In a situation with a great variety of applications, such as in the case of bismuth, it is difficult to provide a reliable estimate of the possible improvement in material efficiency. We follow the general conclusions of Section 7.1.3 and assume that the material efficiency improvement potential of the bismuth applications remaining after maximum substitution is 10%.

Table 7.3.3 Bismuth substitutability in various types of applications.

`Column	1	2	3
	Current distribution of bismuth consumption (2010)	Substitutability, own estimate on the basis of US Geological Survey (2019)	Remaining bismuth consumption after substitution
	% of total consumption		% of total consumption
Fusible alloys, solders, and cartridges	46%	0%	46%
Metallurgical additives	17%	0%	17%
Pharmaceuticals and chemicals	35%	100%	0%
Other applications	2%	10%	2%
Total			65%

7.3.5.5 Dissipation reduction

Dissipation reduction is relevant for the pharmaceutical and chemical applications of bismuth and for a small part of the "other applications," such as pigments and fireworks. However, the pharmaceutical and chemical applications are assumed to have been substituted or abandoned, and the other applications are only a minor part of the total consumption after substitution and material efficiency measures. Hence, we assume that, after the substitution of bismuth in pharmaceuticals and chemicals, dissipation reduction provides almost no potential to further reduce the use of primary bismuth.

7.3.5.6 In-use stock

As a result of all of the bismuth–economizing measures, the in-use stock of bismuth will stop increasing and will reach a new stable level. This means that any further accumulation of bismuth in the in-use stock does not need to be taken into account.

7.3.5.7 Increase in recycling

We consider only the end-of life products remaining after substitution. This concerns fusible alloys, solders, and cartridges (about 70% of bismuth in the remaining end-of-life products), metallurgical additives (about 25%

Table 7.3.4 Different mixes of measures to achieve a sustainable bismuth production rate, while realizing a global bismuth service level that is similar to the bismuth service level in developed countries in 2020. The main measures are presented in the boldfaced rows.

		Sustainability ambition (years of sufficient availability)					
		1000	500			200	
Ultimately available bismuth resources[a]	Mt	20	11	20	20	6	20
Assumed future recovery efficiency[b]	%	70%	70%	70%	70%	70%	60%
Future availability of bismuth resources after increase in recovery efficiency	Mt	23	13	23	23	7	20
Sustainable bismuth production[c]	kt/year	23	26	47	47	35	100
Aspired future bismuth service level	g/capita/year	9	9	9	9	9	9
Assumed future substitution	%	**35%**	**35%**	**35%**	**0%**	**35%**	**0%**
Assumed increase in material efficiency	%	10%	10%	10%	10%	10%	0%
Aspired future bismuth consumption level	g/capita/year	5	5	5	8	5	9
Corresponding annual bismuth consumption	kt/year	53	53	53	81	53	90
Assumed in-use dissipation of remaining products[d]	%	0%	0%	0%	0%	0%	35%
Bismuth in end-of-life products	kt/year	53	53	53	81	53	59
Required bismuth recycling of remaining products	kt/year	29	26	6	34	18	0
Required bismuth recycling rate[e]	%	**56%**	**50%**	**11%**	**42%**	**34%**	**0%**

The assumed maximum substitutability of bismuth in the future is 35%, and the assumed material efficiency increase is 10%. The assumed minimum in-use dissipation rate of remaining products is 0%, and the assumed maximum recyclability of remaining products is 50%.

[a]Low estimate: 6 Mt; high estimate: 20 Mt (see Section 7.3.3).
[b]Current recovery efficiency: 60% (see Section 7.3.4).
[c]It is assumed that the world population stabilizes at 10 billion people.
[d]Current in-use dissipation rate: 35% (see Section 7.3.4).
[e]Current end-of-life recycling rate: 0% (see Section 7.3.4).

of bismuth in the remaining end-of-life products), and other applications (about 5% of bismuth in the remaining end-of-life products). If we assume that a maximum of 50% of bismuth will be recyclable from the first two mentioned applications, the maximum recyclability of bismuth from bismuth-containing end-of-life products remaining after substitution is about 50%.

7.3.6 Conclusions

Using Table 7.3.4, we analyze which mixes of technical measures are necessary to achieve a sustainable production of bismuth in the future, while simultaneously bringing the bismuth service level globally to the level of developed countries in 2020 (8 g/capita/year).

The conclusions are as follows:
- A sustainability ambition of 1000 years is not achievable.
- A sustainability ambition of 500 years is achievable on the condition that the ultimately available bismuth resources are a minimum of 11 Mt of bismuth, in combination with an increase in the recovery efficiency from 60% to 70%, substitution of bismuth of 35%, an improvement in material efficiency of 10%, and an end-of-life recycling rate of 50%. With 20 Mt of ultimately available bismuth resources, the necessary measures are less far reaching.
- A sustainability ambition of 200 years is achievable with ultimately available resources of 6 Mt, in combination with substitution and recycling measures. If 20 Mt of bismuth resources are available, no measures need to be taken compared to the current situation.

References

Avdonen, V.V., Boitsov, V.E., Grigor'ev, V.M., Seminskii, ZhV., Solodov, N.A., 2005. Starostin VI, 2005, Metal Ore Deposits. Triksta, Moscow cited by Krenev et al. (2015).

Chen, Y., Liao, T., Li, G., Chen, B., Shi, X., 2012. Recovery of bismuth and arsenic from copper smelter flue dusts after copper and zinc extraction. Miner. Eng. 39 (2012), 23—28.

Ha, T.K., Kwon, B.H., Park, K.S., Mohapatra, D., 2015. Selective leaching and recovery of bismuth as Bi_2O_3 from copper smelter converter dust. Separ. Purif. Technol. 142, 116—122.

Krenev, V.A., Drobot, N.F., Formichef, S.V., 2015. Processes for the recovery of bismuth from ores and concentrates. Theor. Found. Chem. Eng. 49 (4), 540—544.

Metalpedia, 2018. Available from: http://metalpedia.asianmetal.com/metal/bismuth/bismuth.shtml. (Accessed 6 December 2018).

Mohan, R., 2010. Green bismuth. Nat. Chem. 2. Available from: www.nature.com/naturechemistry.

Mudd, G.M., Yellishetty, M., Reck, B.K., Graedel, T.E., 2014. Quantifying the recoverable resources of companion metals: a preliminary study of Australian mineral resources. Resources 3, 657–671.

Umicore, 2018. Available from: https://www.umicore.com/en/about/elements/bismuth. (Accessed 6 December 2018).

UNEP International Resource Panel, 2011. Recycling Rate of Metals, a Status Report, 2011.

US Geological Survey, 1996–2017. Mineral Commodity Summaries, Bismuth.

US Geological Survey, 2000. 1998 Assessment of undiscovered deposits of gold, silver, copper, lead and zinc in the United States. Circular 1178, 2000.

US Geological Survey, 2005. Bismuth statistics. In: Kelly, T.D., Matos, G.R., comps (Eds.), Historical Statistics for Mineral and Material Commodities in the United States: U.S. Geological Survey Data Series 140. Available from: http://pubs.usgs.gov/ds/2005/140/ . (Accessed 12 December 2018).

US Geological Survey, 2009. Bismuth, Mineral Commodity Summaries, 2009.

US Geological Survey, 2017a. Bismuth Statistics, 2017.

US Geological Survey, 2017b. Mineral Commodity Summaries, Bismuth, January 2017.

US Geological Survey, 2017c. 2015 Minerals Yearbook, Bismuth, February 2017.

US Geological Survey, 2018. Mineral Commodity Summaries, Bismuth, 2018.

US Geological Survey, 2019. Mineral Commodity Summaries, Bismuth, 2019.

Vitkova, M., Ettler, V., Hyks, J., Astrup, T., Kribek, B., 2011. Leaching of metals from copper smelter flue dust (Mufulira Zambian Copperbelt). Appl. Geochem. 26, S263–S266.

Wikipedia, 2018. Bismuth. https://en.wikipedia.org/wiki/Bismuth.

Yao, G.S., 2003. Preparation of zinc sulfate from copper smelter flue dust. Nonferrous Smelting 3, 41–43.

Yu, Z.Z., Pan, X.J., 1997. Preparation of zinc sulfate from copper smelter flue dust, nonferrous metals. Extractive metallurgy 6, 20–24.

CHAPTER 7.4

Boron[3]

7.4.1 Introduction

Boron's most important applications are in different types of glass and in fertilizer as a micronutrient. Boron is on the European Union's list of critical raw materials, which means that boron combines a relatively high supply risk with a relatively high vulnerability to non-supply or low supply in Europe's economy.

In this section, we investigate to what extent the primary boron extraction rate can be made sustainable, while simultaneously bringing boron's global service level to the level of developed countries in 2020. We start by discussing the current boron applications and flows, followed by a systematic evaluation of the possibilities to reduce current boron use.

7.4.2 Properties and applications

Boron is a non-metal, with a melting point of 2076°C and a boiling point of 3927°C. Its density at the melting point is 2.08 g/cm^3. Elemental boron is not found naturally, because boron combines easily with oxygen to form borates. Boron's presence in the Earth's crust is mostly in borate minerals. About half of the world's supply of boron is used in glass products, and the second most important boron application is in fertilizer (see Fig. 7.4.1).

In this section, we present the global end uses as provided by US Geological Survey (2018a,b) (column 1 in Table 7.4.1), completed with additional data from other sources.

7.4.2.1 Glass

Glass represents the most important boron application, with 50% of boron's total global use. Boron compounds improve glass's thermal shock resistance, increase the aqueous durability and chemical resistance, improve the mechanical strength, provide higher resistance to devitrification during

---

[3] This section is based on the publication of Henckens, M.L.C.M., Driessen, P.P.J., Worrell, E., 2015. Toward a sustainable use of primary boron. Approach to a sustainable use of primary resources. Resour. Conserv. Recycling 103, 9—18.

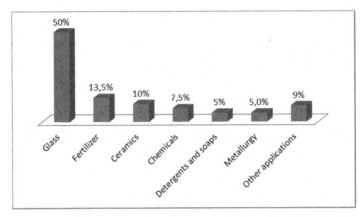

Figure 7.4.1 Global boron end uses (US Geological Survey, 2018b).

processing, lower the glass melting temperature (fluxing agent), improve the refining process and formability, and improve optical properties (Borax, 2019). About 85% of the applications of boron in glass are in fiber glass, which is used for glass wool for insulation and glass fiber (for textiles and reinforcement of plastics; see Table 7.4.2).

A smaller portion of boron application in glass (about 15%) is used for borosilicate glass. The applications of borosilicate glass are shown in Table 7.4.3. Depending on the specific application, borosilicate glass may contain around 5%–20% B_2O_3 (Borax, 2019).

7.4.2.2 Fertilizers

Fertilizers represent the second largest application type of borates, with 13.5% of the global use (see column 1 of Table 7.4.1). Boron is a micronutrient that is primarily used in fruit and seed production (US Geological Survey, 2013). Normal plant leaves typically contain 25–100 ppm boron (US Geological Survey, 2016), whereas the average boron concentration in the Earth's crust is 13 ppm (see Table 2.2). To keep sufficient boron available for crops, it is necessary to supply 1 kg per hectare per year (US Geological Survey, 2016).

7.4.2.3 Ceramics

About 10% of global boron consumption is for ceramics (US Geological Survey, column 1 of Fig. 7.4.1). Borates are applied in ceramic glazes and enamels, increasing chemical, thermal, and wear resistance. Enamel is a smooth, durable, vitreous coating on metal, glass, or ceramics. It is hard,

Table 7.4.1 Boron end uses in the world, the United States, and the EU.

Column	1	2	3	4
	World (US Geological Survey, 2018a,b) 2016 data	USA (US Geological Survey, 2005), 2003 data	EU (European Commission, 2013) Year of data is not indicated	World Statista (2019a,b), 2012 data
Glass	50%	79%	51%	51%
Insulation material (glass wool)	30%[a]	50%		
Glass fiber	13%[a]	22%		
Borosilicate glass	7%	7%		
Fertilizer	13.5%[b]	3%	13%	14%
Ceramics	10%	4%	14%	13%
Detergents and soaps	5%	5%	1%	3%
Flame-retardants	1%[b]	4%	1%	
Other applications	20.5%[b]	5%	20%	19%[c]
Chemicals	7.5%[d]		7%	
Cosmetics, pharmaceuticals and toiletries	2%[d]		2%	
Industrial fluids	2%[d]		2%	
Metallurgical applications	5%[d]		5%	
Construction materials	4%[d]		4%	
Total	100%	100%	100%	100%

[a]Global distribution of boron applications in insulation materials and glass fiber is assumed to be similar to the distribution in the United States in these two applications (column 2).
[b]US Geological Survey (2018a,b) does not provide details on the global use of boron in fertilizer, flame-retardants, and other applications. This is assumed to be the remainder of total global boron use (35%), and it is assumed that these three types of applications have the same relative use as in the EU.
[c]Includes flame-retardants.
[d]US Geological Survey does not provide details on the other applications. Therefore, it is assumed that the other applications have the same relative uses as in the EU.

chemically resistant, and scratch resistant, has long-lasting color fastness, is easy to clean, and cannot burn (Fedak and Baldwin, 2005). Enamels are mostly used on steel (Borax, 2019). A specific application of boron in ceramics is in lightweight armor, in which boron carbide is a key ingredient (US Geological Survey, 2016). The use of borates in the production of ceramic tiles reduces the temperature and energy requirements. Moreover,

Table 7.4.2 Glass fiber applications of boron (Crangle, 2018).

Glass fiber applications	Function
Glass fiber textiles	Resistance against corrosion and heat, high strength
Insulation (glass wool)	Thermal insulation and acoustic insulation. Thermal insulation glass wool contains about 4%—5% of boron oxide to aid melting, to inhibit devitrification, and to improve the aqueous durability (Lyday, 2003)
Glass fiber reinforced plastics (GFRP or FRP), also called E-glass	Alumino-borosilicate glass with less than 1% alkali oxides (Fitzer et al., 2008). Important applications are boats, wind turbine blades, pipes, tanks, and lightweight composite structural components for cars, trucks, trains, and aircraft (Borax, 2019). The boric oxide concentration varies between 0% and 10% B_2O_3, but typically between 6% and 10% boron oxide (Lyday, 2003)

Table 7.4.3 Borosilicate applications (Borax, 2019).

Borosilicate glass applications	Application examples
Heat-resistant glass	Pyrex kitchenware, microwave dishes, laboratory glass
Display screens	LCD screens. This is one of the major boron-consuming areas that have grown recently. Generally, flat-screen glass contains 11%—13% boron oxide. The cover glass of touchscreens of smartphones and tablets consists of borosilicate glass
Lighting glass	Headlights, halogen bulbs, fluorescent tubes
Sealing glasses	Tungsten filament lamps, lamps in street lighting, cathode ray tubes
Neutral glasses	Ampoules and vials for medicine for increased chemical resistance
Cosmetics containers	For chemical resistance
Solar glass	Cover glass and substrate glass for photovoltaic cells, and evacuated solar collector tubes
Glass microspheres	Airport runway reflector systems
Other	Optical glass, prisms, lenses, opal glassware, telescope mirror blanks

borates increase the dry mechanical strength of unfired tiles by 30% to 80% (Lyday, 2003). Ceramic glazes with boron are applied in tiles and tableware (porcelain, china, stoneware, and earthenware) (Borax, 2019). Large consumers of boric oxide in ceramics are glazes for wall and floor tiles (Borax, 2019). Wall tiles contain between 3% and 20% B_2O_3, depending on the firing time and temperature (Borax, 2019). Enamels contain typically 14% B_2O_3.

7.4.2.4 Detergents and soaps

The use of borates in detergents and soaps accounts for 5% of global consumption (US Geological Survey, 2018). Borates are used in detergents and soaps as alkaline buffers, enzyme stabilizers, oxygen-based bleaching agents, water softeners, for improvement of surfactant performance, and for soil removal (Borax, 2019). Sodium perborate, in contact with hot water, produces hydrogen peroxide, a very effective bleaching agent. Modern laundry detergents typically contain 15% sodium perborate.

7.4.2.5 Flame-retardants

Zinc borate is used as a flame-retardant in plastic and rubber applications, pressed boards, paper boards, cellulose-based insulation material, gypsum board, cotton batting in mattresses, and fabrics requiring flame-retardant treatment. It forms a glassy coating that protects the surface (EFRA, 2007). Normally, zinc borate is used in combination with other flame-retardants, such as antimony trioxide, aluminum trioxide, magnesium hydroxide, or red phosphorus (EFRA, 2007).

7.4.2.6 Other applications

Boron is used in a multitude of other products, including:
- Chemicals: in insecticides, wood preservatives, pH buffers, lubricators, in nuclear power plants as a moderator, in semiconductors, airbags, magnets, abrasives, ballistic vests, electrolytic capacitors, lithium–ion batteries, starch adhesives for example for paper bags and corrugated cardboard, paints, coatings, and printing inks. In oil and gas production as, for example, a retarder in concrete, drilling fluid, and profile control. Anhydrous borax is used in gold refining as part of flux formulations to dissolve metal oxides.

- Cosmetics and pharmaceuticals: e.g., in cosmetic creams, skin lotions, hair shampoos, dyes and gels, bath salts, and denture cleaners (Borax, 2019) and pharmaceuticals (eye drops). Boric acid is used as an antiseptic and an antibacterial compound.
- Industrial fluids. According to Borax (2019), borates are used in antifreeze, lubricants, brake fluids, metalworking fluids, water treatment chemicals, corrosion inhibition, freezing point reduction, boiling point elevation, fuel additives for the prevention of pre-ignition, leather tanning as a pH buffer liquid, and photo development solutions (pH buffer).
- Metallurgical applications. Borate fluxes are used to remove oxide impurities from metals. The addition of boric oxide to steel slags prevents 'dusty' slag, resulting in a stable rock-like material (Borax, 2019). About 15 kg of B_2O_3 is used per ton of slag. Ferroboron in steel increases the steel strength.
- Construction materials. Borates and other boron compounds are used in the manufacture of gypsum board and to protect wood from termites, pests, and fungi. Applications are in, for example, solid wood, plywood, engineered wood, and railroad crossties (sleepers) (Borax, 2019).

7.4.2.7 Future developments

Future boron use may increase because of applications in automotive fuel cells (i.e., sodium borohydride), in cars to replace metal parts with fiber glass reinforced plastics, and in batteries (i.e., titanium diboride) (Lyday, 2003).

7.4.3 Production and resources
7.4.3.1 Production

The global production of boron since 1964 is presented in Fig. 7.4.2. On the basis of data from Lyday (2003) and US Geological Survey (2017a, 2017b, 2018a,b, 2019, 2020), boron world production is estimated as follows:

- The content of B_2O_3 in tradable boron oxide ores produced in the United States is about 50%, and similar in other boron-producing countries. As 31% of B_2O_3 is boron, the share of boron in tradable boron concentrates is about 15.5%.

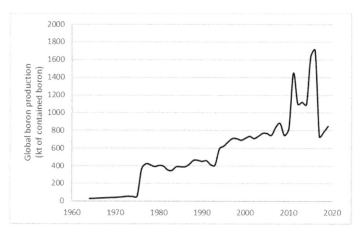

Figure 7.4.2 World production of boron. *(Derived from US Geological Survey, 2017a, 2017b, 2018a,b, 2019, 2020.)*

- The US Geological Survey has not provided production figures for boron in the United States since 2005. However, the annual production of borates in the United States was rather stable between 1975 and 2005, at about 600 ktons average annually, expressed as B_2O_3. On this basis, we have assumed that the annual production of boron (all tradable forms) in the United States after 2005 can be estimated at about $2 \times 600 = 1200$ ktons annually.

The main borate-producing countries and their share in the global boron production are presented in Fig. 7.4.3. Turkey (50% of world borate production in 2019) and the United States (an estimated 24% of world borate production in 2019) are the largest boron producers and also have the largest boron reserves.

7.4.3.2 Resources

The average concentration of boron in the Earth's crust is 13 ppm (see Table 2.2). Boron is usually found in borate minerals, such as tincal. According to US Geological Survey (2018), global boron reserves are 1100 Mt (all tradable forms). This implies a reserve estimate of $0.155 \times 1100 = 170$ Mt of boron.

On the basis of the assumption that the extractable part of a resource is 0.01% of the total amount of that resource in the upper 3 km of the Earth's continental crust, the extractable global resources of boron can be estimated at 1560 Mt of boron, or 2 Gt rounded (see Table 2.2).

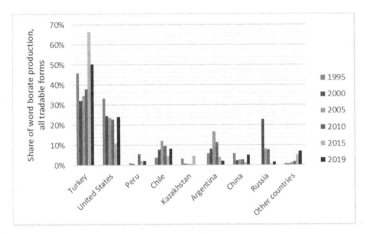

Figure 7.4.3 Main boron-producing countries (US Geological Survey, 1996–2020). The contribution of the United States has been estimated on the basis of an extrapolation of the production until 2006.

On the basis of an extrapolation of the results of an assessment by the US Geological Survey (2000) of deposits of gold, silver, copper, lead, and zinc in the United States in 1998, the global amount of boron resources can be estimated at about 3 Gt (see Table 2.4). We will analyze for the highest estimate (3 Gt) how primary boron production can be made sustainable, while simultaneously realizing a boron service level globally that is equal to boron's service level in developed countries in 2020.

7.4.4 Current stocks and flows

7.4.4.1 Recovery efficiency

According to Özdemir and Kipcak (2010) and Boncukcuoglu et al. (2003), 5% of the extracted boron ends up in waste sludge in Turkey, the country with the highest boron reserves in the world. This means that, at 95%, the boron recovery efficiency is quite high compared to the recovery efficiency of other resources.

7.4.4.2 Dissipation through use

The following boron applications are dissipative: fertilizers, detergents, chemicals, cosmetics, pharmaceuticals, industrial fluids, and metallurgical applications. In total, this concerns 35% (see column 1 of Table 7.4.1). Hence, a relatively important part of boron's use is dissipative.

7.4.4.3 In-use stock

Based on the end uses, we estimate the average lifetime of boron-containing products to be about 15 years (see Table 7.4.4).

Based on an average lifetime of 15 years and the boron production figures, we estimate that the increase in the annual in-use stock of boron is about 30% of the annual boron consumption. This estimate is based on the approach presented in Chapter 7.1.

7.4.4.4 Recycling

According to US Geological Survey (2019), the current recycling of boron is negligible.

7.4.4.5 Current boron flows summarized

Based on the data in this section, Fig. 7.4.4 presents the current anthropogenic boron flows.

Table 7.4.4 Estimation of the average lifetime of boron-containing products.

Application	End use (US Geological Survey, 2018a,b)	Lifetime (years)	Contribution to average lifetime (years)
Insulation material (glass wool)	30%	30	9
Glass fiber	13%	10	1.3
Borosilicate glass	7%	10	0.7
Fertilizer	13.5%	0	0
Ceramics	10%	20	2
Detergents and soaps	5%	0	0
Flame-retardants	1%	10	0.1
Chemicals	7.5%	0	0
Cosmetics, pharmaceuticals, and toiletries	2%	0	0
Industrial fluids	2%	0	0
Metallurgical applications	5%	0	0
Construction materials	4%	30	1.2
Total	100%	14.3	Rounded: 15 years

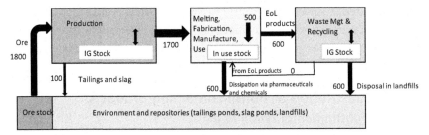

Figure 7.4.4 Simplified global anthropogenic boron cycle for 2017. Flows are in kt. Based on: end-of-life (EoL) recycling rate of 0%, recovery efficiency at the mining stage of 95%, increase of in-use stock of 30%, and in-use dissipation of 35%. The arrow widths are a rough indication of flow magnitudes. IG stock is industrial, commercial, and government stocks.

7.4.5 Sustainable boron flows

7.4.5.1 Introduction

A reduction in the use of primary raw materials can be realized by a mixture of (1) improvements in recovery efficiency, (2) substitution, (3) material efficiency improvement, (4) dissipation reduction, and (5) recycling. The approach will be to investigate these five types of economizing measures in the indicated order. For each of these measures, we investigate the possibilities to reduce the use of primary boron. Boron applications that can be adequately substituted do not need further attention in the framework of improving material efficiency, dissipation reduction, and recycling. The aim is a sustainable boron production rate while simultaneously realizing a global boron service level that is equal to boron's service level in developed countries in 2020, which was estimated at 600 g/capita/year (see Table 5.3).

7.4.5.2 Increase in recovery efficiency

According to Boncukcuoglu et al. (2003), it is possible to recover 90% of the boron in waste sludge at boron production locations in storage ponds, by acid leaching followed by precipitation and crystallization. This would lead to an overall recovery efficiency of 99.5%, which would be very high. Others (Uslu and Arol, 2004; Kavas, 2006; Christogerou et al., 2009) have studied the possible use of borax waste sludge as an additive in the production of red bricks and heavy clay ceramics. While this solves the environmental problems of borax waste sludge and reduces the use of other primary materials, it will not contribute to the recovery and efficient use of boron. As the grade of boron-containing ores will probably further decrease in the future, we assume that the recovery efficiency will remain at the current level of 95%.

7.4.5.3 Substitution of boron-containing products

7.4.5.3.1 Substitution of boron in glass

Glass wool can be substituted by various foams, by rock wool, or by natural fibers. Rock wool has approximately the same properties as glass wool, while insulation foams such as expanded polystyrene, extruded polystyrene, and polyurethane are currently used as an alternative for glass wool. Natural materials such as cellulose and cork can also be used for insulation. Contrary to glass wool and rock wool, synthetic foams and natural materials have the disadvantage that they are inflammable. There are several new developments in this field such as vacuum insulation panels, gas-filled panels, aerogels, and so-called phase change materials. An aerogel is a synthetic porous material derived from a gel, in which the liquid component of the gel has been replaced with a gas. The result is a solid with an extremely low density and low thermal conductivity. For a review of the state of the art in this field, we refer to the publications of Jelle (2011), Dewick and Miozzo (2002), and Papadopoulos (2005). The conclusion is that, in principle, it is possible to replace 100% of glass wool for insulation purposes with alternative materials.

Glass fibers in glass fiber reinforced plastics (GFRPs) may be partly substituted by other fibers such as carbon fiber, aramid fiber, high-modulus PE fiber, quartz fiber, basalt fiber, ceramic fiber, and natural fibers. Each of these fibers has a specific application area. Glass fiber is the oldest and by far the most common reinforcement in applications used to replace heavier metal parts, while carbon fiber is the most used fiber in high-performance applications. The general performance of natural fibers is not yet as high as the performance of glass fibers, in particular because of the hydrophilic nature of natural fibers. However, there has been some development in this area and, according to Faruk et al. (2014), natural fibers have the potential to replace a substantial part of glass fibers in reinforced plastics. The most popular natural fibers are: flax, jute, hemp, sisal, ramie, and kenaf (Faruk et al., 2014), while abaca, bamboo, wheat straw, curaua, and rice husk fibers are gaining interest.

Hence, the substitution of glass fiber with other types of fiber seems possible. We conservatively and provisionally assume that the substitutability of glass fiber with other types of fiber in glass fiber reinforced plastics is between 0% and 50%.

The substitution of *borosilicate glass* by a non-boron-containing material appears to be more difficult. In some applications, glass may be replaced with other materials. However, we have assumed that, for this boron application, the substitutability is zero.

The substitution of boron in glass wool (currently one of the key applications of boron) as an insulation material is not limited by the substitutes' performance, EHS properties, costs, or geological availability, as shown in Table 7.4.5.

7.4.5.3.2 Substitution of boron in ceramics

According to US Geological Survey (2019), boron in enamel may be replaced by phosphates. Furthermore, glass fiber from recycled GFRPs may (partly) replace the use of primary boron in ceramics (López et al., 2012a,b). According to a paper of the European Commission (2010), boron in ceramics is substitutable, but at high costs. We therefore assume that the substitutability of boron in ceramics is between 0% and 50%.

7.4.5.3.3 Substitution of boron in detergents and soaps

Perborate as a bleaching agent in detergents may be replaced by sodium percarbonate, and in soaps by sodium and potassium salts of fatty acids (US Geological Survey, 2019). In their use as an enzyme stabilizer, borates are

Table 7.4.5 Glass wool substitutability assessment.

	Substitutes			
	Rock wool	Foams	Natural fibers	References
Performance	0	0	0	Jelle, 2011; Papadopoulos, 2005
EHS properties	0	−	+	Papadopoulos (2005)
Costs	0	−	−	Papadopoulos (2005)
Geological availability of substitutes	++	0	++	

EHS, Environment, health, and safety.
Performance scale: ++: much better than original; 0: equal to original; −: much lower than original.
EHS scale: ++: much better than original; 0: equal to original; −: much worse than original.
Cost scale: ++: much cheaper than the original; 0: equal to original; −: much more expensive than the original.
Geological availability: ++: much less scarce than the original; 0: equal to original; − much scarcer than the original.

considered not to be substitutable (Risk and Policy Analysts, 2008). Over the period 2003–2008, the use of sodium perborate in detergents in western Europe decreased by around 80%, and was mostly substituted by sodium percarbonate (Risk and Policy Analysts, 2008). Nevertheless, compared to their use in the United States and the rest of the world, the use of perborates in detergents in Europe remains relatively high. It is however the case that sodium perborate performs better than sodium percarbonate in warmer climates (Risk and Policy Analysts, 2008). Table 7.4.6 presents the results of a substitutability assessment of the substitution of perborate in detergents with sodium percarbonate.

For the more general functions as an alkaline buffer, bleaching agent, water softener, surfactant performance improver, and for soil removal, boron-containing detergents and soaps can probably be partly substituted by non-boron-containing products.

Our conclusion is that we may assume prudently that the substitutability of boron in detergents and soaps is between 50% and 100%.

7.4.5.3.4 Substitution of boron in fertilizer
Boron as a micronutrient cannot be substituted.

7.4.5.3.5 Substitution of boron in flame-retardants
There are several alternative flame-retardant systems. Zinc borate is typically used as a synergist for other flame-retardant systems and is hardly used

Table 7.4.6 Substitutability assessment of perborate in detergents.

	Substitute sodium percarbonate	References
Performance	−	Risk and Policy Analysts (2008)
EHS properties	+	Risk and Policy Analysts (2008)
Costs	+	Alibaba (2014)
Geological availability of substitutes	+	

EHS, Environment, health, and safety.
Performance scale: ++: much better than original; 0: equal to original; −: much lower than original.
EHS scale: ++: much better than original; 0: equal to original; −: much worse than original.
Cost scale: ++: much cheaper than the original; 0: equal to original; −: much more expensive than the original.
Geological availability: ++: much less scarce than the original; 0: equal to original; − much scarcer than the original.

as a flame-retardant on its own. Therefore, it is assumed that replacement might be possible, but that it would decrease the performance of the applications in which it is used currently. Therefore, we assume that boron compounds can be substituted only with difficultly in its flame-retardant applications.

7.4.5.3.6 Substitution of boron in other applications

- Chemicals. Some of the boron applications are specific (e.g., lubricants, neutron moderators in nuclear power plants, semiconductors, airbags and magnets, abrasives and ballistic vests, electrolytic capacitors, and gold refining). Other boron applications are less specific and may be easily replaced, such as insecticides and wood preservatives, as a pH buffer, in starch adhesives, paints, coatings, and printing inks. We assume that boron's substitutability in its applications in chemicals is between 0% and 50%.
- Cosmetics and pharmaceuticals. In many cosmetics and pharmaceuticals, boron applications are not indispensable. Therefore, we assume that the substitutability of boron in its applications in cosmetics and pharmaceuticals is between 25% and 50%.
- Industrial fluids. Some of the boron applications in industrial fluids are relatively specific (as a lubricant, and in brake fluids, metal working fluids, and fuel additives). However, others are not specific (such as a pH buffer or corrosion inhibitor). We assume that the substitutability of boron in its applications in industrial fluids is between 25% and 50%.
- Metallurgical applications. The boron applications in steel production are quite specific. We assume that the substitutability of boron in its metallurgical applications is between 0% and 50%.
- Construction materials. Boron's applications in construction materials also are quite specific. It is assumed that boron's substitutability in construction materials is between 0% and 50%.

Based on the above assumptions, we consider three boron substitution scenarios (see Table 7.4.7).

Table 7.4.7 shows that the overall substitutability of primary boron could be as high as 57% of current boron consumption. In all scenarios, the determining factor for total substitutability is the 100% substitutability of glass wool by other insulation materials.

Table 7.4.7 Boron substitution scenarios.

Application	Proportion of applications	Substitutability			
		Scenario 0	Scenario 1	Scenario 2	Scenario 3
Glass applications					
- Glass fiber in glass wool	30%	0%	100%	100%	100%
- Glass fiber in glass fiber reinforced plastics and for high-strength textiles	13%	0%	0%	0%	50%
- Borosilicate glass	7%	0%	0%	0%	0%
Ceramics	10%	0%	0%	0%	50%
Detergents and soaps	5%	0%	50%	100%	100%
Fertilizer	13.5%	0%	0%	0%	0%
Flame-retardants	1%	0%	0%	0%	0%
Other applications					
- Chemicals	7.5%	0%	10%	25%	50%
- Cosmetics, pharmaceuticals	2%	0%	25%	50%	50%
- Industrial fluids	2%	0%	25%	50%	50%
- Metallurgical applications	5%	0%	0%	25%	50%
- Construction materials	4%	0%	0%	25%	50%
Total	100%	0%	34%	41%	57%

7.4.5.4 Material efficiency improvement

In the case of boron material efficiency, improvement will be especially a matter of minimizing use and decreasing losses (e.g., through precision fertilizer application).

Referring to Chapter 7.1, we assume a default material efficiency potential of 10% after the boron substitution potential has been used.

7.4.5.5 Dissipation reduction

Currently, 35% of boron use is dissipative. After the described substitution measures and material efficiency improvement, the dissipative use of boron is about 47% of the remaining boron consumption in scenario 1, 42% in scenario 2, and 50% in scenario 3. Hence, after substitution, the relative share of in-use dissipation of boron in the anthropogenic boron flows will further increase compared to the current situation.

7.4.5.6 In-use stock

As a result of boron-economizing measures and stabilizing global boron use per capita, the in-use stock of boron will stop increasing and reach a stable level.

7.4.5.7 Increase in recycling

Glass wool

Waste glass wool is generated in three ways: construction, renovation, and demolition. It is relatively easy to reuse or recycle glass wool from construction sites, because it is not polluted and easily separable, and some glass wool producers offer a take-back scheme for their products (Väntsi and Kärki, 2014). It is possible to return waste glass wool into the glass wool production process. However, the fine particles can clog the air and oxygen feeding equipment for the cupola furnace. To solve this problem, it is possible to briquet the waste glass wool using binder materials (Väntsi and Kärki, 2014). Construction waste impurities in waste glass wool may prevent recycling in the glass wool production process (Väntsi and Kärki, 2014). State-of-the-art selective demolition and sorting are therefore important for the adequate recycling of glass wool. Modern separation techniques are expected to enable reasonable results to be achieved in the separation of mineral wool from a mix of construction and demolition waste. However, glass wool is still barely recycled as the recycling costs (including the costs for separation and transport) are usually higher than the costs of the primary raw materials. The main challenges for the reuse of waste glass wool are its voluminous character, making transport relatively expensive, and the varying composition and availability. Conservatively, we shall assume a glass wool recycling potential of 10%.

Waste glass wool may be used as a raw material in other products, such as cement and concrete (Shi and Zheng, 2007), and ceramics and tiles, and as an artificial substrate for growing plants (Väntsi and Kärki, 2014). This approach reduces the use of other primary raw materials, but will not contribute to the sustainable use of boron as boron is downcycled in this way and finally dissipated into the environment.

Glass fiber in glass fiber reinforced plastics

The literature describes three potential recycling methods for glass fibers: chemical, thermal, and mechanical. An extensive overview of the possibilities for recycling GFRPs is provided by Gopalraj and Kärki (2020). Asmatulu et al. (2014) compared the methods, suggesting that chemical recycling offers the highest tensile strength of the recycled fibers (98% of the original strength), while mechanical and thermal recycling result in lower quality fibers (75% and 50%—75% of the strength of virgin fibers, respectively). Mechanical recycling consists of cutting and grinding the composite into small pieces, and separating the fibers from the rest of the particles. However, the recovered glass fibers lose between 18% and 30% of their strength, potentially limiting applicability. Beauson et al. (2014), in an investigation of the recyclability of glass fiber from wind turbine blades, confirmed this conclusion. Thermal treatment can encompass pyrolysis and combustion (López et al., 2012a,b; Zheng et al., 2009; Akesson et al., 2012; Asmatulu et al., 2014). The resulting glass fraction cannot be directly reused as glass fiber, but needs reprocessing because of the relatively low-quality mechanical properties. Contrary to this, Mizuguchi et al. (2013) reported on a thermal method to decompose GFRPs yielding the embedded reinforcing fibers in their original form without any noticeable difference between the virgin reinforcing fibers and the recovered ones. In the chemical treatment method, the polymeric matrix is dissolved in organic solvents or strong inorganic acids (see, e.g., Liu et al., 2006). Thermal treatment of GFRPs therefore may also be oriented to produce high-calorific oils and gases, as well as glass for glass–ceramic applications (as glaze) in the building sector. The recycled glass fiber, which can be as much as 99% of the glass fiber in the GFRP, can be used as glaze on tiles and may reduce primary boron use for ceramics glazing (López et al., 2012a,b). Asmatulu et al. (2014) also mention the possibility of "direct structural composite recycling." The idea is that large composite products are cut into smaller pieces that can be directly used in small composite products. García et al. (2014) investigated the addition of GFRP waste to microconcrete. Under specific conditions, the addition can be beneficial for the mechanical compressive and bending strength of the microconcrete. Although the use of other primary materials is prevented in this way, the method will not reduce boron consumption.

Although recycling of glass fiber is still in a developmental stage, research results are promising. Recycling will depend on the costs of recovering and recycling glass fiber-containing products versus the costs of primary boron. However, we estimate that, in practice, the recycling potential of boron from GFRPs is 25% maximum.

Borosilicate glass is recyclable if it is separated from other waste glass. However, borosilicate glass cannot be mixed with other glass for recycling, because of its impact on the viscosity of the melt. Recyclability is therefore assumed to be limited. We shall assume a maximum recycling potential for borosilicate glass of 40%.

Recycling of boron in ceramics

The majority of boron-containing waste ceramics consists of tiles and sanitary ware in construction and demolition waste. The rest is in broken tableware in municipal waste. The waste composition is complex and the overall boron concentration is low. We therefore assume that boron from these waste flows cannot be recycled.

Recycling of boron in other boron applications

The other boron applications are dissipative or are used in quantities and concentrations that are assumed to be too low to be suitable for economical recycling.

Overall boron recycling potential

Taking the above assumptions into consideration, the overall recycling potential of boron from end-of-life products after substitution is about 17% in scenario 1, 18% in scenario 2, and 21% in scenario 3.

7.4.6 Conclusions

The above observations regarding possible substitution, material efficiency improvement, and recycling are summarized in Table 7.4.8.

Table 7.4.8 Different mixes of measures to achieve a sustainable boron production rate, while realizing a global boron service level that is similar to the boron service level in developed countries in 2020. The boldfaced rows include the main (combination of) measures

		Sustainability ambition (years of sufficient availability		
		1000	500	200
Ultimately available boron resources[a]	Gt	3	3	3
Sustainable boron production[b]	Mt/year	3	6	15
Aspired future boron service level	kg/capita/year	0.6	0.6	0.6
Assumed future substitution	**%**	**41%**	**0%**	**0%**
		Substitution scenario 2	No substitution	No substitution
Assumed increase in material efficiency	%	10%	0%	0%
Derived future boron consumption level	kg/capita/year	0.3	0.6	0.6
Corresponding annual boron consumption	Mt/year	3	6	6
In-use dissipation of remaining products[c]	%	42%	35%	35%
Boron in end-of-life products	Mt/year	1.8	4	4
Required boron recycling of remaining products	Mt/year	0.2	0	0
Required boron recycling rate[d]	**%**	**10%**	**0%**	**0%**

It is assumed that the boron recovery efficiency remains at 95% in the future. The assumed maximum substitutability of boron in the future is 57%, and the assumed maximum material efficiency increase is 10%. The minimum in-use dissipation rate and the assumed maximum recyclability of the remaining products depends on the substitution scenario.

[a]Current recovery efficiency: 95%.
[b]It is assumed that the world population stabilizes at 10 billion people.
[c]Current in-use dissipation rate: 35% (see Section 7.4.4).
[d]Current end-of-life recycling rate: 0% (see Section 7.4.4).

The conclusions are:

- A sustainability ambition of 1000 years is achievable for boron on the condition of a substitution level of 41%, a material efficiency improvement of 10%, and an end-of-life recycling rate of 10%.
- Sustainability ambitions of 500 and 200 years are achievable without taking any measures compared to the current situation.

It must be noted that these observations assume a consumption ceiling of 600 g boron per capita per year, which represents the assumed present consumption level of boron in developed countries. However, if boron consumption in developed countries continues to grow beyond 600 g/capita/year, the sustainability ambitions will be more difficult to achieve for boron, in particular seen from the perspective of its high dissipation rate in combination with its low end-of-life recycling potential.

References

Akesson, D., Foltynowicz, Z., Christéen, J., Skrifvars, M., 2012. Microwave pyrolysis as a method of recycling glass fiber from used blades of wind turbines. J. Reinforc. Plast. Compos. 31, 1136—1142.

Alibaba, 2014. Price Sodium Perborate, Price Sodium Percarbonate. www.alibaba.com.

Asmatulu, E., Twomey, J., Overcash, M., 2014. Recycling of fiber-reinforced composites and direct structural composite recycling concept. J. Compos. Mater. 48, 593—608.

Beauson, J., Lilholt, H., Brondsted, P., 2014. Recycling solid residues recovered from glass-fibre-reinforced composites — a review applied to wind turbine blade materials. J. Reinforc. Plast. Compos. 33 (16), 1542—1556.

Boncukcuoğlu, R., Kocakerim, M.M., Kocadağistan, E., Yilmaz, M.T., 2003. Recovery of boron of the sieve reject in the production of borax. Resour. Conserv. Recycl. 37, 147—157.

Borax, 2019. Available from. www.borax.com/applications.

Christogerou, A., Kavas, T., Pontikes, Y., Koyas, S., Tabak, Y., Angelopoulos, G.N., 2009. Use of boron wastes in the production of heavy clay ceramics. Ceram. Int. 35, 447—452.

Crangle Jr., R.D., 2018. Boron, US Geological Survey 2016 Minerals Yearbook.

Dewick, P., Miozzo, M., 2002. Sustainable technologies and the innovation — regulation paradox. Futures 34, 823—840.

EFRA, 2007. The European Flame Retardants Association, Flame Retardants, Frequently Asked Questions.

European Commission, 2010. Critical Raw Materials for the EU, Report and Annex V of the Ad-Hoc Working Group on Defining Critical Raw Materials, 2010. European Commission, Enterprise and Industry.

European Commission, 2013. Critical Raw Materials for the EU, Critical Raw Materials Profiles, Enterprise and Industry.

Faruk, O., Bledzki, A.K., Fink, H.P., Sain, M., 2014. Progress report on natural fiber reinforced composites. Macromol. Mater. Eng. 299, 9—26.

Fedak, D., Baldwin, C., 2005. A comparison of enameled and stainless steel surfaces. In: Proceedings of the 67th Porcelain Enamel Institute Technical Forum, pp. 45—54.

Fitzer, E., Kleinholz, R., Tiesler, H., Stacey, M.H., De Bruyne, R., Lefever, I., Foley, A., Frohs, W., Hauke, T., Heine, M., Jäger, H., Sitter, S., 2008. "Fibers, 5.Synthetic Inorganic" Ullmann's Encyclopedia of Industrial Chemistry. Wiley-VCH Verlag GmbH & Co. KGaA, Weinheim, Germany.

García, D., Vegas, I., Cacho, I., 2014. Mechanical recycling of GFRP as short-fiber reinforcements in micro concrete. Construct. Build. Mater. 64, 293—300.

Gopalraj, S.K., Kärki, T., 2020. A review on the recycling of waste carbon fibre/glass fibre — reinforces composites: fibre recovery, properties and life-cycle analysis. SN Appl. Sci. 2, 423. https://doi.org/10.1007/S42452-020-2195-4.

Jelle, P.B., 2011. Traditional, state-of-the-art and future thermal building insulation materials and solutions — properties, requirements and possibilities. Energy Build. 43, 2549—2563.

Kavas, T., 2006. Use of boron waste as a fluxing agent in production of red mud brick. Build. Environ. 41, 1779—1783.

Liu, Y., Meng, L., Huang, Y., Liu, L., 2006. Method of recovering the fibrous fraction of glass/epoxy composites. J. Reinforc. Plast. Compos. 25, 1525—1533.

López, F.A., Martin, M.I., Alguacil, F.J., Rincón, JMa, Centeno, T.A., Romera, M., 2012a. Thermolysis of fibre glass polyester composite and reutilization of the glass fibre residue to obtain a glass-ceramic material. J. Anal. Appl. Pyrol. 93, 104—112.

López, F.A., Martin, M.I., García-Díaz, I., Rodriguez, O., Alguacil, F.J., Romero, M., 2012b. Recycling of glass fibres from fiberglass polyester waste composite for the manufacture of glass-ceramic materials. J. Environ. Protect. 3, 740—747.

Lyday, P.A., 2003. Boron, US Geological Survey Minerals Yearbook 2003.

Mizuguchi, J., Tsukada, Y., Takahashi, H., 2013. Recovery and characterization of reinforcing fibres from fibre reinforced plastics by thermal activation of oxide semi-conductors. Mater. Trans. 54 (3), 384—391.

Özdemir, M., Kipçak, I., 2010. Recovery of boron from borax sludge of boron industry. Miner. Eng. 23, 685—690.

Papadopoulos, A.M., 2005. State of the art in thermal insulation materials and aims for future developments. Energy Build. 37, 77—86.

Risk & Policy Analysts, 2008. Assessment of the Risk to Consumers from Borates and the Impact of Potential Restrictions on Their Marketing and Use, Final Report Prepared for the European Commission DG Enterprise and Industry.

Shi, C., Zheng, K., 2007. A review of the use of waste glasses in the production of cement and concrete. Resour. Conserv. Recycl. 52, 234—247.

Statista, 2019. Available from: www.statist.com/statistics/449828/world-wide-distribution-of-boron-end-use-by-application. (Accessed October 15 2019).

US Geological Survey, 1996—2020. Mineral Commodity Summaries, Boron.

US Geological Survey, 2000. 1998 Assessment of undiscovered deposits of gold, silver, copper, lead and zinc in the United States. Circular 1178, 2000.

US Geological Survey, 2005. Boron End-Use Statistics.

US Geological Survey, 2013. Mineral Commodity Summaries, Boron.

US Geological Survey, 2016. Mineral Commodity Summaries, Boron.

US Geological Survey, 2017a. Mineral Commodity Summaries, Boron.

US Geological Survey, 2017b. Boron Statistics.

US Geological Survey, 2018a. Mineral Commodity Summaries, Boron.

US Geological Survey, 2018b. Boron, 2016 Minerals Yearbook.

US Geological Survey, 2019. Mineral Commodity Summaries, Boron.

US Geological Survey, 2020. Mineral Commodity Summaries, Boron.

Uslu, T., Arol, A.I., 2004. Use of boron waste as an additive in red bricks. Waste Manag. 24, 217–220.

Väntsi, O., Kärki, T., 2014. Mineral wool waste in Europe: a review of mineral wool waste quantity, quality and current recycling methods. J. Mater. Cycles Waste Manag. 16, 62–72.

Zheng, Y., Shen, Z., Ma, S., Cai, C., Zhao, X., Xing, Y., 2009. A novel approach to recycling of glass fibers from non-metal materials of waste printed circuit boards. J. Hazard Mater. 170, 978–982.

CHAPTER 7.5

Chromium

7.5.1 Introduction

Chromium is essential for the production of stainless steel, and more than 90% of chromium is used for the production of different types of stainless steel. The purpose of this section is to explore whether and under which conditions the extraction rate of chromium can be made sustainable, while simultaneously increasing the global service level of chromium to that in developed countries in 2020.

7.5.2 Properties and applications

Chromium has a melting point of 1907°C and a boiling point of 2672°C. The density of chromium is 7.19 g/cm^3. Elemental chromium is extremely hard and also highly resistant to tarnishing. The main applications of chromium are in the metallurgical sector, as a chemical and refractory material. Refractories are materials that retain strength and form at high temperatures, such as fire bricks. Refractory materials are used, for example, in furnaces and kilns.

The proportion of chromium that is used in the metallurgical sector has grown rapidly over the years. Fig. 7.5.1 shows on the other hand that the proportional use of chromium in refractory materials in the United States was still substantial in the 1930s—1960s, but that it has become practically negligible compared to the rest of the current chromium applications.

In the metallurgical industry, most chromium is alloyed with iron for making stainless steel. The remainder is used to make nonferrous alloys, such as aluminum, cobalt, copper, nickel, and titanium alloys. Stainless steel and other chromium alloys are used in a wide variety of products, durable goods and sectors including tools, kitchen appliances, all types of transportation, electrical equipment, machinery, and the chemical industry.

In 2016, chromium end use in steel making accounted for 96% of consumption in the United States. Super alloys and other end uses of chromium (chemicals and refractories) made up only 4% (US Geological Survey, 2020b). According to a study by Johnson et al. (2006a,b), the worldwide chromium application in steel was about 90%, of which 86% in stainless steel, about 6.5% in chemicals, and 3.5% in refractories.

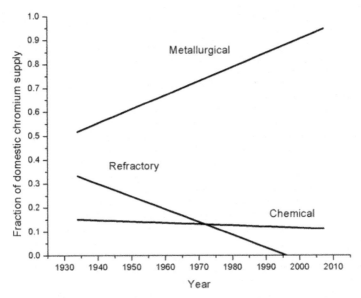

Figure 7.5.1 Trend of distribution of apparent consumption of chromium in the United States among major sectors of application (metallurgical, chemical, and refractory) (Papp, 2004).

There are many types of stainless steel, such as chromium steel with 12%—13% chromium, austenitic steels containing 18% chromium, and heat-resistant steels with 25%—30% chromium (Danish Environmental Protection Agency, 2003). Chromium hardens steel and increases the resistance of steel to corrosion, especially at high temperatures. The chromium content in nonferrous alloys varies. In some alloys, the chromium content is much lower than in stainless steel, for instance, in aluminum (0.04%), copper (0.025%), and titanium (0.49%). In other alloys, the chromium content is higher, such as in cobalt (20%) and nickel (14%) (Papp, 1994).

The second most important use of chromium, though small in comparison to chromium's use in stainless steel and other alloys, is in chemicals. The main applications of chromium-containing chemicals are for tanning leather, for pigments in paint and plastic, and for surface treatment (e.g., chromium plating).

Table 7.5.1 presents the proportional chromium consumption in Denmark in different sectors in 1999.

Table 7.5.1 Consumption of chromium, chromium compounds, and chromium as a trace constituent in Denmark in 1999 (Danish Environmental Protection Agency, 2003).

Metallurgical sector	97%
Iron and steel	97%
Aluminum alloys	0.2%
Copper alloys	0.03%
Other chromium use	2.2%
Tanning	0.8%
Pigments	0.23%
Surface treatment	0.14%
Other	0.91%
Chromium as trace constituent	0.8%
Coal and oil	0.53%
Cement	0.24%
Total	100%

Most leather products are *chromium tanned*. This concerns products such as footwear, gloves, jackets, furniture, bags, wallets, belts, and upholstery for cars. Tanning takes place with a double salt consisting of potassium sulfate and chromium sulfate, and shoe leather contains about 2% chromium (Danish Environmental Protection Agency, 2003).

Of the chromium-based *pigments*, the yellow lead chromate and zinc chromate and the greenish chromium oxide are popular (Danish Environmental Protection Agency, 2003).

Chromium-based *surface treatment* includes chromium plating, hard chromium plating, black chromium plating, zinc passivation, chromating of aluminum, chromic acid pickling of aluminum, chromic acid pickling of plastic, anodizing of aluminum, and chromium passivation after phosphatizing (Danish Environmental Protection Agency, 2003). Decorations, for example of automobiles, are chromium plated, because chromium has a unique combination of the following properties: high specular reflection, high hardness, and high resistance to corrosion.

In the chemical industry, hexavalent chromium is used as an *oxidizer* (in the form of potassium dichromate in concentrated sulfuric acid) and in *catalysts* (Danish Environmental Protection Agency, 2003). In this framework, it is relevant to mention that hexavalent chromium compounds are classified as toxic and carcinogenic.

The third type of application of chromium is in refractories. Today, the iron and steel industry and metal casting sectors use approximately 70% of

Table 7.5.2 The top 10 internationally traded products in 2000 ranked according to chromium content (kt) traded. These products contained about 15% of all chromium entering use.

	Assumed Cr content	Global trade (kt Cr)	% of all chromium entering use in 2000
Passenger cars	0.27%	91	3.1%
Stainless steel kitchen items	18%	88	3.0%
Automotive exhaust systems	9.8%	45	1.5%
Cutlery	11.2%	42	1.4%
Turbo jets and other gas turbine engines	16%	40	1.3%
Pumps	2.70%	40	1.3%
Food processing machines	14.4%	37	1.2%
Tanning chemicals	17%	37	1.2%
Tableware (not cutlery)	12.3%	32	1.1%
Taps, cocks, and valves	0.72%	24	0.8%

Derived from Johnson, J., Schewel, L., Graedel, T.E., 2006. The contemporary in-use chromium cycle. Environ. Sci. Technol. 40, 7060−7069.

all refractories produced (Garbers-Craig, 2008). The use of chromium in refractories is very small compared to the total use of chromium.

Globally, 25% of chromium is used in buildings and infrastructure, 25% in industrial machinery, 15% in transportation, 5% in household appliances and electronics, and 30% in a variety of metal goods and parts (Johnson et al., 2006a,b). Table 7.5.2 gives an impression of some of the important end uses of chromium worldwide. The table illustrates only the top 10 internationally traded products, representing about 15% of all chromium use.

7.5.3 Production and resources

7.5.3.1 Production

The global production of chromium has increased rapidly. Between 1980 and 2015, the annual production increase was an average of about 3.5% (see Fig. 7.5.2). In 2019, the main production countries were South Africa,

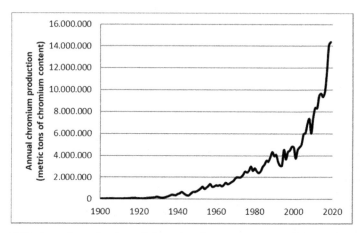

Figure 7.5.2 Annual production of chromium between 1900 and 2019. The production in 2019 was about 14.4 Mt of chromium content. *(Data derived from US Geological Survey, 2017a. Chromium Statistics and US Geological Survey, 2020a. Mineral Commodity Summaries, Chromium, 1996–2020.)*

Turkey, Kazakhstan, and India, which had a combined total of 86% of global chromium production (see Fig. 7.5.3).

The main source of chromium is chromite ore. The beneficiation of chromite ore includes crushing, grinding, and separation. Separation includes gravimetric and magnetic techniques and flotation in heavy media.

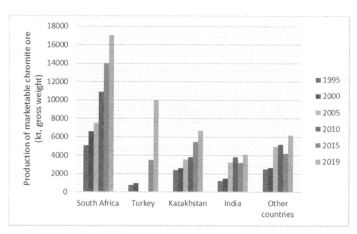

Figure 7.5.3 Main chromite-producing countries between 1995 and 2019. *(Derived from US Geological Survey, 2020a. Mineral Commodity Summaries, Chromium, 2020.)*

The result is a product that is suitable for further processing to ferrochromium (as a raw material for the metallurgical industry), to chromium-based chemicals, or refractories. To produce ferrochromium, chromite ore is fed into an electric arc furnace and heated in the presence of carbon to a temperature of at least 1100–1200°C. The ferrochromium production is energy intensive. The products of this process (high-carbon ferrochromium, low-carbon ferrochromium, and ferrochromium-silicon) are used to add chromium to different metal alloys (Papp, 1994). The chromium content of the different ferrochromium products is between 50% and 70% (ASTM, 2016).

7.5.3.2 Resources

The chromium concentration in the Earth's crust is 110 ppm (see Table 2.2). Chromium does not occur naturally. Although chromium occurs in many types of rocks in small quantities in oxides and silicate minerals, its main production source is chromite. Pure chromite consists of $FeO.Cr_2O_3$ (or $FeCr_2O_4$). However, in nature, Fe^{2+} can be substituted by Mg^{2+} and Cr^{3+} can be substituted by Al^{3+} or Fe^{3+}. Therefore, chromite from mines is an oxide of iron, magnesium, aluminum, and chromium with small amounts of oxides of calcium and silicon. The Cr_2O_3 content in chromite ore can vary between 25% and 65%. The compositions of some typical chromite ores are presented in Table 7.5.3.

In 2000, the chromium content in chromite ore was between 22% and 34%, depending on the production country (Johnson et al., 2006a,b). Mined chromite ore is beneficiated at the mine site to marketable chromite

Table 7.5.3 Chemical composition (weight percent) of some typical chromite ores (Gazulla et al., 2006).

	Cuba	Turkey	Russia	South Africa
Cr_2O_3	30.8	37.1	39.1	44.5
Al_2O_3	29	24.4	17.4	15
MgO	18.9	17.7	16.1	10
FeO equivalent[a]	13	13.8	13.9	24.7
CaO	0.91	0.22	0.65	0.31
SiO_2	5.3	4.3	9.4	3.9
Total	97.91	97.52	96.55	98.41

[a]The amount of iron contained in chromite is presented as being 100% FeO.

ore containing at least 45% Cr_2O_3. The chromium content of pure Cr_2O_3 is 68.4%, which means that marketable chromite ore contains a minimum of 30.8% chromium. In practice, the chromium content of marketable chromite ore is between 30.8% and 38%, with an average of 32.6% between 1994 and 2015 (derived from US Geological Survey 2017 and 2020a).

According to US Geological Survey (2020), chromium reserves, expressed in chromium content, are 175 Mt. According to the same source, the world's identified chromium resources are >3.7 Gt of chromium content and are mainly concentrated in Kazakhstan and South Africa.

If we use the approach that the extractable chromium resources are a maximum of about of 0.01% of the total amount of chromium in the upper 3 km of the Earth's crust (UNEP, 2011; Erickson, 1973; Skinner, 1976), chromium resources are 10 Gt (rounded; see Table 2.3). A third approach for estimating the ultimately extractable chromium resources is to extrapolate the results of assessments of the extractable resources of other elements. US Geological Survey (2000) compared the results of 19 assessments of the estimated total (identified plus undiscovered) deposits of gold, silver, copper, lead, and zinc in the United States in the upper 1 km of the continental crust of the United States (a) with the identified resources of the same elements in the United States that far (b). The a/b ratios were 4.82 for zinc, 3.88 for silver, 2.67 for lead, 2.2 for gold, and 2.12 for copper, (average is 3.14). We apply this average ratio to the most recent US Geological Survey data regarding the identified global resources of chromium (3.7 Gt). Extrapolating the result to the upper 3 km crust, this approach results in globally available chromium resources of 3 x 3.14 x 3.7 = 35 Gt (rounded). In the framework of the analysis of the conditions for the sustainable production of chromium, we use two estimates: 10 Gt (the lower estimate), and 35 Gt (the higher estimate).

7.5.4 Current stocks and flows

7.5.4.1 Recovery efficiency

Losses at the beneficiation stage are estimated at about 14.5% (tailings) and the additional net losses at the stage of further chromite ore processing, in particular the production of ferrochromium, are 11.5% (slag). Therefore, the total recovery efficiency of chromium at the mining and production stages is 74% (Johnson et al., 2006a,b).

7.5.4.2 Dissipation through use

For the described types of applications of chromium, there is practically no in-use dissipation of chromium. Chromium-containing chemicals such as oxidizing agents, pigments, and tanning liquids, end up in landfills or incineration plants through the disposal of end-of-life (EoL) products such as leather and painted objects.

7.5.4.3 In-use stock

Lifetimes of chromium-containing products are presented in Table 7.5.4. According to Gerst and Graedel (2008), the average lifetime of chromium-containing products is about 25 years. Related to the amount of chromium compounds produced and flowing to fabrication and manufacturing, the chromium flow into the in-use stock in 2000 was about 57% globally (Johnson et al., 2006a,b).

Related to end use alone, this was 78%. Hence, the amount of chromium in end-of-life products in 2000 was only about 22% of the annual consumption globally (Johnson et al., 2006a,b). This is why the recycled content of chromium in consumer products is quite low.

7.5.4.4 Recycling

Chromium is primarily recycled as part of stainless steel scrap, and small amounts of chromium are recycled in super alloys (US Geological Survey, 2004). The recycled chromium content as part of the consumption in the United States between 1995 and 2019 is presented in Fig. 7.5.4. In that period, the recycled chromium content was generally between 20% and 40%, with an average of 32%. In 2000, the global chromium end-of-life recycling rate was 34% (Johnson et al., 2006a,b).

Table 7.5.4 Lifetimes of chromium-containing products (Oda et al., 2010).

End use	Mean lifetime (years)
Buildings and construction	28.9
Infrastructure	34.5
Industrial machinery	30.0
Electrical and electronic equipment	12.1
Household appliances	20.3
Passenger cars	9.7
Trucks	11.3
Other transportation	40.0
Containers	30.0

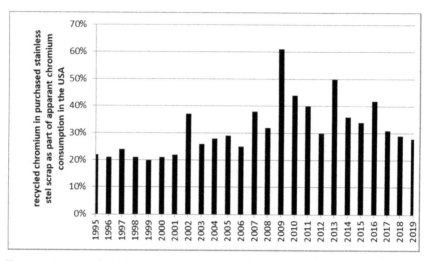

Figure 7.5.4 Recycled chromium content as a proportion of apparent chromium consumption in the United States (US Geological Survey 2020a). About half of the recycled chromium is old scrap and the other half of it is prompt scrap and home scrap.

7.5.4.5 Downcycling

When a chromium-containing alloy such as stainless steel is recycled to the carbon steel cycle or a metal cycle other than stainless steel, chromium loses its functionality. In fact, this *downcycled* chromium cannot be used anymore in stainless steel or in another specific chromium application. In 2000, it was estimated that, globally, about 35% of chromium in end-of-life products and industrial waste was downcycled in this way (Johnson et al., 2006a,b). It is assumed that downcycling mainly concerns chromium waste generated during fabrication and manufacture, but it also partly concerns chromium in end-of-life products (Reck and Gordon, 2008). The recycling rate of chromium from austenitic stainless steels is considerably higher than the recycling rate of chromium from nonaustenitic stainless steels (Daigo et al., 2010; Oda et al., 2010; Nakajima et al., 2013). The reason for this is that austenitic stainless steels are nonferromagnetic, whereas nonaustenitic steels are ferromagnetic. Hence, separating austenitic stainless steel from carbon steels is relatively easy by applying magnetic separation. According to Daigo et al. (2010), the recycling rate of chromium from austenitic stainless steel is more than 75%, whereas the recycling of chromium from ferritic stainless steel is less than 40%. Ferritic stainless steel contains, apart from iron and carbon, only chromium, whereas austenitic stainless steel also contains molybdenum and/or nickel.

7.5.4.6 Disposal

The disposal of chromium occurs with industrial waste, hazardous waste, waste from electronic and electrical equipment, municipal solid waste, end-of-life vehicles, construction and demolition waste, sewage sludge, and industrial machinery and large transportation means, such as trains, ships, and airplanes. Chromium concentrations in some of these flows are presented in Table 7.5.5. Two thirds of the chromium that is finally disposed of originates from industrial and hazardous waste, while only one third of chromium-containing waste is from consumer products. A large part of the industrial waste is produced in the fabrication of stainless steel, and includes stainless steel slag and dust (Johnson et al., 2006a,b; see Fig. 7.5.5).

7.5.4.7 Current chromium flows summarized

On the basis of the figures above, Fig. 7.5.6 presents the estimated global chromium flows and stocks in 2019.

The conclusion is that about half of all chromium contained in chromite ore is currently lost in tailings and slags, disposal sites, and through the downcycling of chromium to other metal cycles.

7.5.5 Sustainable chromium flows

7.5.5.1 Introduction

In this section, we explore how the use of primary chromium can be made sustainable. The sustainable use of chromium means that there remains enough chromium to keep providing chromium for a world population of 10 billion people for an agreed period of time at a chromium service level

Table 7.5.5 Chromium content in different waste flows in 2000.

Waste flow	Approximate chromium content in 2000 (mg/kg)	Source
WEEE scrap	730	Weigand et al. (2003)
Municipal solid waste	210	Weigand et al. (2003)
End-of-life vehicles	2700	Staudinger and Keoleian (2001)
Construction and demolition waste	150	Brunner and Stampfli (1993)

WEEE is waste from electronic and electrical equipment.

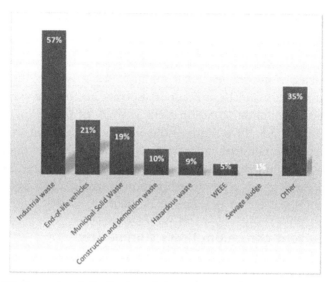

Figure 7.5.5 Proportional distribution of chromium in different waste flows in 2000 (Johnson et al., 2006a,b).

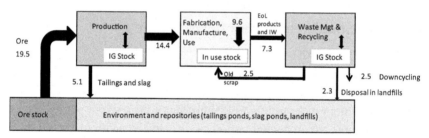

Figure 7.5.6 Simplified scheme of estimated chromium flows in 2019. The indicated amounts are expressed in Mt. Based on: a chromium recovery efficiency of 74%; in-use dissipation of 0%; increase of in-use stocks is 57% of consumption; old scrap is 35% of end-of-life products plus industrial waste; downcycling is 35%. These assumptions are derived from Johnson et al. (2006a,b). The arrow widths indicate the relative volume of the flows. IG stock is industry and government stocks, and IW is industrial waste.

equal to the level in developed countries in 2020. The current chromium consumption in developed countries is about 8 kg per capita per year (see Table 5.3).

7.5.5.2 Increase in recovery efficiency

The recovery efficiency of chromium at the mining and production stages, including the production of ferrochromium, is currently 74% (see Section 7.5.4). We assume that this can be improved to 80% in the future. Though a further improvement in the chromium recovery efficiency might be technically possible, we assume that this will not be feasible financially.

7.5.5.3 Substitution of chromium-containing products

There is no substitute for chromite ore in the production of ferrochromium, chromium chemicals, and chromium refractories. There is also no substitute for chromium in stainless steels, nor for chromium in super alloys (US Geological Survey, 2020a; Oakdene Hollins Fraunhofer, 2013; Johnson et al., 2006a,b).

Hence, we assume that chromium's substitutability is 0%.

7.5.5.4 Material efficiency improvement

We assume that future technological developments will enable a material efficiency increase of 10% (see Section 7.1).

7.5.5.5 Increase in recycling

The increase in chromium recycling regards both recycling at the fabrication and manufacturing stage and the recycling of chromium from end-of-life products. Improving chromium recycling is the remaining approach to decrease the use of primary chromium to a lower level, while simultaneously increasing the chromium service level of a world population of 10 billion people to the service level in developed countries in 2020.

The sorting of steel scrap can be improved by better identifying and specifying the material composition of scrap, and the mixing of alloys with carbon steel must be avoided. The selective dismantling of parts with a high content of alloying elements from end-of-life vehicles is one way to more easily recover chromium from nonaustenitic stainless steels. Using sensor-based scrap sorting processes, such as X-ray transmission and X-ray fluorescence sorters, it is possible to separate high chromium and high molybdenum content scrap from other steel scrap (Nakajima et al., 2013; Mesina et al., 2007). According to a study by Nakamura et al. (2017), a sorting process that separates carbon steel from stainless steel and other alloy steels from carbon steels is able to retain more than 70% of chromium and nickel for functional use over a period of 100 years. However, in a scenario in which only austenitic stainless steel is separated from the rest, less than 30% of chromium and nickel can be retained for functional use over a period of 100 years. The conclusion is that design for recycling, in which the different alloy steels in a product receive clear electronic IDs, will improve the selective sorting of stainless steel and other chromium-containing alloy steels from a scrap mix and will keep stainless steel away from the carbon steel cycle. This is crucial for increasing the end-of-life recycling rate of chromium. Considering the above observations, we assume a maximum chromium recycling rate from end-of-life products and industrial waste of 70%.

7.5.6 Conclusions

In Table 7.5.6, we calculate the required increase in the chromium recycling rate to achieve the sustainable use of primary chromium, based on various assumptions concerning the available resources and sustainability horizons.

The conclusions are as follows:
- A sustainability ambition of 1000 years is only achievable with >20 Gt of ultimately available chromium resources combined with maximum recycling. With ultimately available chromium resources of 35 Gt, a sustainability ambition of 1000 years is possible with 57% end-of-life recycling.
- A sustainability ambition of 500 years is achievable with 10 Gt of ultimately available chromium resources in combination with maximum recycling. With >30 Gt of ultimately available chromium resources, a sustainability ambition of 500 years is possible with no additional measures compared to now.
- A sustainability ambition of 200 years is achievable with ultimately available chromium resources of 10 Gt with an end-of-life recycling rate of 52%. With >12 Gt of ultimately available chromium resources, a sustainability period of 200 years is possible with no additional measures compared to now.

Table 7.5.6 Different mixes of measures to achieve a sustainable chromium production rate, while realizing a global chromium service level that is similar to the chromium service level in developed countries in 2020. The main measures are presented in boldfaced rows.

		Sustainability ambition (years of sufficient availability)					
		1000		500		200	
Ultimately available chromium resources[a]	Gt	20	35	10	30	10	12
Assumed future recovery efficiency[b]	%	80%	74%	80%	74%	80%	74%
Future availability of chromium resources	Gt	21.5	34.1	10.9	29.6	10.5	11.9
Sustainable chromium production[c]	Mt/ year	21.5	34.1	21.9	59.2	52.6	59.4
Aspired future chromium service level	kg/ cap/ year	8	8	8	8	8	8
Assumed future substitution	%	0%	0%	0%	0%	0%	0%
Assumed increase in material efficiency	**%**	**10%**	**0%**	**10%**	**0%**	**0%**	**0%**
Aspired future chromium consumption level	kg/ cap/ year	7.2	8	7.2	8	8	8
Corresponding annual chromium consumption	Mt/ year	72	80	72	80	80	80
Assumed dissipation[d]	%	0%	0%	0%	0%	0%	0%
Accumulation in-use stocks	%	0%	0%	0%	0%	0%	0%
Chromium in end-of-life products	Mt/ year	72	80	72	59	53	59
Required chromium recycling	Mt/ year	51	46	50	21	27	21
Required chromium recycling rate[e]	**%**	**70%**	**57%**	**70%**	**35%**	**52%**	**35%**

The assumed maximum recovery efficiency in the future is 80%, and the assumed maximum substitutability of chromium in the future is 0%. The assumed material efficiency increase is 10%, and the assumed minimum in-use dissipation rate is 0%. The assumed maximum recyclability of end-of-life products and industrial waste in the future is 70%.

[a] Low estimate: 10 Gt; high estimate: 35 Gt (see Section 7.5.3).
[b] Current recovery efficiency: 74% (see Section 7.5.4).
[c] It is assumed that the world population stabilizes at 10 billion people.
[d] Current in-use dissipation rate: 0% (see Section 7.5.4).
[e] Current recycling rate of end-of-life products and industrial waste: 35% (see Section 7.5.4).

References

ASTM, 2016. Standard Specification for Ferrochrome — Silicon, ASTM A482/A482M-11, 2016.

Brunner, P.H., Stampfli, D., 1993. The material balances of a construction waste sorting plant. Waste Manag. Res. 11, 27—48.

Daigo, I., Matsuno, Y., Adachi, Y., 2010. Substance flow analysis of chromium and nickel in the material flow of stainless steel in Japan. Resour. Conserv. Recycl. 54, 851—863.

Danish environmental protection Agency, 2003. Mass flow analysis of chromium and chromium compounds. Environmental. Project No 793.

Erickson, R.L., 1973. Crustal occurrence of elements, mineral reserves and resources. U. S. Geol. Surv. Prof. Pap. 820, 21—25.

Garbers-Craig, A.M., 2008. How cool are refractory materials? J.Southern Afr. Inst. Mining Metall. 106, 1—16.

Gazulla, M.F., Barba, A., Gómez, M.P., Orduña, M., 2006. Chemical characterization of chromites by XRF spectrometry. Geostand. Geoanal. Res. 30 (3), 237—243.

Gerst, M.D., Graedel, T.E., 2008. In-use stocks of metals: status and implications. Environ. Sci. Technol. 42 (19), 7038—7045.

Johnson, J., Schewel, L., Graedel, T.E., 2006a. The contemporary anthropogenic chromium cycle. Environ. Sci. Technol. 40, 7060—7069.

Johnson, J., Schewel, L., Graedel, T.E., 2006b. The contemporary anthropogenic chromium cycle, supplemental information. Environ. Sci. Technol. 40, 7060—7069.

Mesina, M.B., De Jong, T.P.R., Dalmijn, W.L., 2007. Automatic sorting of scrap metals with a combined electromagnetic and dual energy X-ray transmission sensor. Int. J. Miner. Process. 82, 222—232.

Nakajima, K., Ohno, H., Kondo, Y., Matsubae, K., Takeda, O., Miki, T., Nakamura, S., Nagasaka, T., 2013. Simultaneous material flow analysis of nickel, chromium, and molybdenum used in alloy steel by means of input-output analysis. Environ. Sci. Technol. 47, 4653—4660.

Nakamura, S., Kondo, Y., Nakajima, K., Ohmo, H., Pauliuk, S., 2017. Quantifying recycling and losses of Cr and Ni in steel throughout multiple life cycles using MaTrace-alloy. Environ. Sci. Technol. 51, 9459—9476.

Oda, T., Daigo, I., Matsuno, Y., Adachi, Y., 2010. Substance flow and stock of chromium associated with cyclic use of steel in Japan. ISIJ Int. 50 (2), 314—323.

Oakdene, H.F., 2013. Study on Critical Raw Materials at EU Level, Final Report for DG Enterprise and Industry.

Papp, J.F., 2004. Chromium use by market in the United States. In: Proceedings Tenth International Ferroalloys Congress, INFACON X: "Transformation through Technology", 1—4 February, 2004, Cape Town, South Africa.

Papp, 1994. chromium life cycle study, US Bureau of Mines. Information Circular 9411.

Reck, B.K., Gordon, R.B., July 2008. Nickel and chromium cycles: stocks and flows project Part IV. JOM 55—59.

Skinner, B.J., 1976. A second iron age ahead? Am. Sci. 64, 158—169.

Staudinger, J., Keoleian, Ga, 2001. Management of End-of-Life Vehicles (ELVs) in the US, Report No CSS01-01, University of Michigan, Ann Arbor, MI.

UNEP, 2011. International Panel on Sustainable Resources Management, 2011, Estimating Long-Run Geological Stocks of Metals, International Panel on Sustainable Resources Management, Working Group on Geological Stocks of Metals, Working Paper.

US Geological Survey, 2000. 1998 Assessment of undiscovered deposits of gold, silver, copper, lead and zinc in the United States. Circular 1178.

US Geological Survey, 2004. Chromium Recycling in the United States in 1998, US Geological Survey Circular 1196-C.

US Geological Survey, 2017. Chromium Statistics, January, 19, 2017.
US Geological Survey, 2020a. Mineral Commodity Summaries, Chromium, 1996—2020.
US Geological Survey, 2020b. 2016 Minerals Yearbook, Chromium, January 2020.
Weigand, H., Fripan, J., Przybilla, L., Marb, C., 2003. Composition and contaminant loads of household waste in Bavaria, Germany. In: Proceedings of the Nineth International Waste Management and Landfill Symposium, Sardinia, Italy, October 6—10, 2003.

CHAPTER 7.6

Copper[4]

7.6.1 Introduction

Copper is a crucial metal for applications such as the generation, transmission, and distribution of electricity, and for electric motors. In this section, we investigate how to achieve the sustainable extraction of copper while simultaneously realizing a copper service level globally at a level that is equal to the service level of copper in developed countries in 2020.

7.6.2 Properties and applications

Copper is strong, ductile, and resistant to creeping. Copper's density is 8.96 g/cm^3, its melting point is 1083°C, and its boiling point is 2570°C.

Copper's properties make this metal particularly appropriate for industry and households. It is a very good conductor of heat and electricity, resists corrosion and is antimicrobial. Copper is an essential micronutrient for life on Earth, including for humans, animals, and plants.

Copper has a large array of applications (see Fig. 7.6.1, Fig. 7.6.2 and Table 7.6.1).

A laptop contains in the order of 0.1 kg of copper, a car 20 kg, a typical modern home contains in the order of 100 kg copper, and a wind turbine in the order of 5000 kg (British Geological Survey, 2007; International Copper Study Group, 2017; European Copper Institute, 2018).

Copper is used in many alloys, for example:

Copper-zinc alloys: brass with 55%—70% copper and 30%—45% zinc.

Copper-tin alloys: bronze with 70%—90% copper and 10%—30% tin.

Copper-nickel alloys: for example, 75% copper and 25% nickel, used for coins (Fig. 7.6.2).

---

[4] This section is based on material from the publication of Henckens, M.L.C.M., Worrell, E., 20120. Reviewing the availability of copper and nickel for future generations. The balance between production growth, sustainability and recycling rates. J. Cleaner Prod. 264. 2020, 121460.

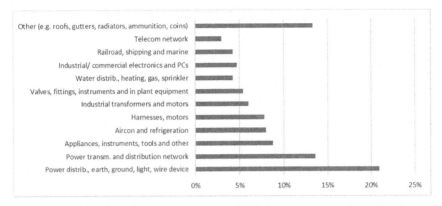

Figure 7.6.1 Global end uses of copper in 2017 (based on data from the Copper Alliance, 2019).

Figure 7.6.2 Copper-nickel alloy is used for coins. *(Photo by Theo Henckens.)*

7.6.3 Production and resources

7.6.3.1 Production

Copper use is growing rapidly. Since 1965, annual copper production has doubled every 25 years, with a 2.8% annual increase or a fourfold increase in the 50 years after 1965. If the average production growth rate in the period between 1980 and 2015 is extrapolated to the year 2100, copper production in 2100 will be 11 times higher than in 2018. Although the real production growth rate of copper might be slower (or higher), the obvious question is whether such growth rates are sustainable. Copper may become so expensive for future generations that the services provided by it will be hardly attainable for poor nations and poor people (Fig. 7.6.3).

Table 7.6.1 Subdivision of copper end uses (based on data from the Copper Alliance, 2019).

Building construction		Plumbing	Water distribution, heating, gas, sprinkler	4%	28%
		Building plant	Air conditioning	1%	
		Architecture	Roofs, gutters, flashing, decoration	1%	
		Communications	Wiring in buildings	1%	
		Electrical power	Power distribution, earthing, lights, wire device	21%	
Infrastructure		Power utility	Power transmission, distribution network	14%	17%
		Telecommunications	Telecom network	3%	
Equipment manufacture	Industrial	Electrical	Industrial transformers, motors	6%	11%
		Non-electrical	Valves, fittings, instruments, in-plant equipment	5%	
	Transport	Automotive electrical	Harnesses, motors	8%	13%
		Automotive non-electrical	Radiators, tubing	1%	
		Other transport	Railroad, shipping, marine	4%	
	Other equipment	Consumer and general products	Appliances, instruments, tools, other	9%	32%

Table 7.6.1 Subdivision of copper end uses (based on data from the Copper Alliance, 2019).—cont'd

		Cooling	Air conditioning, refrigeration	8%	
		Electronic	Industrial/ commercial electronics, PCs	5%	
		Diverse	Ammunition, clothing, coins, other	10%	

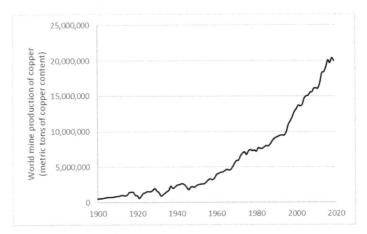

Figure 7.6.3 World copper production from 1900 to 2019 (US Geological Survey, 2017, 2018, 2019, 2020).

World copper extraction in 2019 was about 20 million metric tons. The main copper-producing countries were Chile, Peru, and China. Chile is also the country with the highest copper reserves (see Fig. 7.6.4).

The growth in copper consumption is unequaled. While it is nearly stable in developed countries, growth rates are very high in industrializing countries such as China and India. China's copper consumption in 1995 was 10% of the world consumption. By 2014, China was consuming about half of the world's copper production, although some of this copper was exported again in copper-containing products. In the United States, the apparent copper consumption remained stable at a little more than two million tons per year (see Fig. 7.6.5).

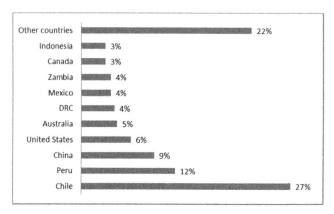

Figure 7.6.4 Top 10 copper-producing countries in 2017 (US Geological Survey, 2018).

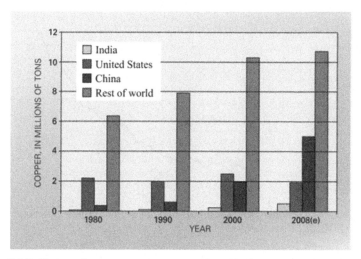

Figure 7.6.5 Changes in apparent copper consumption between 1980 and 2008 (US Geological Survey, 2009). From left to right: India, US, China, rest of the world (US Geological Survey, 2009).

The transition to a low-carbon economy will require an additional amount of copper on top of the amount of copper required if no such transition occurs. The creation of a low-carbon energy system capable of supplying the world's electricity needs in 2050 will require two years of current global copper production (Hertwich et al., 2015). However, even if this is achieved, total global copper demand will still be mainly determined by an increase in the global economy and wealth (per capita GDP and total

GDP). Without the energy transition, copper demand in 2050 is estimated to be 3 times higher than in 2000; with the energy transition, it is estimated to be 3.3 times higher (De Koning et al., 2018).

7.6.3.2 Resources

Of the different copper deposit types, porphyry copper deposits are currently the main source of copper, at about 60% (US Geological Survey, 2013). Copper ore is extracted in three ways: open-pit mining, underground mining, and in situ leaching. The ore is crushed and ground, followed by flotation or leaching to separate the gangue from the copper-rich mineral (Fig. 7.6.6).

The total amount of copper in the Earth's crust is enormous. The average copper concentration in the Earth's crust is 50 ppm.[5] Hence, the upper 3 km of the Earth's continental crust contain about 60,000 billion

Figure 7.6.6 Bingham Canyon open-pit copper mine. This is one of the largest man-made excavations in the world. It has been in operation for more than 100 years and has produced approximately 15 million tons of copper. *(Photo retrieved from https://nl. wikipedia.org/wiki/Bingham_Canyon_Mine on 13-1-2019.)*

---

[5] The scientific literature provides different values for the crustal abundance of elements. We have taken the average value of seven sources: McLennan (upper crustal abundance), 2001; Darling, 2007; Barbalace, 2007, Webelements, 2007; Jefferson Lab, 2007; Wedepohl, 1995; Rudnick and Fountain, 1995.

metric tons of copper. However, in practice, only a very small fraction of this amount can be economically extracted. Currently, most copper mines operate with minimum copper concentrations of between 0.4% and 0.8%. These concentrations are a factor 80−160 higher than the average crustal abundance of copper. As copper mining and production technology continue to develop, the extraction of lower grade copper and from deeper mines may become feasible. Different estimates of the copper resources can be found in the literature:

- 5600 Mt (identified plus undiscovered resources) (US Geological Survey, 2018);
- 6400 Mt (Singer, 2017);
- >1781 Mt (Northy et al., 2014);
- 2459 Mt (Schodde, 2010);
- 3035 Mt (Mudd and Jowitt, 2018);
- 2800 Mt (Sverdrup et al., 2014).

According to Kesler and Wilkinson (2008), copper resources in the upper 3 km of the Earth's crust could even be as much as 25 Gt (see Section 2.4).

A generic approach is to assume that the upper limit for the extractable amount of a resource is 0.01% of that resource in the mineable part of the Earth's continental crust (UNEP, 2011; Erickson, 1973; Skinner, 1976). Based on this approach, the ultimately available amount of copper in the upper 3 km of the Earth's continental crust is 6 Gt.

A thorough assessment of US copper resources in 1998 indicated 540,000 kt of copper in the upper 1 km of the continental crust in the United States (290,000 kt identified and 660,000 kt undiscovered) (US Geological Survey, 2000). This means that the ratio between total estimated copper resources and identified copper resources in the United States in 1998 was 540,000/290,000 = 2.12. For the estimation of the ultimately available copper resources globally, we apply this ratio of 2.12 to the most recent data on identified global copper resources (2.1 Gt). This results in 10 Gt of ultimately available copper resources in the upper 3 km of the Earth's continental crust.

In the framework of the analysis of conditions for the sustainable production of copper, we consider a low estimate of 6 Gt and a high estimate of 25 Gt for the ultimately available copper resources.

The above amounts do not include copper on ocean bottoms in deep-sea nodules and submarine massive sulfides. Global copper resources in nodules have been estimated at 700 Mt in nodules and 30 million tons in submarine massive sulfides (Hannington et al., 2011). Though these amounts are not negligible, they are only in the order of 10% of the identified and nonidentified resources in the upper 1 km of the Earth's continental crust. Furthermore, responsible exploitation of these resources is challenging from an environmental protection point of view.

The grade-tonnage distribution of copper in the Earth's crust is illustrated in Fig. 7.6.7. There are vast amounts of copper in common rock, but they are of a very low grade and, moreover, occur in a different, less easily extractable matrix. The copper ores mined currently are sulfide and oxide minerals, whereas in crustal rock copper usually occurs in a silicate matrix. The energy required to separate copper from a silicate matrix is 100−1000 times higher than the current cost of primary copper production (Skinner, 1987; Gordon et al., 2016).

Extracted copper ore grades were 10%−20% until late in the 19th century, and decreased to 2%−3% in the early part of the 20th century. Since about 1950, extracted copper ore grades have declined below 1%. Currently, the extraction of copper ores with grades as low as 0.4% is already quite common (Kerr, 2014; see Fig. 7.6.8).

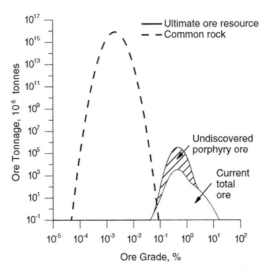

Figure 7.6.7 Grade-tonnage density curves for copper in the Earth's crust. The "current total ore" area represents the currently known copper ore resources (British Geological Survey, 2007; Gerst, 2008).

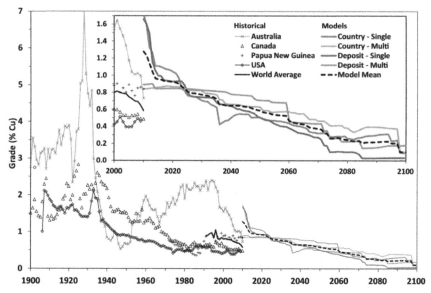

Figure 7.6.8 Historical copper ore grades for selected countries to 2010 and projected ore grades for 2010 onwards (Northey et al., 2014).

The gradual depletion of higher ore grades is characteristic of all mineral resources. It is not the amount of a mineral in the Earth's crust as such that may lead to a scarcity problem, but the effort needed to extract the mineral in terms of energy use, water use, waste generation, and landscape degradation. Energy use in relation to copper ore grade is presented in Fig. 3.7. Roughly speaking, halving copper ore grades will double the energy requirements for copper production. However, energy efficiency increases may save 30% of the energy used currently for copper extraction (Elshkaki et al., 2016). Taking this energy-saving potential into account, energy demand for copper production may, nevertheless, increase to 1% of total energy demand by 2050 or even reach 2.4% of total energy demand in 2050 compared to only 0.3% currently (Elshkaki et al., 2016). It is expected that the CO_2 footprint of copper production will at least triple by 2050 (Kuipers et al., 2018).

7.6.4 Current stocks and flows

7.6.4.1 Recovery efficiency

Production losses of copper through tailings and slag are about 16% (Glöser et al., 2013; Lifset et al., 2002).

7.6.4.2 Dissipative use

The three main routes of in-use dissipation of copper are via food, chemicals, and corrosion. Additionally, some copper is dissipated during use via discarded copper-containing products, such as staples. Humans need to ingest 1.5—3 mg of copper per day with food, and animals and plants need copper too (e.g., pigs need about 12 mg/kg fodder). Chemicals containing copper include wood preservatives, fungicides, pigments, and antifouling paint for ships. Dissipation is estimated at 2% of copper consumption (Glöser et al., 2013).

7.6.4.3 In-use stock

The average expected lifetime for copper products in 2010 was slightly over 25 years (Glöser et al., 2013). This is based on the average lifetimes of different types of copper applications (see Table 7.6.2).

The global in-use stock of copper was estimated to be 57 kg per capita in 2000 (Gerst, 2009) and to range between 9 kg per capita in Africa, to 160 kg per capita in Europe in 2014 (Soulier et al., 2018), and 206 kg per capita in North America in 2000.

The current global annual build-up of copper in anthropogenic stocks is estimated at about 1.4 kg per capita (Glöser et al., 2013). The estimate for the annual build-up of in-use stocks in western Europe was about 1.5 kg per capita in 2014 (Soulier et al., 2018). The annual build-up of copper in the anthropogenic stocks slows down as a country becomes more developed. Fig. 7.1.3 shows the slowdown and even decrease in copper use in more developed countries. Countries like India and China are currently building up large amounts of copper in anthropogenic stocks via major infrastructural projects. Extrapolation leads to the conclusion that the global

Table 7.6.2 Copper in-use residence times (Spatari et al., 2005).

Copper end use	Residence time (years)
Plumbing	40
Wiring	25
Built-in appliances	17
Industrial EEE[a]	20
Consumer EEE	10
Infrastructure	50
Motor vehicles	10
Other transport	30

[a]Electrical and electronic equipment.

in-use amount of copper must have been around 600 Mt in 2017, which is 30 times more than annual copper production today and represents as much as about 85% of all copper mined since 1900. The 20th century accounted for 90% of all copper mined and put into service throughout human history: 70% of this in just the last 70 years and 50% in the last 25 years (Lifset et al., 2002). Currently, the build-up of in-use stocks of copper is estimated at about 44% of the annual consumption of copper globally (Glöser et al., 2010). This proportion will gradually decrease until it reaches zero, when there is equilibrium between copper consumption and the amount of copper in end-of-life products (Glöser et al., 2013).

7.6.4.4 Recycling from end-of-life products

It requires 85% less energy to produce copper through recycling than to produce copper from copper ore (European Copper Institute, 2018).

The copper collection rates from different end-of-life products and the copper recovery rates vary (see Table 7.6.3).

In developed countries, nearly one third of copper usage is currently recycled copper (International Copper Study Group, 2017; UNEP, 2011), but globally this is only one fifth of total copper usage (Elshkaki et al., 2016). Estimated copper recycling rates vary for different parts of the world. In western Europe, about 80% of the copper in end-of-life products is collected for recycling. Of this amount, about 80% is recycled into the copper loop (Soulier et al., 2018). The remaining 20% of the copper

Table 7.6.3 Copper content, copper collection rate, and copper recovery rate from different waste streams in 1999 in western Europe (Ruhrberg, 2006).

	Copper content (%)	Copper collection for recovery (%)	Copper recovery (%)
Construction and demolition waste	0.3	78	91
Industrial electrical equipment waste	24	64	95
Industrial non-electrical equipment waste	0.14	98	78
End-of-life vehicles	1.4	67	61
Waste from electric and electronic equipment	7.3	50	85
Municipal solid waste	0.05−0.2	52	62

Comparable figures are provided by other authors (Graedel et al., 2004).

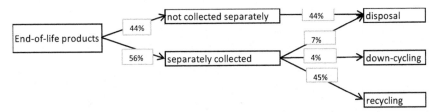

Figure 7.6.9 The fate of copper in copper-containing end-of-life products. This figure assumes 45% copper recycling from end-of-life products globally.

collected in copper-containing end-of-life products goes to non-copper metal loops, such as aluminum and steel (about one third) and to disposal sites (about two thirds) (Ruhrberg, 2006). The resulting recycling rate of copper from end-of-life products in western Europe is estimated to be between 48% (Ruhrberg, 2006; Bertram et al., 2002) and 65% (Ruhrberg, 2006). However, there are also lower estimates than 48%. The end-of-life recycling rate in Australia is estimated at 56% and takes into consideration that about 20% of the copper collected is not recycled into the copper loop (Van Beers et al., 2007). The end-of-life copper recycling rate is estimated to be 42% in North America (Spatari et al., 2005) and 43% in the United States (Goonan, 2004). Worldwide, the end-of-life recycling rate of copper is estimated at 45% (Glöser et al., 2013). Assuming a global end-of-life recycling rate of 45%, and that this corresponds with 80% of copper in separately collected end-of-life copper products, then the other 20% (or about 11% of the initial amount of copper in end-of-life products) goes to downcycling and disposal. Thereof, one third or 4% is downcycled and two thirds or 7% is disposed of. The total amount of separately collected copper is $10/8 \times 45\% = 56\%$, which means that 44% of copper is not separately collected and is disposed of in landfills and waste incinerators. Also taking into consideration the disposed part of separately collected copper, the conclusion is reached that 55% of copper is currently disposed of in landfills and waste incinerators globally, or is downcycled in the steel or aluminum cycle (see Fig. 7.6.9).

7.6.4.5 Downcycling

In 2006, about 4% of the copper in end-of-life products went to the steel and aluminum loops (Ruhrberg, 2006). In the steel loop, this was mainly from scrapped cars and large household appliances such as washing machines (Ruhrberg, 2006). In the iron and aluminum smelting processes,

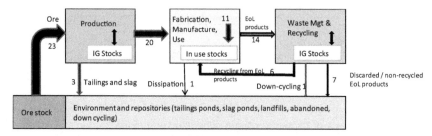

Figure 7.6.10 Simplified global copper flows for the year 2017, extrapolated from 2010 estimates concerning these flows (Glöser, 2013). Flows are in million metric tons. The EoL (end-of-life) recycling rate is 45%, tailings and slags are 16% of ore, accumulation in anthropogenic stocks is 44%, and dissipation is 2%. The arrow widths are a rough indication of flow magnitudes. IG stocks are industrial, commercial, and government stocks.

copper remains in the liquid phase together with iron or aluminum. However, copper is considered to be an undesired impurity in steel and aluminum. Steel properties, for example for machine and automobile construction, are negatively impacted by copper. In steel, the copper content may not be higher than a few tenths of a percent [0.25%−0.5% according to Ruhrberg (2006) and 0.1%−0.3% according to Ayres et al. (2002)], and in aluminum, the copper content may not be higher than 2.5% (Ruhrberg, 2006). However, the removal of copper traces from secondary steel is difficult (Ayres et al., 2002), so it is in the interest of steel and aluminum manufacturers to keep copper out of their raw materials as much as possible, and to limit copper downcycling.

7.6.4.6 Current copper flows summarized

Fig. 7.6.10 presents current copper flows in a simplified way. It is striking that a circular copper economy is still far away, as more than 50% of copper is not reused. Characteristic of the current situation is also the important build-up of copper in the in-use stock: the output from the usage phase is only 60% of the input.

7.6.5 Sustainable copper flows

7.6.5.1 Introduction

For the different assumptions of copper resource availability, we investigate whether and under which conditions a sustainable copper production rate can be achieved while increasing the current service level of copper for an average person in the world to the average copper service level in developed countries in 2020. Copper consumption per capita in developed countries is estimated at about 10 kg per person per year (see Table 5.3).

7.6.5.2 Increase in recovery efficiency

The current copper recovery efficiency is 84% (Lifset et al., 2002; Glöser et al., 2013). Given that ore grades will decrease, it will be difficult to increase the recovery efficiency substantially. However, we assume that the current copper recovery rate of 84% will remain feasible in the future.

7.6.5.3 Substitution

Copper in power cables, electrical equipment, automobile radiators, and cooling and refrigeration tubes can be replaced by aluminum. In heat exchangers, copper can be substituted by titanium and steel, and in telecommunications by optical fibers. In water pipes, drainpipes, and plumbing fixtures, copper can be replaced by plastic (US Geological Survey, 2018). However, substitution has until now been minimal, because copper's properties are often superior to the properties of its substitutes. In 2015 and 2016, net copper substitution was about 1%, mainly in cables and wiring (International Copper Association, 2017; Dewison, 2016).

Refined copper is currently much less energy intensive than aluminum. However, at an ore grade of around 0.1%, this advantage for copper compared to aluminum will disappear (Ayres et al., 2002). For high-voltage overhead transmission lines, aluminum is now preferred to copper because it weighs less, although this advantage is less relevant for underground lines. Because overhead transmission lines are considered ugly and because of (unverified) concerns about the health impacts of overhead transmission lines, high-voltage transmission may occur more through underground copper lines than through overhead aluminum lines in the future (Ayres et al., 2002).

Copper is preferred to aluminum for interior wiring as aluminum has caused overheating problems and fires. In telecommunication, copper wires are replaced by glass fiber, because the performance per kilogram is much higher. Copper can also be replaced in roofing. In some applications, copper seems virtually irreplaceable, such as for the local distribution of electric power, household wiring, motor-generator windings, electronic circuitry, and for some kinds of heat exchangers (Ayres et al., 2002).

We assume a maximum substitution of 10%, mainly in roofs, gutters, radiators, ammunition, water distribution, chemicals, and other applications not related to electricity generation, transmission, and distribution.

7.6.5.4 Material efficiency improvement

Apart from recycling, an important option for improving material efficiency is to increase product lifetimes. The average product lifetime for copper is currently just over 25 years (Glöser et al., 2013), with plumbing and electricity infrastructure having lifetimes of 40 and 50 years, respectively (Spatari et al., 2005). In general, we assume that it is possible to increase the material efficiency by about 10% (see Chapter 7.1).

7.6.5.5 Dissipation reduction

The in-use-dissipation of copper is currently about 2%. The only way to reduce dissipation is to ban certain copper applications, such as in chemicals (wood preservatives, fungicides, pigments, and antifouling paints). The corrosion of copper/bronze products and the use of copper as a micronutrient cannot be avoided. We assume that a minimum copper in-use dissipation rate of 1% is achievable, with this reduction being made possible by the substitution of copper in certain applications.

7.6.5.6 In-use stocks

Future in-use stocks of copper will depend on the worldwide economic growth rate, urbanization, population growth, decline in average household size, and environmental awareness. In-use copper stocks in 2100 are estimated at 1.4–2.1 billion tons (Gerst, 2009). With 10 billion world citizens, this comes to 140–210 kg per capita, which is still less than the current figure for North America. Eventually, equilibrium will be reached and anthropogenic stocks will stop increasing. At this point, the amount of copper entering the usage phase will be equal to the amount of copper in end-of-life products.

7.6.5.7 Decrease in downcycling

The amount of copper downcycled to the iron and steel loop is currently estimated at about 4% of the copper in end-of-life products (Ruhrberg, 2006) (Fig. 7.6.9). We assume that downcycling can be reduced by 50% as a result of the improved separation and sorting of copper-containing end-of-life products.

7.6.5.8 Decrease in disposal in landfills and increase in recycling

Currently, about half of the copper in end-of-life products ends up in landfills, partly in the ash from waste incineration plants. This is a substantial amount. There are three parallel ways to reduce the amount of copper that is lost for recycling in this way: (1) by increasing the fraction of separately collected copper-containing end-of-life products, (2) by increasing the fraction of recycled copper from the separately collected copper-containing fraction, and (3) by recovery of copper from landfills and incinerators. The main incentive for increasing copper recycling will be a higher price for primary copper, making copper recycling from products with a low copper content financially more attractive. This can be achieved by promoting a better recycling-oriented design and by subsidizing recycled copper, possibly in combination with the taxation of primary copper and prohibiting disposal to landfills or the incineration of copper-containing products.

Copper in end-of-life products occurs in seven types of waste: municipal solid waste, construction and demolition waste, industrial waste, hazardous waste, waste from electrical and electronic equipment, end-of-life vehicles, and sewage sludge (Graedel et al., 2004). For the year 1994, it was estimated that WEEE (waste from electronic and electrical equipment) and end-of-life vehicles together contained 70% of all the discarded copper (Graedel et al., 2004). Globally, the end-of-life copper recycling rate is 45% (Glöser et al., 2013). The lower the copper content in an end-of-life product, the more complicated and less rewarding it is to recycle the copper.

Copper concentrations in fly ash from municipal solid waste incinerators vary between about 200 mg/kg and 11,000 mg/kg, with an average of about 1200 mg/kg (Jung et al., 2004; Lam et al., 2010). Copper concentrations in the bottom ash from such incinerators vary between 80 mg/kg and 13,000 mg/kg, with an average of almost 3000 mg/kg (Jung et al., 2004; Lam et al., 2010). Currently, grades of extracted copper ore are as low as 4000 mg/kg. This means that bottom ash from municipal solid waste incinerators may become interesting as a future source of secondary copper. Simultaneously, by removing heavy metals such as copper from municipal solid waste incinerator bottom ash, its potential utility in other applications increases because of the decreased leaching of heavy metals. On the other hand, improved end-of-life recycling will decrease the amount of copper in municipal solid waste, making copper extraction from bottom ash less attractive.

Table 7.6.4 Different mixes of measures to achieve a sustainable copper production rate, while realizing a global copper service level that is similar to the copper service level in developed countries in 2020. The boldfaced rows include the main (combinatiuons of) measures

		Sustainability ambition (years of sufficient availability)				
		1000	500		200	
Available copper resources[a,b]	Gt	25	12	22	6	11
Sustainable copper production	Mt/year	25	25	45	30	56
Aspired future copper service level globally	kg/capita/year	10	10	10	10	10
Assumed future substitution	%	**10%**	**10%**	**10%**	**10%**	**0%**
Assumed increase in material efficiency	%	10%	10%	10%	10%	0%
Derived future copper consumption level	kg/capita/year	8.1	8.1	8.1	8.1	10.0
Corresponding annual copper consumption	Mt/year	81	81	81	81	100
Assumed in-use dissipation of remaining products[c]	%	1%	1%	1%	1%	2%
Copper in end-of-life products	Mt/year	80	80	80	80	98
Required copper recycling of remaining products	Mt/year	56	56	36	51	44
Required copper recycling rate[d,e]	%	**70%**	**70%**	**45%**	**64%**	**45%**

The assumed maximum recovery efficiency in the future is 84%, and the assumed maximum substitutability of copper in the future is 10%. The assumed material efficiency increase is 10% and the assumed minimum in-use dissipation rate of remaining products is 1%. The assumed maximum recyclability of remaining products in the future is 70%.
[a]Low estimate: 6 Gt; high estimate: 25 Gt (see Section 7.6.3.2).
[b]Current recovery efficiency: 84% (see Section 7.6.4).
[c]Current in-use dissipation rate: 2% (see Section 7.6.4).
[d]Current end-of-life recycling rate: 45% (see Section 7.6.4).
[e]It is assumed that the world population stabilizes at 10 billion people.

It can be expected that the more the copper price increases in the future, due to a combination of increasing demand for copper and the higher extraction costs resulting from a further decrease in copper ore grades and deeper mining, the more popular copper recycling will become. The question is whether this market mechanism will be sufficient and timely enough to prevent future generations from being confronted with rocketing copper prices.

Overall, we think that it will be a challenge to increase the end-of-life recycling rate of copper to more than 70%.

7.6.6 Conclusions

Table 7.6.4 presents different mixes of measures that will be required to achieve a sustainable copper production rate while simultaneously realizing a global copper service level that is equal to the copper service level in developed countries in 2020.

The conclusions are as follows:

- A sustainability ambition of 1000 years is only achievable with 25 Gt of ultimately available copper resources in combination with 10% substitution, 10% material efficiency improvement, and 70% end-of life recycling.

- A sustainability ambition of 500 years is only achievable with >12 Gt of ultimately available copper resources in combination with 10% substitution, 10% material efficiency improvement, and 70% end-of-life recycling. If the ultimately available copper resources are higher, the required measures will be less.

- A sustainability ambition of 200 years is achievable with the lowest considered amount of ultimately available resources (6 Gt) if substitution is 10%, material efficiency improvement is 10%, and the end-of-life recycling rate is 64%. If the ultimately available copper resources are >11 Gt, no extra measures are required.

References

Ayres, R.U., Ayres, L.W., Rade, I., 2002. The Life Cycle of Copper, its Co-products and By-Products. International Institute for Environment and Development.

Barbalace, K., 2007. Periodic Table of Elements. Retrieved from: http://environmentalchemistry.com/yogi/periodic. (Accessed 14 April 2007).

Bertram, M., Graedel, T.E., Rechberger, H., Spatari, S., 2002. The contemporary European copper cycle: waste management subsystem. Ecol. Econ. 42, 43–57.

British Geological Survey, 2007. Copper. www.mineralsuk.com.

Copper Alliance, 2019. Global 2018 Semis End Use Data Set. Retrieved on. https://copperalliance.org/trends-and-innovations/data-set/. (Accessed 14 January 2019).

Darling, D., 2007. Elements, Terrestrial Abundance. http://www.daviddarling.info/encyclopedia/E/elterr.html.

De Koning, A., Kleijn, R., Huppes, G., Sprecher, B., Van Engelen, G., 2018. Metal supply constraints for a low-carbon economy. Resour. Conserv. Recycl. 129, 202–208.

Dewison, P., 2016. Realities of Substitution in the Global Copper Market, Metals Research and Consulting.

Elshkaki, A., Graedel, T.E., Ciacci, L., Reck, B.K., 2016. Copper demand, supply, and associated energy use to 2050. Global Environ. Change 39, 3050315.

European Copper Institute, 2018. Copper Recycling. http://copperalliance.eu/about-copper/recycling.

Gerst, M., 2009. Linking material flow analysis and resource policy via future scenarios of in-use-stock: an example for copper. Environ. Sci. Technol. 43 (16), 6320−6325.

Glöser, S., Soulier, M., Espinoza, L.A.T., 2013. Dynamic analysis of global copper flows. Global stocks, Postconsumer material flows, recycling indicators, and uncertainty evaluation. Environ. Sci. Technol. 47, 6564−6572.

Goonan, T.G., 2004. Copper recycling in the United State in 2004. US Geol. Survey Circ. 1196-X.

Gordon, R.B., Koopmans, J.J., Nordhaus, W.B., Skinner, B.J., 2016. Towards a New Iron Age. Harvard University Press, Cambridge MA cited by Elshkaki et al (1987).

Graedel, T.E., Van Beers, D., Bertram, M., Fuse, K., Gordon, R.B., Gritsinin, A., Kapur, A., Klee, R.J., Lifset, R.J., Memon, L., Rechberger, H., Spatari, S., Vexler, D., 2004. Multilevel cycle of anthropogenic copper. Environ. Sci. Technol 38, 1242−1252.

Hannington, M., Jamieson, J., Monecke, T., Petersen, S., Beaulieu, S., 2011. The abundance of seafloor massive sulfide deposits. Geology 39 (12), 1155−1158.

Hertwich, E.G., Gibon, T., Bouman, E.A., Arvesen, A., Suh, S., Heath, G.A., Bergesen, J.D., Ramirez, A., Vega, M.I., Shi, L., 2015. Integrated life-cycle assessment of electricity supply scenarios confirms global environmental benefit of low-carbon technologies. Proc. Natl. Acad. Sci. U.S.A. 112, 6277−6282. Cited by De Koning et al. 2018.

International Copper Association, 2017. Global Copper Substitution and Regulatory Trends.

International Copper Study group, 2017. The World Copper Factbook.

Jefferson Lab, 2007. It's Elemental, the Periodic Table of Elements. http://education.jlab.org/itselemental/index.html.

Jung, C.H., Matsuto, T., Tanaka, N., Okada, T., 2004. Metal distribution in incineration residues of municipal solid waste (MSW) in Japan. Waste Manag. 24, 381−391.

Kerr, R.A., 2014. The coming copper peak. Science 343 (6172), 722−724.

Kesler, W.E., Wilkinson, B.H., 2008. Earth's copper resources estimated from tectonic diffusion of porphyry copper deposits. Geology 255−258.

Kuipers, K.J.J., Van Oers, L.F.C.M., Verboon, M., Van der Voet, E., 2018. Assessing environmental implications associated with global copper demand and supply scenarios from 2010 to 2050. Global Environ. Change 49, 106−115.

Lam, C.H.K., Ip, A.W.M., Barford, J.P., McKay, G., 2010. Use of incineration MSW ash: a review. Sustainability 2, 1943−1968.

Lifset, R.J., Gordon, R.B., Graedel, T.E., Spatari, S., Bertram, M., 2002. Where has all the copper gone: the stocks and flows project, Part 1. JOM 21−26.

McLennan, S.M., 2001. Relationships between the trace element composition of sedimentary rocks and upper continental crust. G-cubed 2. April 20, paper number 2000GC000109.

Mudd, G.M., Jowitt, S.M., 2018. Growing global copper resources, reserves and production; discovery is not the only control on supply. Econ. Ecol. 113 (6), 1235−1267.

Northey, S., Mohr, S., Mudd, G.M., Weng, Z., Giurco, D., 2014. Modelling future copper ore grade decline based on a detailed assessment of copper resources and mining. Resour. Conserv. Recycl. 83, 190−201. Data from Mudd, G.M., Weng, Z., Base Metals, In: Letcher, T.M., Scott, J.L. (Eds.), Materials for a sustainable future, London Royal Society of Chemistry, 2012, 11−59 and from Mudd, G., Weng, Z., Jowitt, S.M., 2013 A detailed assessment of global Cu reserve and resource trends and worldwide Cu endowments. Econ. Geol. 108 (5), 1163−1183.

Rudnick, R.L., Fountain, D.M., 1995. Nature and composition of the continental crust, a lower crustal perspective. Rev. Geophys. 267−309.

Ruhrberg, M., 2006. Assessing the recycling efficiency of copper from end-of-life products in Western Europe, Resources. Conserv. Recycl. 48, 141−165.

Singer, D.A., 2017. Future copper resources. Ore Geol. Rev. 86, 271−279.

Schodde, R.C., 2010. The key drivers behind resource growth: an analysis of the copper industry over the last 100 years. In: Presentation to the Mineral Economics & Management Society (MEMS). Session at the 2010 SME Annual Conference.

Skinner, B.J., 1987. Supplies of geochemically scarce metals. In: MacLaren, D.J., Skinner, B.J. (Eds.), Resources and World Development: 305−325. John Wiley and Sons, Chichester, UK. Cited by Elskakhi et al. (2016).

Soulier, M., Glöser-Chahoud, S., Goldman, D., Espinoza, L.A.T., 2018. Dynamic analysis of European copper flows, Resources. Conserv. Recycl. 129, 143−152.

Spatari, S., Bertram, M., Gordon, R.B., Henderson, K., Graedel, T.E., 2005. Twentieth-century copper stocks and flows in North America: a dynamic analysis. Ecol. Econ. 54, 37−51.

Sverdrup, H.U., Ragnarsdottir, K.V., Koca, D., 2014. On modelling the global copper mining rates, market supply, copper price and the end of copper reserves. Resour. Conserv. Recycl. 87, 158−174.

UNEP International Resource Panel, 2011. Decoupling Natural Resource Use and Environmental Impacts from Economic Growth. UNEP.

US Geological Survey, 2009. Copper − A Metal for Ages, Fact Sheet 2009-3031.

US Geological Survey, 2000. 1998 Assessment of undiscovered deposits of gold, silver, copper, lead and zinc in the United States. Circular 1178, 2000.

US Geological Survey, 2014. Estimate of Undiscovered Copper Resources of the World, 2013, Fact Sheet 2014-3004.

US Geological Survey, 2018. Mineral Commodity Summaries, Copper, January 2018.

US Geological Survey, 2019. Mineral Commodity Summaries, Copper, January 2019.

US Geological Survey, 2020. Mineral Commodity Summaries, Copper, January 2020.

Van Beers, D., Kapur, A., Graedel, T., 2007. Copper and zinc recycling in Australia: potential quantities and policy options. J. Clean. Prod. 15, 862−877.

Webelements, 2007. Abundance in Earth's Crust. http://www.webelements.com/webelements/properties/tekst/image-flah/abund-crust.html.

Wedepohl, K.H., 1995. The composition of the continental crust. Geochimica et Cosmochimica 59 (7), 1217−1232.

CHAPTER 7.7

Gold

7.7.1 Introduction

Gold is massively used for jewelry and adornment, and as an investment in gold bars and coins. Gold is also an important metal for modern technology, and is used in printed circuit boards (PCBs) in computers, smartphones, and other electronic hardware.

The purpose of this section is to investigate what, if anything, needs to be done to make the current service level of gold in wealthy countries sustainable at a worldwide scale. Following a description of the properties and applications of gold, we investigate the current mass flows of gold. Then, based on data on the total available resources of gold, we analyze whether or not the current extraction rate of gold is sustainable and, if not, what is required to make the extraction rate sustainable, while simultaneously extending the service level of gold in the world as a whole to the service level of gold in developed countries in 2020.

7.7.2 Properties and applications

Gold has a density of 19.3 g/cm^3, a melting point of $1064°C$, and a boiling point of $2970°C$. The most important properties of gold are that it does not react with oxygen under normal conditions, that it is very corrosion-resistant, ductile, and malleable, and that it has high conductivity for heat and electricity. To increase gold's strength, it is alloyed with other metals such as silver, copper, and zinc. The purity of gold is expressed in karats, to indicate the proportion of solid gold in an alloy based on a total of 24 parts (pure gold is therefore 24-karat gold). Gold of 8 karats therefore contains one third gold and two thirds other materials. Depending on the amount and type of metal additions, gold alloys have different colors. Rose gold of 18 karats contains 75% gold, 22.25% copper, and 2.75% silver. White gold of 18 karats contains 75% gold, and 25% palladium or platinum. The non-gold part of white 18-karat gold may also be composed of 10% palladium, 10% nickel, and 5% zinc.

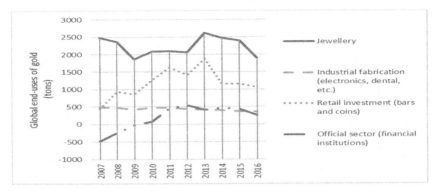

Figure 7.7.1 End uses of gold between 2007 and 2016 (tons). *(Derived from Thomson Reuters, 2017. GFMS Gold Survey 2017.)*

At present, gold is used in jewelry and as a decorative material (about 50%), for retail investment in bars and coins (about 30%), in technological applications (about 10%), and by financial institutions (about 10%) (Thomson Reuters, 2017; see Fig. 7.7.1). Gold applications between 2007 and 2016 are presented in Fig. 7.7.1. Technological applications consist of electronic and electrical applications (about 70%), dental and medical services (about 10%), and other industrial applications (about 20%) (Thomson Reuters, 2017; see Table 7.7.1). Individual buyers can buy gold bars directly or through gold exchange traded funds. Gold use for technological purposes was fairly stable in the period 2007 to 2016, but retail investment showed large fluctuations, as this type of gold use seems to be dependent on fluctuations in the gold price, which had a peak in 2012. In electronic applications, gold is indispensable in electronic circuitry, connectors, switch

Table 7.7.1 Gold use in technological applications between 2007 and 2016 (tons) (Thomson Reuters, 2017).

	2007	2008	2009	2010	2011	2012	2013	2014	2015	2016
Electronics	345	334	295	346	343	307	300	290	258	254
Dental and medical	58	56	53	48	43	39	36	34	32	30
Other technological	89	89	79	86	85	83	85	79	76	70
Total technological applications	492	479	427	480	471	429	421	403	366	354

and relay contacts, soldered joints, connecting wires, and strips. Other technological applications include catalytic converters (US Geological Survey, 2005), and some of the high-temperature brazing alloys used in the manufacture of aircraft turbine engines are gold alloys (US Geological Survey, 2005). Gold is also used on or in glass for reflecting building heat inward and solar heat outward (US Geological Survey, 2005).

As a decorative material, gold is used, for example, in (very) thin layers on inexpensive jewelry, wrist watches, pens, cutlery, eye glasses, bathroom fixtures, gold-plated silver wound silk thread for expensive wedding saris in India, and gold leaf (US Geological Survey, 2005).

For medical purposes, gold is used in medication for facial paralysis and rheumatoid arthritis in a weak solution of gold salts (e.g., aurothioglucose; US Geological Survey, 2005).

7.7.3 Production and resources

7.7.3.1 Production

Gold is mainly found as "*native*" gold, meaning that it occurs alone and not in combination with other elements (Kesler and Simon, 2017). A small quantity of gold is mined as a by-product of porphyry copper, although some individual copper mines produce a significant quantity of gold, like the Grasberg copper mine in Irian Jaya (Indonesia) (Kesler and Simon, 2017). The two main methods of gold recovery from ores are amalgamation (selectively dissolving gold with mercury) and the cyanide process, selectively dissolving gold in a cyanide solution. The cyanide process is used for 90% of gold production and the amalgamation process for 10% (Kesler and Simon, 2017). These processes are applied on gold ore after crushing the ore and separation of the gold-containing particles from the gangue. Despite declining ore grades, annual gold production has increased by a factor of six since 1900. In 2016, gold was produced in 90 countries (US Geological Survey, 2016). The main gold-producing countries are China, Australia, Russia, the United States, Canada, Peru, and South Africa, in that order. These countries are responsible for about half of the global gold production. In the 1970s, South Africa still produced the large majority of the world's gold (about 1000 tons annually alone). However, South Africa's gold production declined because of decreasing ore grades. The other main gold-producing countries could not fully compensate for the decrease in South Africa's gold production and the share of countries with a lower gold production increased (see Fig. 7.7.2).

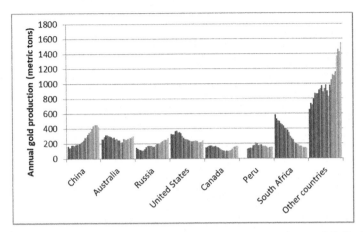

Figure 7.7.2 Main gold-producing countries between 1994 and 2017 (US Geological Survey, 2019). Total gold production in 2018 was 3280 metric tons.

With a production of 3000 tons per year, and a unit value of about 42 million US$/ton (per 21-3-2019), the total value of gold produced annually is about 130 billion US$. Gold therefore ranks along with copper (about 140 billion US$ in 2018[6]) and aluminum (about 110 billion US$ in 2018[7]) in terms of total production value.

The production of gold has increased quickly (see Fig. 7.7.3). Between 1980 and 2015, gold production increased by 2.7% annually. This growth percentage will not be maintained, and the metal consumption of a country will decouple from its GDP growth from a certain level of GDP per capita (Halada et al., 2008). In Japan, gold consumption stabilized at about 3 g/cap/year from an annual GDP of about US$25,000/cap (Halada et al., 2008).

7.7.3.2 Resources

The average abundance of gold in the Earth's crust is 2.5 ppb (see Table 2.2). Extracted gold ore grades have declined greatly in the last century: from an average of about 10−15 g/ton (10,000−15,000 ppb) a century ago to less than 3 g/ton (3000 ppb) now (in the United States, Australia, South Africa, Brazil) (Mudd, 2007). The gold content of porphyry copper ores is between about 0.1 and 1.3 g per ton (between 100 and

---

[6] 21,000,000 tons, 6.52 USD/kg. Data from US Geological Survey, 2019b.
[7] 60,000,000 tons, 1.89 USD/kg. Data from US Geological Survey, 2019b.

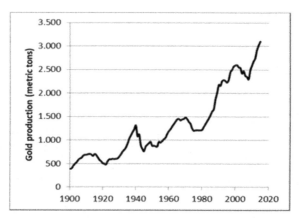

Figure 7.7.3 Global gold production between 1900 and 2017 (US Geological Survey, 2017). Gold production was 3260 tons in 2018 (US Geological Survey, 2019).

1300 ppb) (Kesler and Simon, 2017). The declining ore grade points to a substantial increase in the resource intensity of gold production in the near future (Mudd, 2007). The resource intensity of the production of a mineral is the amount of water, energy, and chemicals consumed, the amount of greenhouse gases emitted, and the amount of waste rock produced per unit of the mineral produced.

In 2017, global gold reserves were estimated at 54 kt (US Geological Survey, 2019b). The gold reserve base in 2009 was 100 kt (US Geological Survey, 2009). According to the tectonic diffusion approach of Wilkinson and Kesler (2010), mineable ore deposits to a depth of 3 km are 113 kt.

The countries with the most important gold reserves are presented in Fig. 7.7.4.

A generic approach is to assume that the upper limit for the extractable amount of a resource is 0.01% of that resource in the mineable part of the Earth's continental crust (UNEP, 2011; Erickson, 1973; Skinner, 1976). Based on this approach, the ultimately available amount of gold in the upper 3 km of the Earth's continental crust is 300 kt (see Table 2.2).

A thorough assessment of US gold resources in the US continental crust to a depth of 1 km in 1998 indicated 33,000 tons of gold in the United States (15,000 tons identified resources and 18,000 tons undiscovered resources) (US Geological Survey, 2000). The ratio between the estimate of total gold resources in the United States in 1998 (33,000 tons) and the

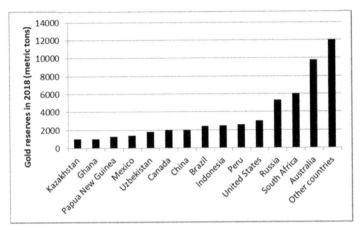

Figure 7.7.4 Gold reserves in 2018 (US Geological Survey, 2019).

identified resources in the United States identified that far (15,000 tons) is therefore 2.2. To estimate the ultimate gold resources in the world, we apply this ratio of 2.2 to the world's identified gold resources. We use recent data on the identified gold resources instead of data from 1998, because the world outside the United States has been explored to a lesser extent than the United States. However, the US Geological Survey does not provide a recent estimate of the identified global gold resources. Therefore, we use a proxy, as follows. For 14 metals (copper, iron, lead, lithium, magnesium, mercury, molybdenum, nickel, platinum group metals, rhenium, strontium, titanium, vanadium, and zinc), the average ratio between identified resources as provided by US Geological Survey in 2018 and the reserve base as provided by US Geological Survey in 2009 is 2.8. The global reserve base of gold in 2009 was 100,000 tons (US Geological Survey, 2009). Hence, the proxy for the identified gold resources is 2.8 × 100,000 tons = 280,000 tons. Applying the ratio 2.2 to this amount, and assuming extraction to a depth of 3 km in the Earth's continental crust rather than to 1 km, results in an estimate for the ultimate exploitable gold resources in the world of 3 × 2.2 × 280,000, which can be rounded to about 2 Mt.

We therefore have two quite different numbers for the ultimately available gold resources: 300 and 2000 kt. We will take both figures into consideration.

7.7.4 Current stocks and flows

7.7.4.1 Recovery efficiency in mining and production

The cyanide process results in a 90% recovery of gold (Hylander et al., 2007), while the amalgamation process is less efficient (Hylander et al., 2007).

7.7.4.2 Dissipative use

The dissipative use of gold is very small, and only takes place in some medical applications.

7.7.4.3 In-use stock

Much of the newly extracted gold is added to the in-use stock. The total historical gold production up to the end of 2017 is estimated at about 187,000 tons. This is based on an extrapolation of 1999 data from Butterman and Amey (2005) and more recent data from Thomson Reuters (2017). Estimates of how much gold has been buried in landfills and graves (or is otherwise unaccounted for) vary between 1% and 15% (Butterman and Amey, 2005). Table 7.7.2 presents an estimate of the main components of global above-ground stocks of gold. In Table 7.7.2, the irretrievable amount of gold in landfills and graves is estimated at less than 2%.

Gold is a good example of how, with good management, very little metal needs to be lost. Because of its high price, few people throw gold away. This means that, over a period of 6000 years, only a few percent of all

Table 7.7.2 Estimate of above-ground stocks of world gold in tons, at the end of 2017.

Total gold mined through 2017	187,000	
Irretrievable losses	3000	
Total in above-ground stocks or in-use	184,000	
Composition of above-ground stocks		
Gold bars (official reserves and private investment)	72,000	40%
Jewelry	89,000	48%
Other fabricated products still in-use or in-stock	23,000	12%
Total available above-ground stocks	184,000	100%

Derived from Thomson Reuters, 2017. GFMS Gold Survey 2017.

the gold mined has been irretrievably lost. However, because about 10% of the gold is used in technological applications, mostly in the electronics sector, the risk of irretrievable gold loss is increasing due to the fact that many people are not aware of the presence of gold in the printed circuit board of their smartphones, and that the collection and recycling of end-of-life products from the electrical and electronics sector is still deficient, especially in less developed countries.

7.7.4.4 Current recycling from end-of-life products

Most old gold scrap is from jewelry and other nontechnological gold applications. In the United States, this was 92% in 1998. On a global scale, we expect that this proportion will be even higher, because the recycling of gold from technological applications is not yet very developed. We therefore assume that, in 2015, 95% of old gold scrap was from non-technological gold applications. Because of its extremely high value, gold is recycled from gold-containing end-of-life products at a very high rate (>90%). In a typical year, about 600 tons of gold is recovered from old scrap (Butterman and Amey, 2005). This is about half of the total gold scrap (Thomson Reuters, 2017), and the rest consists of new scrap. Total scrap counted for about one third of the total gold supply to the market in the period 2007 to 2016 (Thomson Reuters, 2017; see Fig. 7.7.5). Gold consumption in 2015 was about 4400 ton.

Data on the end-of-life recycling rate of gold-containing products are not available. In practice, only gold from jewelry, coins, and technological applications, such as electronics and dental applications, is recycled.

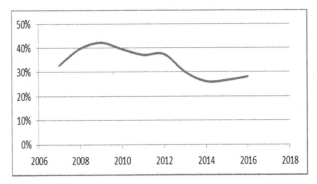

Figure 7.7.5 Recycled gold as a part of total gold supply. *(Derived from Thomson Reuters, (2017). GFMS Gold Survey 2017.)*

The end-of-life recycling rate of gold from jewelry and coins is close to 100%, while the recyclability of gold from applications in the electronics and electrical sector is high, but more complicated because gold is only a relatively small portion of the product.

The largest part of the gold that is used in electronical and electrical applications is in PCBs. According to Holgersson et al. (2018), the gold content in PCBs in mobile phones is about 1 g/kg or 1 million ppb. This is a far higher gold concentration than in ores currently mined. In fact, the gold content is the main reason for recycling mobile phones, as the value of the gold contained in PCBs is higher than the value of all the other materials together. In 2012, the value of the gold in 1000 kg of average PCBs was about 15,000 US$; more than 70% of the value of all the contained metals (Wang and Gaustad, 2012; Cucchiella et al., 2015). This means that 1000 kg of mobile phones can contain 300−350 g of gold (Sarath et al., 2015). With a gold price of 42 US$ per gram (21-3-2019), the value of gold in 1 ton of mobile phones is 12,000−14,000 US$. Assuming an average weight per mobile phone of 150 g, this means that the gold value in an average mobile phone is about 1.8 US$.

Old scrap consists of two main parts: nontechnological applications, mostly jewelry, and technological applications. In the United States, scrap from technological applications was 8% of old scrap in 1998, and scrap from nontechnological applications was 92% (Amey, 1998). Worldwide, the nontechnological part of old scrap will probably be higher. For the flow scheme in Fig. 7.7.6 we assume a figure of 95%. In the United States, the overall recycling efficiency of gold from collected old scrap including jewelry and electronic scrap was 96% in 1996 (Amey, 1998). However, the collection rate of WEEE is currently only 25% at a global scale (Baldé et al., 2017). This does not mean that 75% of gold-containing WEEE is discarded, as a substantial part of end-of-life electrical and electronic products is traded from developed countries to developing countries and starts a second life (Ongondo et al., 2011). Another part, especially small WEEE such as mobile phones, laptops, and computers, is stocked at home and still available for future collection. There are no data available regarding the part of WEEE that is exported or still in-stock in the waste management phase. From the flow scheme presented in Fig. 7.7.6, it can be derived that, in 2015, about three fourths of the non-collected WEEE was still in-stock or had been exported for reuse, and one fourth had been disposed of in landfills or waste incineration plants.

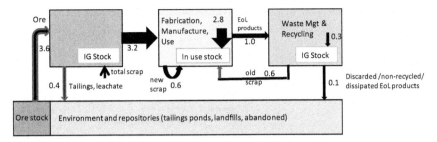

Figure 7.7.6 Simplified global gold flow scheme for the year 2015. Flows are in thousands of metric tons. The arrow widths are a rough indication of flow magnitudes. IG stock is industrial, commercial, and government stocks.

Table 7.7.3 shows that the current gold recycling rate of end-of-life technological gold applications is only 7%. Gold in dental applications is partly recycled by dentists, and we are not aware whether gold teeth are removed from deceased persons.

7.7.4.5 Gold discards to the environment

In Fig. 7.7.6, the loss of gold in end-of-life products to the environment, mainly via landfills and the bottom ash and fly ash of waste incineration plants, is assumed to be about 3% of the annual gold production. This is based on historic gold losses, which are estimated at less than 2% (see Table 7.7.2). However, the current losses will be somewhat higher, due to the 10% of gold that is used for technological applications. Most of the gold currently lost is through disposal of non–separated electronic waste.

7.7.4.6 Current gold flows summarized

The above data are summarized in the simplified anthropogenic gold flow scheme presented in Fig. 7.7.6. The assumptions made for the mass balancing in Fig. 7.7.6 are included in Table 7.7.3.

7.7.5 Sustainable gold flows

For different assumptions of gold resource availability, we investigate under which conditions the gold extraction rate can be made sustainable, while realizing a global service level of gold for technological applications that is equal to the gold service level for technological applications in developed

Table 7.7.3 Gold flows in 2015. Flows are in thousands of metric tons.

Points of departure		Derived		
Gold production in 2015 (1) (kt)	3.209			
Recovery efficiency (5)	90%	Mined ore (kt)	3.57	Kt
		Tailings and leachate	0.36	Kt
Net hedging supply (1) (kt)	0.021			
Total scrap in 2015(1)				
	1,172	Consumption in 2015	4.4	Kt
		of which for technological use (1)	0.37	Kt
		of which for non-technological use	3.99	Kt
		Surplus in 2015	0.04	Kt
Old scrap, see section 7.6.4 (2) (kt)	0.6			
		New scrap	0.57	Kt
EoL technological gold products (3) (kt)	0.43	Old scrap from non- technological applications	0.57	Kt
Non-technological EoL as part of total EoL (derived from 4)	95%	Old scrap from technological applications	0.03	Kt
		EoL recycling rate technological applications	7%	
		Non recycled gold from technological applications	0.4	Kt
Discarded (as part of annual gold production) (derived from (1)) 3%		Discarded	0.1	Kt
		Stocked in waste management stage	0.3	Kt
		EoL products	1	Kt
		of which end-of-life technology products	0.43	Kt
		of which end of life jewelry and coins	0.57	Kt
		To in-use stock	2.83	Kt

(1) Thomson Reuters, 2017, (2) Butterman and Amey, 2005, (3) average of annually consumed technological products in period between 2007 and 2016; see section 7.7.2 and (1), (4) Amey (1998), (5) Hylander et al (2007)

countries in 2020. Gold is an unusual metal in the sense that only 10% is used for technological purposes and 90% for decorative and monetary reasons. Moreover, a very large part of the gold that was mined in the past is still available. Hence, for gold's service level we focus on its technological applications.

The estimated gold consumption per capita in developed countries was about 2 g/capita/year in 2020 (see Table 5.3). About 10% of this was for technological applications, or 0.2 g/capita/year.

The possibilities for improving recovery efficiency at the mining stage, substitution, material efficiency improvement, and dissipation are limited in the case of gold. However, the end–of–life recycling rate of gold from industrial products can be substantially improved, in particular by improving the collection rate of waste from electrical and electronic equipment (WEEE). Currently, the end–of–life recycling rate of gold from technological products is estimated at only 7% (see Table 7.7.3). For the future, we assume a maximum end–of–life recycling rate of gold from technological products of 50%, which will be quite a challenge. Table 7.7.4 presents different combinations of measures required to achieve sustainable

Table 7.7.4 Different mixes of measures to achieve a sustainable gold production rate, while realizing a global gold service level for technological applications that is similar to the gold service level in developed countries in 2020. The boldfaced rows include the main measures.

		Sustainability ambition (years of sufficient availability)					
		1000		500		200	
Ultimately available gold resources[a,b]	Mt	1.0	1.9	0.5	0.9	0.3	0.4
Sustainable gold production	kt/ year	1.0	1.9	1.0	1.9	1.5	1.9
Aspired future gold service level for technological applications	g/ capita/ year	0.2	0.2	0.2	0.2	0.2	0.2
Assumed future substitution	%	**0%**	**0%**	**0%**	**0%**	**0%**	**0%**
Assumed increase in material efficiency	%	0%	0%	0%	0%	0%	0%
Derived future gold consumption level for technological applications	g/ capita/ year	0.2	0.2	0.2	0.2	0.2	0.2
Corresponding annual gold consumption	kt/ year	2	2	2	2	2	2
Assumed in-use dissipation of remaining products[c]	%	0%	0%	0%	0%	0%	0%
Gold in end-of-life products of technological applications	kt/ year	2	2	2	2	2	2
Required gold recycling from technological applications	kt/ year	1.0	0.1	1.0	0.1	0.5	0.1
Required gold recycling rate for keeping gold available for technological applications at the level of developed countries in 2020[d,e]	%	**50%**	**7%**	**50%**	**7%**	**25%**	**7%**

The assumed maximum recovery efficiency in the future remains 90%, and the assumed maximum substitutability of gold in technological applications in the future is 0%. The assumed material efficiency increase in technological applications is 0%, and the assumed in-use dissipation rate of gold from technological applications remains 0%. The assumed future maximum gold recyclability from technological products is 50%.

[a]Low estimate: 0.3 Mt; high estimate: 2 Mt (see Section 7.7.3.2).
[b]Current and assumed future recovery efficiency: 90% (see Section 7.7.4).
[c]Current in-use dissipation rate: 0% (see Section 7.7.4).
[d]Current end-of-life recycling rate from technological applications: 7% (see Section 7.7.4).
[e]It is assumed that the world population stabilizes at 10 billion people.

gold production, while simultaneously realizing a service level of gold for technological applications that is equal to the service level of gold in developed countries in 2020.

7.7.6 Conclusions

The conclusions are as follows:

- A sustainability ambition of 1000 years can only be achieved with >1 Mt of ultimately available gold resources combined with 50% end-of-life recycling of gold from technological products. With ultimately available gold resources of >1.9 Mt, no measures will be necessary to have sufficient gold available for technological applications. Extra recycling or additional resources will cause gold to become available for nontechnological applications.
- A sustainability ambition of 500 years can only be achieved with 0.5 Mt of ultimately available gold resources. With ultimately available gold resources of >0.9 Mt, no measures will be necessary to have sufficient gold available for technological applications. Extra recycling or additional resources will cause gold to become available for nontechnological applications.
- A sustainability ambition of 200 years can be achieved with 0.3 Mt of ultimately available gold resources and 25% recycling of gold from technological applications. To the extent that the ultimately available gold resources are > 0.4 Mt, gold will become available for nontechnological applications.

References

Amey, E.B., 1998. Gold Recycling in the United States in 1998, US Geological Survey Circular 1196-A.

Baldé, C.P., Forti, V., Gray, V., Kuehr, R., Stegmann, P., 2017. The Global E-Waste Monitor 2017. United Nations University, International Telecommunication Union, International Solid Waste Association. ISBN 978-92-808-9054-9.

Butterman, W.C., Amey, E.B., 2005. Mineral Commodity Profiles, Gold, US Geological Survey Open-File Report 02-303.

Cucchiella, F., D'Adamo, I., Lenny Kok, S.C., Rosa, P., 2015. Recycling of WEEEs: An economic assessment of present and future e-waste streams. Renew. Sustain. Ener. Rev. 51, 263–272.

Halada, K., Shimada, M., Ijima, K., 2008. Decoupling status of metal consumption from economic growth. Mater. Trans. 49 (3), 411–418.

Holgersson, S., Steenari, B.-M., Björkman, M., Cullbrand, K., 2018. Analysis of the metal content of small-sized Waste Electric and Electronic Equipment (WEEE) printed circuit boards — part 1: internet Routers, mobile phones and smartphones. Resour. Conserv. Recycl. 133, 300—308.

Hylander, L.D., Plath, D., Miranda, C.R., Lücke, S., Öhlander, J., Rivera, A.T.F., 2007. Comparison of different gold recovery methods with regard to pollution control and efficiency. Clean 35 (1), 52—61.

Kesler, S.E., Simon, A.C., 2017. Mineral Resources, Economics and the Environment, second ed. Cambridge University Press.

Mudd, G.M., 2007. Global trends in gold mining: towards quantifying environmental and resource sustainability? Resour. Pol. 32, 42—56.

Ongondo, F.O., Williams, I.D., Cherrett, T.J., 2011. How are WEEE doing? A global review of the management of electrical and electronic wastes. Waste Manag. 31 (2011), 714—730.

Sarath, P., Bonda, S., Mohanty, S., Nayak, S.K., 2015. Mobile phone management and recycling: views and trends. Waste Manag. 46, 536—545.

Thomson Reuters, 2017. GFMS Gold Survey 2017.

US Geological Survey, 1998. Gold, Mineral Commodities Summaries.

US Geological Survey, 2000. 1998 Assessment of Undiscovered Deposits of Gold, Silver, Copper, Lead and Zinc in the United States.

US Geological Survey, 2005. Mineral Commodity Profiles — Gold, Open-File Report 02-303.

US Geological Survey, 2009. Mineral Commodity Summaries, Gold.

US Geological Survey, 2016. Gold, Minerals Yearbook 2016.

US Geological Survey, 2017. Gold Statistics.

US Geological Survey, 2019. Mineral Commodity Summaries, Gold, January 1996—February 2019.

Wang, X., Gaustad, G., 2012. Prioritizing material recovery for end-of-life printed circuit boards. Waste Manag. 32, 1903—1913.

Wilkinson, B.H., Kesler, S.E., 2010. Tectonic-diffusion estimate of orogenic gold resources. Econ. Geol. 105, 1321—1334.

CHAPTER 7.8

Indium

7.8.1 Introduction

A thin film of indium tin oxide (ITO) containing about 75% indium is an essential part of every flat screen and touchscreen today. Indium is therefore very important for the current information society. The European Union and other countries consider indium to be a critical raw material, because it is essential for modern technology, and because more than 70% of refined indium is produced in only two countries (China and South Korea). According to Goe and Gaustad (2014), indium is the fourth most critical raw material (of 17 considered metals) for use in photovoltaics in the United States. This is due to a combination of high demand, the difficulty of recycling and substitution, import reliance, constrained secondary production, the importance of clean energy, and the geological scarcity of indium.

The purpose of this section is to investigate the conditions for the sustainable extraction of primary indium. The sustainable extraction of indium is defined as a extraction that can be sustained for an agreed period of time that provides indium to a world population of 10 billion people at a service level that is equal to the indium service level in developed countries in 2020, for an affordable price. In this book, we investigate three sustainability periods: 200, 500, and 1000 years.

7.8.2 Properties and applications

An important property of indium is its low melting point (156.6°C), which is lower than the melting point of tin. Indium is also a very soft metal, making indium suitable for alloys with a low melting point. The density of indium at normal temperature is about 7.3 g/cm^3. Indium's first industrial use was in bearings in military aircraft motors in the 1940s. For several decades, indium continued to be used mainly in solders, other alloys, and dental applications (amalgam), but from the 1990s its use in flat panel displays, architectural glass, and LEDs increased. Over the last decade, indium's use in solar panels and thermal interface materials has also started to increase.

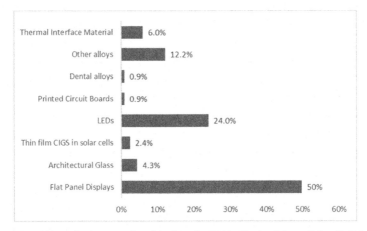

Figure 7.8.1 Uses of primary refined indium in 2011. *(Derived from Licht, C., Peiró, L.T., Villalba, G., 2015, Global substance flow analysis of gallium, germanium and indium: Quantification of extraction, uses, and dissipative losses within their anthropogenic cycles. J. Ind. Ecol. 2015, 890–903.)*

Currently, refined indium is used for four main groups of applications in semifinished products: ITO, semiconductors, alloys, and thermal interface materials. Fig. 7.8.1 presents an overview of the indium applications in 2011. It is clear that flat panel displays are the most important indium application.

Indium tin oxide

ITO is a mixture of 90% indium oxide (In_2O_3) and 10% tin oxide (SnO_2). The indium content of ITO is about 75%. ITO combines electrical conductivity, transparency, and heat reflection, making it suitable as a transparent and conductive coating on flat panel displays and liquid crystal displays (LCDs) in TVs, laptops, watches, and portable phones, to convert data from an electrical to an optical form. The ITO film thickness varies between 100 and 500 nm (Yoshimura et al., 2013), which means that the amount of ITO per display is very low. ITO is also used as a coating on glass due to the combination of heat reflection and electrical conductivity. In this application, ITO is used on automobile and aircraft windshield glasses for demisting and deicing.

ITO is also used for the production of CIGS (copper-indium-gallium-selenide), which is applied in thin-film solar cells. In 2016, CIGS films accounted for 2% of global solar cell energy production (US Geological Survey, 2019b). Thin-film CIGS solar cells are less efficient than crystalline silicon solar cells, but they are less expensive to produce and more flexible (US Geological Survey, 2010a), and the use of thin-film CIGS in photovoltaics is increasing. The thickness of the indium-containing layer on CIGS solar cells is between 1000 and 3000 nm (Zimmermann and Gössling-Reisemann, 2014), and the average amount of indium used in CIGS solar panels is 16.5 kg/MW (Zimmermann and Gössling-Reisemann, 2014).

Semiconductors

High-purity indium is used in semiconductors, for instance in LEDs, photovoltaic cells, and optoelectronic devices such as laser diodes, photodetectors, and infrared detectors. In this application, indium, often in combination with gallium and zinc, is included in compounds such as oxides, phosphides, antimonides, and arsenides (US Geological Survey, 2005, 2019b). Indium phosphide lasers are used in fiber-optic networks with connections to 3G, 4G, and 5G wireless antennas (US Geological Survey, 2020). Indium is also used in batteries as a substitute to mercury to prevent zinc from corroding (US Geological Survey, 2010b).

Alloys

Indium is used to lower the melting point of alloys. Low melting point alloys are used in fusible alloys, for instance in indoor fire sprinkler systems, holding agents, and solders. Indium is also used in alloys with precious metals (gold and palladium) for dental applications (US Geological Survey, 2005). Furthermore, indium is used in lead-free solders in combination with silver and tin.

Thermal interface materials

Indium silver alloys or pure indium foil are used as thermal interface materials in electronics for effective heat transfer to a heat sink.

Future developments

In general, global demand for electronic products (with indium) is expected to follow GDP development. Faster (than GDP) growth is expected of the application of indium in solar cells, because of the energy transition to fossil-free electricity production. However, models of the demand for indium for solar panels show large divergences. Stamp et al. (2014)

estimated the cumulative demand for refined indium for CIGS solar panels between 2010 and 2050 to be a minimum of about 300 tons and a maximum of 260,000 tons in the most far-reaching scenario. Elshkaki and Graedel (2013) estimated the cumulative indium demand for PV solar technology between 2010 and 2050 at about 10,000 tons. These demand figures are very divergent and the uncertainty is high. In 2011, the use of refined primary indium for CIGS solar cells was about 13 tons (derived from Licht et al., 2015). The transition to lead-free solders also will probably increase indium demand for application in solders faster than GDP. Another growth market for indium could be its application in Li-ion batteries as a coating on lithium electrodes (Santhosha et al., 2019).

7.8.3 Production and resources

7.8.3.1 Production

In 2019, 71% of the production of refined indium was concentrated in just two countries: China (40%) and South Korea (31%) (see Fig. 7.8.2). It is therefore no coincidence that demand for refined indium in these two countries is also high. A considerable part of the world's electronic consumer products is manufactured in China and South Korea. China is also the largest zinc-producing country, with about 33% of the global zinc production in 2019. This is relevant because, at present, 95% of indium is recovered from zinc sulfide ores. The rest is mainly recovered from copper and tin ores (Werner et al., 2015). Despite the high contribution to zinc

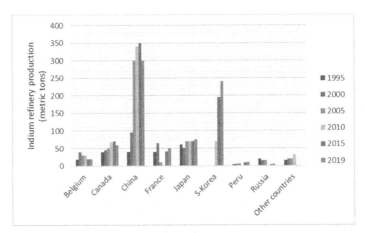

Figure 7.8.2 Refined indium-producing countries between 1995 and 2019 (US Geological Survey, 1996–2020).

production and indium refining, China has transitioned from an indium-exporting country to an indium-importing country (Ylä-Mella and Pongrácz, 2016). Indium refining takes place partly in countries without zinc production, for instance in Belgium, France, Japan, and South Korea. On the other hand, some important zinc-producing countries have little or no indium-refining capacity, such as the United States, Australia, India, Bolivia, and Mexico. Of these, Australia and Bolivia are important exporters of zinc concentrate for the purpose of indium refining (European Commission, 2013; US Geological Survey, 2019a; US Geological Survey, 1996–2020).

7.8.3.2 Resources

The average indium concentration in the upper continental crust is 0.24 ppm (see Table 2.3). The US Geological Survey estimated the reserve base of indium at 16,000 tons in 2008 (US Geological Survey, 2008), and no new figures have been provided by the US Geological Survey regarding the reserve base or the identified resources of indium since this time. According to the Indium Corporation of America, global indium resources are about 50,000 tons (Moss et al., 2011). Indium is mainly extracted from zinc ore. Hence, the availability of indium will be more or less concurrent with the availability of zinc (see Fig. 7.8.3).

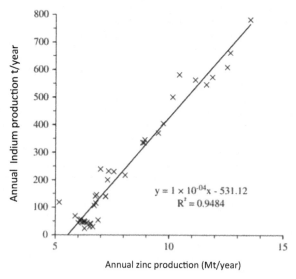

$$y = 1 \times 10^{-04}x - 531.12$$
$$R^2 = 0.9484$$

Figure 7.8.3 Indium production as related to zinc production. *(Derived from Werner, T.T., Mudd, G.M., Jowitt, S.M., 2015. Indium: key issues in assessing mineral resources and long-term supply for recycling. B. Appl. Earth Sci. https://doi.org/10.1179/1743275815Y. 0000000007.)*

The indium content of zinc deposits, from which indium is recovered, is between 1 and 100 ppm (US Geological Survey, 2020). According to data from Werner et al. (2015), the indium grade in 10 well documented deposits, most being zinc mines, was between 5 and 65 ppm, with an average of about 10 ppm. If we can relate this to the estimated ultimately available amount of zinc (30,000 Mt; see Chapter 2), and a recovery efficiency of zinc from zinc ore of 80%, this results in indium resources of 375,000 tons or 400 kt (rounded).

A generic approach for estimating the ultimately extractable resources of an element is to extrapolate the results of assessments of the extractable resources of other elements. US Geological Survey (2000) compared the results of 19 assessments of the estimated total (identified plus undiscovered) deposits of gold, silver, copper, lead, and zinc in the United States in the upper 1 km of the continental crust of the United States (a) with the identified resources of the same elements in the United States that far (b). The ratios a/b were 4.82 for zinc, 3.88 for silver, 2.67 for lead, 2.2 for gold, and 2.12 for copper, with an average of 3.14. The US Geological Survey does not provide data regarding the identified indium resources. Hence, we estimated these by applying the average ratio between the identified resources and the reserve base of 14 raw materials, which is 2.8 (see Table 2.5). Extrapolating this result to the upper 3 km crust, this approach results in globally available indium resources of $3 \times 3.14 \times 2.8 \times 16 = 400$ kt (rounded). This is the same result as above.

If we use the approach that the extractable indium resources are a maximum of about 0.01% of the total amount of indium in the upper 3 km of the Earth's crust (UNEP, 2011; Erickson, 1973; Skinner, 1976), indium resources could be as much as 30 Mt (rounded).

We include both results for the ultimately available indium resources (0.4 and 30 Mt) in our observations regarding the sustainable use of indium.

7.8.4 Indium as a companion metal: impact on availability and price

Between 1950 and 2015, the market price of indium per kg was about 200 times higher than the market price of zinc, though with a higher volatility (see Fig. 7.8.4). This means that the total value of zinc produced in 2018

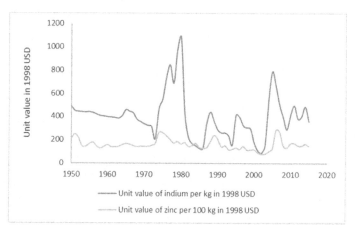

Figure 7.8.4 Market price development of zinc and indium between 1950 and 2015 (US Geological Survey 2017a,b).

(13 Mt) was about 20,000/200 = 100 times as much as the value of indium produced in the same operation. The conclusion is that indium extraction remains a minor business for zinc mining companies: about 1% of their revenues as long as zinc and indium maintain the same price ratio as today. Zinc mining companies will not be inclined to increase zinc mining to increase their indium revenues.

However, between 1980 and 2015, the annual increase in zinc production was 2.2%, whereas the annual increase in refined indium production in the same period was 8.1%. This may have been realized by increasing the indium recovery efficiency and/or extracting indium from historical zinc residues and from historical indium-refining operations (see Section 7.8.5).

According to Lokanc et al. (2015), producers will stop producing indium at prices below US$100 per kg, and they can cover the variable indium production costs at prices between US$150 and US$300 per kg (in 2011 US dollars). Prices higher than US$350 (2011 US$) are attractive for indium producers, but they will not be able to increase their production quickly, as that would require an increase in zinc production and/or an increase in the recovery efficiency of indium, requiring relatively high investments. Hence, indium production is fairly inflexible, which explains the volatility of the indium price.

7.8.5 Current stocks and flows

Table 7.8.1 presents the production and uses of refined indium in 2011. The numbers are based on data from Licht et al. (2015). In 2011, refined indium production was 662 tons, extracted from zinc ore (64%) and residue stocks (36%) (Licht et al., 2015). Of the indium production in 2011, 12 tons of indium were stockpiled, and 650 tons were utilized (Licht et al., 2015).

The following conclusions can be drawn.

7.8.5.1 Recovery efficiency

The recovery efficiency of the extraction of indium from zinc ore is relatively low. In 2011, only 35% of the indium contained in the extracted zinc ore was recovered (derived from Licht et al., 2015). However, the recovery efficiency of indium from the ore in which it is contained is increasing quickly: in 2004, the indium recovery efficiency was only about 10% (Yoshimura et al., 2013). According to Werner et al. (2015), indium production per unit of zinc has increased about six-fold in the period between 1988 and 2012.

7.8.5.2 Losses and recycling

During fabrication and manufacturing, 62% of all the indium used is lost as industrial waste. Of these losses, about 85% are generated in the fabrication and manufacturing of ITO-based applications, such as flat panel displays, and about 15% during the fabrication and manufacturing of applications with semiconductors. In 2011, the indium in flat panel displays represented 16% of the indium in consumer products. About 50% of primary indium and 71% of primary plus secondary indium were lost in connection with the production of flat panel displays. Hence, with regard to indium use, the production of flat panel displays really is an indium–wasting process. The losses are especially large during the "sputtering" of ITO on displays, glass surfaces, and solar cells. According to the data of Licht et al. (2015), only 10% of the ITO is successfully deposited on the substrate, about 70% of the total ITO used in the sputtering process can be recovered, and about 20% is lost on the surfaces of tools and chamber walls. Hence, the sputtering process is still very inefficient. The further processing of the substrate is also quite inefficient, as only 30% of the indium on the substrate (or 5% of the ITO originally used) ends up in consumer products. The rest goes to waste

Table 7.8.1 Indium production and use in 2011 (metric tons).

| Raw materials | | | | Semifinished products | | | Consumer products | | | | | |
Primary indium	Secondary indium	Total								Primary + secondary indium use	Primary indium use	Indium in consumer products
364	1138	1502	For the fabrication of:	ITO	1502	81%	For manufacturing of	Flat panel displays	40	2.7%	71%	50%	16%
								Arch. glass	3.5	0.2%	6.3%	4%	1.4%
								CIGS	1.6	0.1%	2.9%	2%	0.7%
								Losses	300	20%			
								Recycling	1157	77%			
								Total	1502	100%			
156	77	233		Semiconductors	233	12%	For manufacturing of:	CIGS	1	0%	0.2%	0.4%	0.4%
								LEDs etc.	69	30%	12%	24.0%	28%
								Losses	55	24%			
								Recycling	108	46%			
								Total	233	100%			

	Alloys	For manufacturing of:	PCBs					
				6	7%	0.3%	0.9%	2.4%
91			Dental alloy	6	7%	0.3%	0.9%	2.4%
			Other alloy	79	87%	4.2%	12.2%	32%
			Total	91	100%			
39	Thermal interface material	For manufacturing of:	Thermal interface material	39		2.1%	6.0%	16%
650	Total		Total	1865		100%	100%	100%

			Alloys				
91	1215	1865		91	5%		
39			Thermal interface material	39	2%		
650			Total	1865	100%		

Derived from Licht, C., Peiró, L.T. Villalba, G., 2015. Global substance flow analysis of gallium, germanium and indium: Quantification of extraction, uses and dissipative losses within their anthropogenic cycles. J. Ind. Ecol. 2015, 890—903.

streams or is recycled. Finally, only about 12% of the primary indium used for ITO production is included in consumer products. The production of LEDs is also wasteful, as only about 30% of the semiconductors produced ends up in consumer products (Licht et al., 2015).

Another problem is that end-of life products are not recycled to recover indium. Globally, the separated collection rate of waste from electronic and electrical equipment (WEEE) is still quite limited, if it takes place at all. According to Baldé et al. (2017), it is about 25%, although other scholars indicate an end-of-life WEEE recycling rate of 15% to 20% worldwide (Yang et al., 2017). Moreover, the recovery of materials from WEEE scrap is focused on precious metals (such as gold, silver, PGMs), and metals like indium are still hardly recovered from WEEE, or not at all. This is not because it is not technically possible, but because it is not profitable.

7.8.5.3 Downcycling

The indium used in architectural glass and dental applications may be part of a recycling process, not for the sake of recycling indium, but for the recycling of glass and precious metals. This indium is downcycled. In 2011, this affected about 4% of the indium in end-of-life products (Licht et al., 2015).

7.8.5.4 Disposal

Indium is mainly used in electronic and electrical equipment. As only a relatively small part of end-of-life electronic and electrical equipment is separately collected, and as the focus is on recycling precious metals (gold, silver, PGMs, copper) from WEEE, the indium in end-of-life products, except the above-mentioned small quantity of downcycled indium, ends up in landfills and incinerators.

7.8.5.5 Dissipative use

There is no in-use dissipation of indium. All indium–containing products are discarded after use, not during use.

7.8.5.6 In-use stock

The average lifetime of products that contain indium is estimated at about 10 years. The actual usage time will probably be some years less (Yoshmura et al., 2013); however, electronic products are often kept for a while after they have been replaced by a new product, or they start a second life in a developing country. Based on the approach explained in Chapter 7.1, it is estimated that the in-use stock of indium is currently growing every year by about 50% of the net indium consumption in that year.

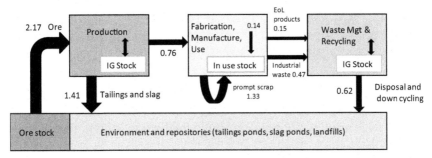

Figure 7.8.5 Simplified global indium cycle for 2019. The amounts are in kt. Assumptions made are: indium recovery efficiency of 35%, annual inflow into in-use stock of 50% of indium in consumer products, and end-of life recycling rate of 0%. The amount of refined indium produced is based on data from US Geological Survey, 2020. The prompt scrap flow and the amount of industrial waste are based on data from Licht et al. (2015). IG stock is industry and government stocks. The arrow widths are a rough indication of the volume of the respective flows.

7.8.5.7 Current indium flows summarized

Fig. 7.8.5 extrapolates the above flow data to the situation in 2019.

7.8.6 Sustainable indium flows

7.8.6.1 Introduction

Here we analyze under which conditions the extraction rate of indium can be made sustainable, while simultaneously increasing the global indium service level to the same level as in developed countries in 2020. The indium consumption in developed countries in 2020 is estimated at 0.4 g/capita/year (see Table 5.3).

There are four opportunities: (1) increasing the indium recovery rate from indium-containing ore, (2) substitution, (3) increasing material efficiency, and (4) reducing the loss of indium during the fabrication and manufacturing stage of ITO-containing products. We explore the possibilities and limitations of these approaches below. The recycling of indium from end-of life products is not considered a feasible option for increasing the sustainable use of indium.

7.8.6.2 Increase in recovery efficiency

The recovery efficiency of indium from zinc ore has increased rapidly over the years; it increased six-fold between 1988 and 2012, and from 10% in 2004 to 35% in 2011 (see Section 7.8.5). The latter figure means that the annual amount of indium contained in ore that ends up in tailings and slag is about 10 times as high as the amount of indium annually put into final products (see Fig. 7.8.5). There is therefore a world of opportunity for further increasing indium recovery efficiency. We assume that it will be possible to increase the indium recovery efficiency to 70% in the future.

7.8.6.3 Substitution of indium-containing products

Price volatility and supply concerns mean that intensive research is ongoing to develop alternatives to indium and ITO (Ylä-Mella and Pongrácz, 2016). In principle, indium can be replaced in many of its applications, although often at the expense of quality. With regard to substitution, the focus is on the most important indium applications, such as the different applications of ITO.

According to US Geological Survey, 2020, ITO coatings in LCDs can be replaced by antimony tin oxide coatings or zinc oxide nanopowder. A disadvantage to this is that antimony and tin are also geologically scarce raw materials. Aluminum-doped zinc oxide can replace indium in flat panel displays (Chen and Graedel, 2012), and carbon nanotube coatings could be an alternative to ITO in flexible displays, solar cells, and touchscreens. In flexible displays and organic light-emitting diodes (OLEDs), ITO can be replaced by poly(3,4-ethylenedioxythiophene). Graphene can replace ITO electrodes in solar cells (US Geological Survey, 2020). According to Graedel et al. (2015), there are not yet any good substitutes for use in ITO film coatings.

In its critical raw material profiles, the European Commission (EC, 2013) provides substitutability scores for different applications of indium (see Table 7.8.2).

We assume a maximum substitutability of indium of 50%.

Table 7.8.2 Substitutability scores of indium applications (European Commission, 2013).

Use	Substitutability score
Flat panel displays	1
Photovoltaics	0.7
Semiconductors	0.7
Solders	0.7
Batteries	0.3
Alloys	0.3
Thermal interface materials	0.3

The meaning of the scores is as follows: 1, not substitutable; 0.7, substitutable at high cost and/or loss of performance; 0.3, substitutable at low cost.

7.8.6.4 Material efficiency improvement

We have assumed that it will be possible to increase the material efficiency by about 10% (see Chapter 7.1).

7.8.6.5 Reduction in losses

The amounts of indium in end-of-life consumer products are very small, and indium is often contained in complex matrices. According to Buchert et al. (2012), 1000 kg of laptop displays, computer monitors, and TVs contain a total of 174 g of indium, or 174 ppm. According to the same authors, a typical white LED contains 0.029 mg of indium. The product design of electronic consumer goods is versatile, and due to further miniaturization in combination with technological improvements, the amount of indium per product is very low. Nevertheless, research is ongoing with regard to the recovery of indium from its main applications in LCD screens and solar cells. This research concentrates on the separation of the indium-containing layer from flat screens and solar cells by chemical leaching and/ or by mechanical or thermal treatment. However, the very small quantity of indium per device will make it difficult to make indium recycling from end-of-life consumer products financially viable, unless the indium price increases substantially and/or indium recycling is subsidized or advantaged in other ways. According to Lokanc et al. (2015), indium market prices would need to consistently exceed 700 US$ (2011 level) to make the recovery of indium from end-of-life consumer products profitable.

For the short term, it makes more sense to focus indium recycling on recovery from the industrial waste generated during the fabrication of semifinished products and the manufacturing of consumer products.

As can be seen in Fig. 7.8.5, the amount of indium in industrial waste in 2011 was about three times higher than the amount of indium contained in end-of-life consumer products. A total of 85% of industrial indium-containing waste is generated by the sputtering process and the manufacture of flat panel displays and other products, and 15% by the fabrication of semiconductors and the manufacturing of LEDs (Licht et al., 2015). The indium concentration in this industrial waste is probably higher than the indium concentration in end-of-life consumer products and can be collected more easily, as it is produced in a limited number of ITO-producing factories and flat panel display-manufacturing companies.

We assume a maximum recyclability of indium from industrial waste of 50%.

7.8.7 Conclusions

Table 7.8.3 presents different mixes of measures that are necessary to achieve a sustainable extraction rate of indium while simultaneously realizing a service level of indium that is equal to the service level of indium in developed countries in 2020.

The conclusions are as follows:

- A sustainability ambition of 1000 years is only achievable if the ultimately available indium resources are >0.6 Mt in combination with a recovery efficiency of 70%, substitution of 50%, and a recycling rate of indium from industrial waste of 50%. If the ultimately available indium resources are >4 Mt, no measures are needed.
- A sustainability ambition of 500 years is achievable with ultimately available indium resources of 0.4 Mt, plus substantial increases in the recovery efficiency and recycling of indium from industrial waste in combination with substitution. If the ultimately available indium resources are >2 Mt, no measures are required.
- A sustainability ambition of 200 years is achievable with 0.4 Mt of ultimately available indium resources and an increase in the recovery efficiency to 70%. If the ultimately available indium resources are >0.8 Mt, no measures are required from the viewpoint of a sustainable indium production rate.

Table 7.8.3 Different mixes of measures to achieve a sustainable indium production rate, while realizing a global indium service level that is similar to the indium service level in developed countries in 2020. The boldfaced rows include the main (combinations of) measures.

Sustainability ambition (Years of sufficient availability)		1000		500			200		
Ultimately available indium resources[a]	Mt	0.6	4.0	0.4	0.4	2	0.4	0.4	0.8
Assumed future recovery efficiency[b]	**%**	**70%**	**35%**	**70%**	**70%**	**35%**	**70%**	**35%**	**35%**
Future availability of indium resources	Mt	1.1	4.0	0.8	1.6	2.0	0.8	0.4	0.8
Sustainable indium production[c]	kt/year	1.1	4.0	1.6	4.0	4.0	4.0	2.0	4.0
Aspired future indium service level	g/capita/year	0.4	0.4	0.4	0.4	0.4	0.4	0.4	0.4
Assumed future substitution	**%**	**50%**	**0%**	**50%**	**28%**	**0%**	**0%**	**10%**	**0%**
Assumed increase in material efficiency	%	10%	0%	10%	10%	0%	0%	10%	0%
Aspired future indium consumption level	g/capita/year	0.2	0.4	0.2	0.3	0.4	0.4	0.3	0.4
Corresponding annual indium consumption	kt/year	1.8	4.0	1.8	2.6	4.0	4.0	3.2	4.0
Assumed in-use dissipation[c]	%	0%	0%	0%	0%	0%	0%	0%	0%
Indium in end-of-life products and industrial waste	kt/year	1.8	4.0	1.8	2.6	4.0	4.0	3.2	4.0
Of which industrial waste	%	76%	76%	76%	76%	76%	76%	76%	76%
Required indium recycling from industrial waste	kt/year	0.7	0.00	0.20	0.99	0.00	0.0	1.2	0.0
Required indium recycling rate from industrial waste[d]	**%**	**50%**	**0%**	**15%**	**50%**	**0%**	**0%**	**50%**	**0%**

The assumed maximum recovery efficiency in the future is 70%, and the assumed maximum substitutability of indium in the future is 50%. The assumed material efficiency increase is 10%, and the assumed minimum in-use dissipation rate of remaining products is 0%. The assumed maximum recyclability of indium from industrial waste in the future is 50%. It is assumed that the world population stabilizes at 10 billion people.
[a]Low estimate: 0.4 Mt; high estimate: 30 Mt (see Section 7.8.3.2).
[b]Current recovery efficiency: 35% (see Section 7.8.5).
[c]Current in-use dissipation rate: 0% (see Section 7.8.5).
[d]Current end-of-life recycling rate: 0% (see Section 7.8.5).

It is uncertain whether future indium demand will stop growing at the level of developed countries in 2020. Indium is mainly used in high-tech applications with strong growth potential. An increase in indium consumption above 400 mg/cap/year at a global level would require additional measures to the measures in Table 7.8.3 to keep indium use sustainable, if the ultimately available indium amount is limited to 0.4 Mt.

References

Baldé, C.P., Forti, V., Gray, V., Kuehr, R., Stegman, P., 2017. The Global E-Waste Monitor 2017. United Nations University, International Communication Union, International Solid Waste Association. ISBN: 978-92-808-9054-9.

Buchert, M., Manhart, A., Bleher, D., Pingel, D., 2012. Recycling Critical Raw Materials from Waste Electronic Equipment. Öko-Institut e.V., Freiburg, Germany.

Chen, W.Q., Graedel, T.E., 2012. Anthropogenic cycles of the elements: a critical review. Environ. Sci. Technol. 46 (16), 8574–8586.

Elshkaki, A., Graedel, T.E., 2013. Dynamic analysis of the global metal flows and stocks in electricity generation technologies. J. Clean. Prod. 59, 260–273.

Erickson, R.L., 1973. Crustal occurrence of elements, mineral reserves and resources. U. S. Geol. Surv. Prof. Pap. 820, 21–25.

European Commission, 2013. Report on Critical Raw Materials for the EU, Critical Raw Materials Profiles. DG Enterprise and Industry.

Goe, M., Gaustad, G., 2014. Identifying critical raw materials for photovoltaics in the US: a multi-metric approach. Appl. Energy 123, 387–396.

Graedel, T.E., Harper, E.M., Nassar, N.T., Reck, B.K., 2015. On the materials basis of modern society. Proc. Natl. Acad. Sci. U.S.A. 112, 6295–6300.

Licht, C., Peiró, L.T., Villalba, G., 2015. Global substance flows analysis of gallium, germanium and indium, quantification of extraction, uses and dissipative losses within their anthropogenic cycles. J. Ind. Ecol. 890–903.

Lokanc, M., Eggert, R., Redlinger, M., 2015. The Availability of Indium: The Present, Medium Term, and Long Term. NREL.

Moss, R.L., Kara, H., Willis, P., Kooroshy, J., 2011. Critical Metals in Strategic Energy Technologies. Assessing Rare Metals as Supply-Chain Bottlenecks in Low-Carbon Energy Technologies, JRD Scientific and Technical Reports. European Commission Joint Research Centre, Netherlands.

Santhosha, A.L., Medenbach, L., Buchheim, J.R., Adelhelm, P., 2019. The indium-lithium electrode in solid-state lithium-ion batteries: phase formation, redox potentials, and interface stability. Chem. Eur. 2 (67), 524–529.

Skinner, B.J., 1976. A second iron age ahead? Am. Sci. 64, 158–169.

Stamp, A., Wäger, P.A., Hellweg, S., 2014. Linking energy scenarios with metal demand modeling – the case of indium in CIGS solar cells, Resources. Conserv. Recycl. 93, 156–167.

UNEP, 2011. International Panel on Sustainable Resources Management, Estimating Long-Run Geological Stocks of Metals, Working Group on Geological Stocks of Metals. Working Paper.

US Geological Survey, 1996–2020. Mineral Commodity Summaries, Indium.

US Geological Survey, 2000. 1998 Assessment of undiscovered deposits of gold, silver, copper, lead and zincin the United States. Circular 1178, 2000.

US Geological Survey, 2005. Indium, Mineral Commodity Profile, Open-File Report 2004-1300.

US Geological Survey, 2008. Mineral Commodity Summaries, Indium.

US Geological Survey, 2010a. Byproduct mineral commodities used for the production of photovoltaic cells. Circular 1365.

US Geological Survey, 2010b. Minerals Yearbook, 2008, Metals and Minerals.

US Geological Survey, 2017c. Zinc Statistics.

US Geological Survey, 2017d. Indium Statistics.

US Geological Survey, 2019a. Mineral Commodity Summaries, Zinc.

US Geological Survey, 2019b. 2016 Minerals Yearbook, Indium.

US Geological Survey, 2020. Mineral Commodity Summaries, Indium.

Werner, T.T., Mudd, G.M., Jowitt, S.M., 2015. Indium: key issues in assessing mineral resources and long-term supply for recycling. Appl. Earth Sci. https://doi.org/10.1179/1743275815Y.0000000007.

Yang, C., Tan, Q., Liu, L., Dong, Q., Li, J., 2017. Recycling tin from electronic waste: a problem that needs more attention. ACS Sustain. Chem. Eng. 5, 9586—9598.

Ylä-Mella, J., Pongrácz, E., 2016. Drivers and constraints of critical materials recycling: the case of indium. Resources 5, 34. https://doi.org/10.3390/resources5040034.

Yoshimura, A., Daigo, I., Matsuno, Y., 2013. Global substance flow analysis of indium. Mater. Trans. 54 (1), 102—109.

Zimmermann, T., Gössling-Reisemann, S., 2014. Recycling potentials of critical metals — analyzing secondary flows from selected applications. Resources 3, 291—318.

CHAPTER 7.9

Molybdenum[8]

7.9.1 Introduction

Molybdenum is an important element for the infrastructure of modern society. More than 80% of molybdenum is applied in high-quality steels to improve a range of characteristics, such as hardenability and the ability to withstand high temperatures, seawater, and corrosive chemicals. Molybdenum is an essential metal in the framework of the transition to fossil-free power generation envisioned in the 2015 Paris Climate Agreement. According to Kleijn et al. (2011), the transition to global non-fossil fuel energy generation would itself require almost as much molybdenum as the total current annual molybdenum production. Wind power is the most iron- and steel-intensive of all non-fossil fuel power generation methods, and molybdenum is an important alloying agent used to strengthen steel construction and reduce its weight. Molybdenum is also used to manufacture high-performance gear steels for wind turbines (International Molybdenum Association, 2011). Molybdenum also plays an important role in thin-film PV systems, as one of the metals in the back electrodes of a thin-film solar panel (International Molybdenum Association, 2013). Ensuring the continuing availability of molybdenum is therefore very important for society and even more so for future generations.

The European Union considers molybdenum to be a critical raw material, because a large part of molybdenum is produced in just three countries outside of the European Union (supply risk), and because molybdenum is an essential material for the European economy, while it is difficult to substitute molybdenum and also the end-of-life recycling rate is low (vulnerability).

We investigate whether the molybdenum extraction rate can be made sustainable, while simultaneously increasing molybdenum's global service level to the service level of molybdenum in developed countries in 2020.

---

[8] This section is based on the publication of Henckens, M.L.C.M., Driessen, P.P.J., Worrell, E., 2018. Molybdenum resources: their depletion and safeguarding for future generations. Resour. Conserv. Recycling 134, 61–69.

7.9.2 Properties and applications

Molybdenum has a density of 10.28 g/cm^3 and its melting point is 2623°C. Its main application is to improve various characteristics of its alloys. A small part of molybdenum is used in a variety of chemicals. Molybdenum's applications worldwide are presented in Fig. 7.9.1.

7.9.2.1 Molybdenum in grade alloy steels and irons

Molybdenum is used in alloy steel and iron to improve a variety of characteristics, such as hardenability, and resistance to high temperatures, seawater, and chemicals. Industrial sectors where these steels are used include automotive, shipbuilding, aircraft, energy, chemical, and offshore.

The molybdenum content is usually between 0.2% and 0.5%, and seldom exceeds 1% (International Molybdenum Association, 2015).

7.9.2.2 Molybdenum in stainless steels

Molybdenum improves the corrosion resistance of stainless steels, especially in chloride-containing liquids such as seawater.

Stainless steels are categorized according to three types: ferritic, austenitic, and duplex steels. These three main categories of stainless steel contain some 150 different stainless steel grades (Battrum, 2008). Most of the stainless steel market concerns non-molybdenum-containing grades,

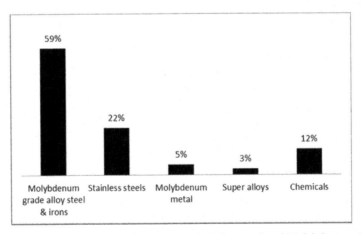

Figure 7.9.1 Molybdenum's applications in 2012 (International Molybdenum Association, 2015).

and only 10% of the stainless steels (about 15 different stainless steel grades) are really relevant to molybdenum (Battrum, 2008). Austenitic stainless steel is the main steel used for building and construction, household applications, and industrial applications (International Molybdenum Association, 2015a). Of these austenitic steels, "316" stainless steel is the major molybdenum-bearing grade, with a molybdenum content of 2.1%. However, the market share of this steel grade is only 7% (Battrum, 2008). According to these data and the data on the global use of 316 stainless steel (about 30 million tons per year), the application of molybdenum in 316 stainless steel covers about 20% of global molybdenum use. "Lean duplex" stainless steels contain about 0.3% molybdenum. According to Battrum (2008), these grades are gradually replacing 304 stainless steel, which occupies half of the stainless steel market and does not contain molybdenum. High-quality stainless steels used in corrosive environments may contain up to 3%–6% of molybdenum. Examples are stainless steel grades such as 254 SMO (6% molybdenum) and SAF 2507 duplex stainless steel (3%–5% molybdenum), both used in desalination plants and in oil and gas coolers in offshore oil and gas exploitation.

7.9.2.3 Molybdenum metal

Molybdenum metal is utilized in many applications, such as high-temperature heating elements, rotating X-ray anodes used in clinical diagnostics, and glass melting furnace electrodes. For specialized applications, molybdenum metal is alloyed with other metals such as tungsten, copper, and rhenium (International Molybdenum Association, 2015).

7.9.2.4 Molybdenum super alloys

Molybdenum super alloys are low in iron content but high in nickel and chromium content (about 40%–70% and 20%–30%, respectively). These alloys fall into two basic classes: corrosion-resistant alloys (5.5%–28.5% molybdenum) and high-temperature alloys (2%–15% molybdenum) (International Molybdenum Association, 2015).

7.9.2.5 Molybdenum in chemicals

The main applications are in desulfurization catalysts, pigments, corrosion inhibitors, smoke suppressants, lubricants, and in agriculture as a micro-nutrient (International Molybdenum Association, 2015).

7.9.3 Production and resources

7.9.3.1 Production

World molybdenum production over time is presented in Fig. 7.9.2. In 1900, global molybdenum production was only 10 metric tons; in 1950, it had increased to 14,500 metric tons (US Geological Survey, 2017c), and by 2018 production had increased to 300,000 metric tons (US Geological Survey, 2019). Molybdenum production growth is however slowing over time. Between 1900 and 2018, annual production growth was 7.0%, between 1950 and 2018 it was 4.6%, and between 1980 and 2018 it was 2.7%.

The most important molybdenum-producing countries are China, Chile, and the United States with, 45%, 18%, and 15% of world production respectively in 2019 (US Geological Survey, 2020) (Fig. 7.9.3).

7.9.3.2 Resources

Available data on molybdenum resources are presented in Table 7.9.1.

Though not negligible, seafloor resources of molybdenum (in manganese nodules and seafloor massive sulfides) do not seem to be high enough to fundamentally change the overall picture (Glasby, 2000; Hein et al., 2013; Sharma, 2011; Hannington et al., 2011).

In the framework of the evaluation of necessary measures for achieving a sustainable production rate of molybdenum, we consider a range of ultimately available molybdenum resources of 50 Mt to 200 Mt.

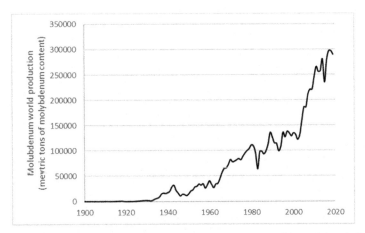

Figure 7.9.2 World molybdenum production between 1900 and 2019 (t/year). Derived from US Geological Survey (2017c, 2018, 2019, 2020).

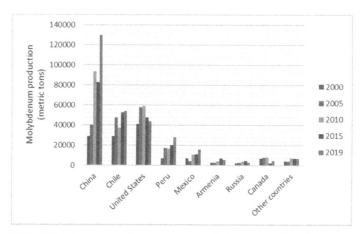

Figure 7.9.3 Molybdenum-producing countries (US Geological Survey, 2002, 2007, 2012, 2017b, 2020).

Table 7.9.1 Estimates of the extractable global amount of molybdenum.

Source	Specification	Estimated amount of molybdenum resources (Mt)
US Geological Survey (2017), identified resources	Identified resources	19.4
Based on UNEP, 2011	"Not unreasonable upper limit" of extractable global molybdenum resources, based on 0.01% of the total amount of molybdenum in the upper 1 km of the Earth's crust. We assume that, in the future, extraction will be economically feasible to a depth of 3 km	180
Extrapolation of the assessment of the US Geological Survey (2000), see Chapter 2.3	Identified plus undiscovered resources, based on an assessment of the ratio between identified and undiscovered resources for five metals in the United States	200

Table 7.9.1 Estimates of the extractable global amount of molybdenum.—cont'd

Source	Specification	Estimated amount of molybdenum resources (Mt)
Sverdrup et al. (2015)	Ultimate recovery rate	46
Own calculation of maximum recoverable amount of molybdenum (Henckens et al., 2018)	Ultimately recoverable amount of molybdenum, based on the estimated recoverable amount of molybdenum in porphyry copper resources	180

Derived from Henckens, M.L.C.M., Driessen, P.P.J., Worrell, E., 2018. Molybdenum resources: their depletion and safeguarding for future generations. Resour. Conserv. Recycl. 134, 61–69.

7.9.4 Current stocks and flows

The available scientific literature on in-use molybdenum flows is scanty, therefore we have made estimates on the basis of data on comparable other mineral resources and our own estimate. This approach provides a first indication which should be refined by future analyses.

7.9.4.1 Recovery efficiency in mining and production

A total of 55%–60% of molybdenum is produced as a by-product or co-product of porphyry copper ore extraction. Porphyry copper ore accounts for 60% of global copper production (US Geological Survey, 2014a). In 2010, global copper production was about 16 Mt (US Geological Survey, 2012). Hence, the potential amount of molybdenum as a by-product or co-product of copper porphyry ore was about 290 kt in 2010, assuming a molybdenum content of 0.03% in porphyry copper ore (Habashi, 1997, cited by Ayres and Peiró, 2013). However, actual molybdenum production as a by-product of copper was 133 kt in 2010, representing a recovery rate of 45% of the potential output, which is close to the recovery rate mentioned by Ayres and Peiró (2013). The amount of 133 kt molybdenum corresponds to 55%–60% of total molybdenum production, including molybdenum from primary molybdenum mines. Using data on a number of other metals (Table 7.9.2), we assume a molybdenum recovery rate from primary molybdenum ore of about 80%. This leads to the conclusion that the overall recovery rate of molybdenum at production must be around 60% currently, which suggests there is room for improvement.

Table 7.9.2 Metal content in ore, tailings, and slag in the production of seven metals.

Metal	Year	Metal content in tailings (kt)	Metal content in slag (kt)	Total metal content in tailings + slag (kt)	Total metal content in ore (kt)	Remaining in tailings + slag (% of total ore)	Source
Cu	1994	1300	—	1300	10,710	12%	Graedel et al. (2004)
Zn	1994	1030	330	1360	7800	17%	Graedel et al. (2005)
Pb	2000	530	230	760	3500	22%	Mao et al. (2008)
Ag	1997	4	0.4	4.4	20.2	22%	Johnson et al. (2005)
Ni	2000	167	74	241	1338	18%	Reck et al. (2008)
Fe	2000	92,000	28,000	120,000	694,000	17%	Wang et al. (2007)
Cr	2000	740	640	1380	5140	26%	Johnson et al. (2006)

7.9.4.2 Dissipative use

Dissipation of molybdenum mainly occurs through its use in chemicals, which covers about 12% of total molybdenum end use. The main applications of molybdenum in chemicals are in desulfurization catalysts, pigments, corrosion inhibitors, smoke suppressants, lubricants, and as a micronutrient in fertilizer (International Molybdenum Association, 2015). According to Nakajima et al. (2007), 36% of molybdenum in chemicals was used as a catalyst and was recycled in 2004 in Japan. The other types of molybdenum-containing chemicals (e.g., pigments, corrosion inhibitors, smoke suppressants, lubricants, and fertilizer) are used in more or less dissipative ways. Thus, based on the Japanese situation, the dissipation of molybdenum from chemicals can be estimated at 64% of 12%, therefore about 8% of total molybdenum use. The additional dissipation of molybdenum from other applications will be relatively low, because of the type of application in metals and metal alloys. The wear of railroads can be mentioned, but the volume involved is small.

7.9.4.3 In-use stock

We found no figures specifically on in-use stocks of molybdenum, or its accumulation. According to Chen and Graedel (2012), no global-level cycle of molybdenum had been derived prior to 2012. According to Blossom (2002), the lifetime of molybdenum-containing products is between 10 and 60 years, but with an average of 20 years. Assuming an average lifetime of 20 years and a dissipation rate of 8% of the molybdenum consumption, application of the formula in Chapter 7.1 results in a net flow into the in-use stock of 50% of the consumption in that year.

The further increase of the molybdenum in-use stock will decline gradually to the extent that the global per capita molybdenum consumption level will catch up with the per capita level of molybdenum consumption in developed countries at present, assuming that this level does not grow much further.

Similar to other metals, molybdenum's per capita accumulation in a country's in-use stock differs, depending on the pace of that country's industrialization. Finally, the in-use stock will become the average molybdenum lifetime × the annual molybdenum consumption.

Although the further accumulation of molybdenum in the in-use stocks of developed countries has slowed down in recent decades, it may increase again as a result of molybdenum being used in a fossil-free energy supply scenario (in wind turbines and solar cells). In industrializing countries, in-use stocks of molybdenum are just beginning to increase and are doing so relatively rapidly.

7.9.4.4 Recycling from end-of-life products

About 88% of molybdenum is used in various alloys or as molybdenum metal (see Fig. 7.9.1). According to Blossom (2002), the old scrap molybdenum recycling efficiency (molybdenum recycled from old scrap divided by total molybdenum in old scrap produced) in the United States was 30% in 1998. This figure for old scrap molybdenum recycling efficiency is also used by UNEP International Resource Panel (2011).

Table 7.9.3 presents a comparison of the end-of-life recycling rates globally with the end-of-life recycling rates in North America for seven metals. For copper, silver, and iron, the global end-of-life recycling rate is higher globally than in North America, but for zinc, lead, nickel, and chromium, it is lower. Steel alloys are recycled for the purpose of steel recycling. Hence, molybdenum recycling from molybdenum–steel alloys is relatively complex. We therefore assume that molybdenum recycling

Table 7.9.3 End-of-life (EoL) recycling rates (RR) for the world compared to end-of-life recycling rates in North America.

Metal	Year	Global EoL RR (%)	EoL RR in North America (%)	(1)/(2)	Source
		(1)	(2)		
Zinc	2010	32%	36%	0.89	Meylan and Reck (2017)
Lead	2000	65%	72%	0.90	Mao et al. (2008)
Copper	1994	53%	50%	1.06	Graedel et al. (2004)
Silver	1997	71%	43%	1.65	Johnson et al. (2005)
Nickel	2000	70%	78%	0.90	Reck et al. (2008)
Iron	2000	67%	64%	1.05	Wang et al. (2007)
Chromium	2000	19%	55%	0.35	Johnson et al. (2006)
Average				0.97	

globally will be lower than in North America, as is the case for zinc, lead, nickel, and chromium. As a first approach, we have assumed that the global end-of-life recycling rate is three fourths of the end-of-life recycling rate of molybdenum in the United States. This is the average proportion of the global recycling rates of zinc, lead, nickel, and chromium compared to the recycling rates of these four metals in North America. Given the 30% end-of-life recycling rate of molybdenum in the United States, this results in a global end-of-life recycling rate of 20% (rounded), assuming that recycling in the United States is representative of recycling in North America. This overall molybdenum end-of-life recycling rate can be achieved by a combination of, for example, 5% molybdenum recycling from molybdenum grade alloy steel and iron, 30% molybdenum recycling from stainless steels, 80% from end-of-life molybdenum metal, 80% from end-of-life super alloys, and 0% from chemicals.

The main reason that the amount of secondary molybdenum is so low is that old scrap containing molybdenum is normally purchased and recycled not for the sake of molybdenum, but for other metals, mostly iron. Steel scrap is processed in electric arc furnaces, so that molybdenum and other elements such as nickel, cobalt, tungsten, and copper remain unintentionally in the molten steel (Nakajima et al., 2011). In this way, a substantial part of the molybdenum is downcycled and "lost" in diluted form in various lower quality types of steel. The recycling efficiency of molybdenum is not expected to increase significantly as long as cheaper alternatives are available in the form of relatively cheap primary molybdenum. It is assumed that most molybdenum contained in grade alloy steel and iron (about 80%) is downcycled. Based on the global end-of-life recycling rate of iron (67%, see Table 7.9.3), this is about 50% of molybdenum in end-of-life products.

7.9.4.5 Current molybdenum flows summarized

Fig. 7.9.4 presents a simplified global in-use molybdenum cycle for 2018. The flows are based on the data given in the previous sections.

7.9.5 Sustainable molybdenum flows

7.9.5.1 Introduction

In this section, we investigate under which conditions the global service level of molybdenum can be increased to the level of molybdenum's

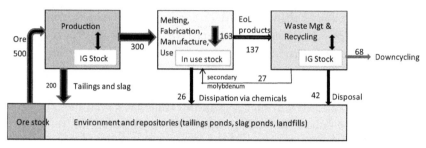

Figure 7.9.4 Simplified global in-use molybdenum cycle for 2018. Flows are in kt. End-of-life (EoL) recycling rate = 20% of the amount contained in end-of-life products; downcycling = 50% of molybdenum in end-of-life products; tailings and slag is 40% of ore; build-up in in-use stock = 50% of molybdenum use; dissipation = 8% of molybdenum use. World mine production of molybdenum in 2018 was 300 kt. The arrow widths are a rough indication of flow magnitudes. IG stock = industrial, commercial, and government stocks.

services in developed countries in 2020, while also achieving a sustainable molybdenum extraction rate. The molybdenum consumption in developed countries in 2020 is estimated at 200 g/capita/year (see Table 5.3).

7.9.5.2 Increase in recovery efficiency at the mining stage

Based on data regarding other metals, we assume that it will be possible to increase molybdenum recovery as a by-product of copper production from the current 45%–50% to 80% in the future. Molybdenum recovery from primary molybdenum mines is assumed to remain at 80%. For the combination of molybdenum as a by-product of copper production plus molybdenum from dedicated molybdenum mines, this will result in an assumed *overall* increase in molybdenum recovery during production, from the current 60% to 80% in the future. However, the actual recovery rate will depend on the molybdenum market price.

7.9.5.3 Substitution of molybdenum

There is little scientific literature on the substitutability of molybdenum. According to the US Geological Survey (2017b), potential substitutes for molybdenum in some of its applications include:
- Chromium, vanadium, niobium (columbium), and boron in alloy steels,
- Tungsten in tool steels,

- Graphite, tungsten, and tantalum for refractory materials in high-temperature electric furnaces,
- Chrome-orange, cadmium-red, and organic-orange pigments for molybdenum orange.

However, most of the substitutes are geologically scarce themselves. In general, according to the US Geological Survey (2017b), there is little substitutability for molybdenum in its major application as an alloying element in steels and cast irons. Oakdene (2013) have also concluded that, in most applications, molybdenum is not substitutable. Based on the above considerations, we assume that the maximum substitutability of molybdenum is 10%.

7.9.5.4 Material efficiency improvement

We assume a conservative (default) material efficiency potential of 10%, ignoring the impact of increased recycling of materials from end-of-life products and of an increase in molybdenum recovery when the raw material is being produced (see Chapter 7.1).

7.9.5.5 Dissipation reduction

The dissipative use of molybdenum is linked to its use in catalysts, pigments, corrosion inhibitors, smoke suppressants, lubricants, and fertilizer. The dissipation is partly direct (use as fertilizer), and partly indirect (via landfills and incinerators). The only possibility for reducing this dissipation is to reduce the use of molybdenum applications in chemicals. This might be possible for some applications, such as the more general use of molybdenum in pigments, as a corrosion inhibitor, and as a smoke suppressant. Nevertheless, we prudently assume that the current dissipation of molybdenum (8%) cannot be reduced, unless a ban is imposed on the respective applications.

7.9.5.6 In-use stock

Once molybdenum consumption stabilizes, the in-use stock of molybdenum will stabilize too. The amount will depend on the lifetime of molybdenum-containing products and will be equal to the annual consumption × lifetime.

Table 7.9.4 Different mixes of measures to achieve a sustainable molybdenum production rate, while realizing a global molybdenum service level that is similar to the molybdenum service level in developed countries in 2020. The boldfaced rows include the main (combinations of) measures.

		Sustainability ambition (years of sufficient availability)		
		500	200	
Ultimately available molybdenum resources[a]	Mt	200	131	200
Assumed future recovery efficiency[b]	**%**	**80%**	**80%**	**79%**
Future availability of molybdenum resources	Mt	267	175	263
Sustainable molybdenum production	kt/year	533	873	1317
Aspired future molybdenum service level	g/capita/ year	200	200	200
Assumed future substitution	**%**	**10%**	**10%**	**10%**
Assumed increase in material efficiency	%	10%	10%	10%
Derived future molybdenum consumption level[d]	g/capita/ year	162	162	162
Corresponding annual molybdenum consumption	kt/year	1620	1620	1620
Assumed in-use dissipation of remaining products[c]	%	8%	8%	8%
Molybdenum in end-of-life products	kt/year	1490	1490	1490
Required molybdenum recycling of remaining products	kt/year	1087	747	303
Required molybdenum recycling rate[d]	**%**	**73%**	**50%**	**20%**

The assumed maximum recovery efficiency in the future is 80%, and the assumed maximum substitutability of molybdenum in the future is 10%. The assumed material efficiency increase is 10%, and the assumed minimum in-use dissipation rate of the remaining products is 8%. The assumed maximum recyclability of end-of-life molybdenum in the future is 50%.
[a]Low estimate: 50 Mt; high estimate: 200 Mt (see Section 7.9.3.2).
[b]Current recovery efficiency: 60% (see Section 7.9.4.1).
[c]Current in-use dissipation rate: 8% (see Section 7.9.4.2).
[d]Current end-of-life recycling rate: 20% (see Section 7.9.4.4).
[e]It is assumed that the world population stabilizes at 10 billion people.

7.9.5.7 Increase in recycling

Molybdenum's major application (about 60%) is in grade alloy steel and iron in concentrations of only between 0.2% and 0.5%. Molybdenum-oriented recycling of these alloys will require a major restructuring of iron and steel recycling. The recycling of molybdenum from its applications in stainless steels, molybdenum metal, and super alloys will be easier, because molybdenum concentrations in these instances are much higher. However, this affects only about 30% of molybdenum applications. We assume a maximum recyclability of end-of-life molybdenum of 50% in the future, increasing from 20% currently. This will be a major challenge.

7.9.6 Conclusions

Table 7.9.4 presents the necessary measures to achieve a sustainable molybdenum extraction rate, with a molybdenum service level for 10 billion people that is equal to the molybdenum service level in developed countries in 2020.

The conclusions are as follows:

- Sustainability ambitions of 1000 and 500 years are not achievable for molybdenum.
- A sustainability level of 200 years is only achievable for molybdenum with ultimately available molybdenum resources of >131 Mt combined with an increase in the recovery efficiency from 60% to 80%, 10% substitution, a 10% material efficiency increase, and an increase in the end-of-life recycling rate from 20% to 50%. With ultimately available resources of 200 Mt, the necessary measures would be to increase the molybdenum recovery efficiency from 60% to 79%, 10% substitution, and a 10% material efficiency increase. These measures would enable keeping the end-of-life recycling rate at the present 20%.

References

Ayres, R.U., Peiró, L.T., 2013. Material efficiency: rare and critical metals. Phil. Trans. R. Soc. A 371, 20110563. https://doi.org/10.1098/rsta.2011.0563.

Battrum, D., 2008. Stainless Steel and Molybdenum. http://www.thompsoncreekmetals.com/i/pdf/Molybdenum_Stainless_Steel.pdf.

Blossom, J.W., 2002. Molybdenum Recycling in the United States in 1998, US Geological Survey, Open File Report 02-165.

Chen, W., Graedel, T.E., 2012. In-use cycles of the elements: a critical review. Environ. Sci. Technol. 46, 8574–8586.

Glasby, G.P., 2000. Lessons learned from deep-sea mining. Science 289 (5479), 551–553. July 28.

Graedel, T.E., Van Beers, D., Bertram, M., Fuse, K., Gordon, R.B., Gritsinen, A., Kapur, A., Klee, R.J., Lifset, R.J., Memon, L., Rechberger, H., Spatari, S., Vexler, D., 2004. Multilevel cycle of in-use copper. Environ. Sci. Technol. 38 (4), 1242—1252.

Graedel, T.E., Van Beers, D., Bertram, M., Fuse, K., Gordon, R.B., Gritsinen, A., Harper, E.M., Kapur, A., Klee, R.J., Lifset, R., Memon, L., Spatari, S., 2005. The multilevel cycle of in-use zinc. J. Ind. Ecol. 9 (3), 67—90.

Habashi, F. (Ed.), 1997. Handbook of Extractive Metallurgy. Wiley-VCH, New York.

Hannington, M., Jamieson, J., Monecke, T., Petersen, S., Beaulieu, S., 2011. The abundance of seafloor massive sulfide deposits. Geology 1155—1158.

Hein, J.R., Koschinsky, A., Mizell, K., Conrad, T., 2013. Deep-ocean mineral deposits as a source of critical metals for high — and green — technology applications: comparison with land based resources. Ore Geol. Rev. 51, 1—14.

Henckens, M.L.C.M., Driessen, P.P.J., Worrell, E., 2018. Molybdenum resources: their depletion and safeguarding for future generations. Resour. Conserv. Recycl. 134, 61—69.

International Molybdenum Association (IMOA), 2011. Molybdenum in Iron and Steels for Clean and Green Power Generation. IMOA.

International Molybdenum Association (IMOA), 2013. Molybdenum in Power Generation, Thin Film Photovoltaic Solar Panels, IMOA/07/13.

International Molybdenum Association (IMOA), 2015. Molybdenum Uses, Molybdenum Metal & Alloys. http://www.imoa.info.2015.

Johnson, J., Jirikowic, J., Bertram, M., Van Beers, D., Gordon, R.B., Henderson, K., Klee, R.J., Lanzano, T., Lifset, R., Oetjen, L., Graedel, T.E., 2005. Contemporary in-use silver cycle: a multilevel analysis. Environ. Sci. Technol. 39, 4655—4665.

Johnson, J., Schewel, L., Graedel, T.E., 2006. The contemporary in-use chromium cycle, Environment. Sci. Technol. 40, 7060—7069.

Kleijn, R., Van der Voet, E., Kramer, G.J., Van Oers, L., Van der Giesen, C., 2011. Metal requirements of low-carbon power generation. Energy 36 (9), 5640—5648.

Mao, J.S., Dong, J., Graedel, T.E., 2008. The multilevel cycle of in-use lead, results and discussion, Resources. Conserv. Recycl. 52, 1050—1057.

Meylan, G., Reck, B.K., 2017. The in-use cycle of zinc: status quo and perspectives. Resour. Conserv. Recycl. 123, 1—10.

Nakajima, K., Yokoyama, K., Matsuno, Y., Nagasaka, T., 2007. Substance flow analysis of molybdenum associated with iron and steel flow in Japanese economy. ISIJ Int. 47 (3), 510—515.

Nakajima, O., Takeda, O., Miki, T., Matsubae, K., Nagasaka, T., 2011. Thermodynamic analysis for the controllability of elements in the recycling process of metals. Environ. Sci. Technol. 45, 4929—4936.

Oakdene, H.F., 2013. Study on Critical Raw Materials at EU Level, a Report for DG Enterprise and Industry of the European Commission.

Reck, B.K., Müller, D.B., Rostowski, K., Graedel, T.E., 2008. In-use nickel cycle: insights into use trade and recycling. Environ. Sci. Technol. 42, 3394—3400.

Sharma, R., 2011. Deep-sea mining: economic, technical, technological and environmental considerations for sustainable development. Mar. Technol. Soc. J. 45 (5), 28—41.

Sverdrup, H.U., Ragnarsdottir, K.V., Koca, D., 2015. An assessment of metal supply sustainability as an input to policy: security of supply extraction rates, stocks-in-use, recycling, and risks of scarcity. J. Clean. Prod. https://doi.org/10.2016/j.jclepro.2015.06.085.

UNEP International Panel on Sustainable Resource Management, 2011a. Working Group on Geological Stocks of Metals. Working Paper.

UNEP International Resource Panel, 2011b. Recycling Rates of Metals, a Status Report.

US Geological Survey, 2000. 1998 Assessment of undiscovered deposits of gold, silver, copper, lead and zinc in the United States. Circular 1178, 2000.

US Geological Survey, 2002. Mineral Commodity Summaries, Molybdenum, January 2002.

US Geological Survey, 2007. Mineral Commodity Summaries, Molybdenum, January 2007.

US Geological Survey, 2012. Mineral Commodity Summaries, Molybdenum, January 2012.

US Geological Survey, 2012. Mineral Commodity Summaries, Copper, January 2012.

US Geological Survey, 2014. Estimate of Undiscovered Copper Resources of the World, 2013. Factsheet 2014-3004, January 2014.

US Geological Survey, 2017b. Mineral Commodity Summaries, Molybdenum, January 2017.

US Geological Survey, 2017c. Molybdenum Statistics, January 2017.

US Geological Survey, 2018. Mineral Commodity Summaries, Molybdenum, January 2018.

US Geological Survey, 2019. Mineral Commodity Summaries, Molybdenum, January 2019.

US Geological Survey, 2020. Mineral Commodity Summaries, Molybdenum, January 2020.

Wang, T., Müller, D.B., Graedel, T.E., 2007. Forging the in-use iron cycle, Environment. Sci. Technol. 41, 5120—5129.

CHAPTER 7.10

Nickel[9]

7.10.1 Introduction

Nickel is essential for the production of stainless steel. Nickel makes stainless steel and other alloys stronger and better able to withstand high temperatures and corrosive environments.

The purpose of this chapter is to investigate whether and under which conditions the extraction rate of nickel can be made sustainable, while simultaneously increasing the global service level of nickel to the service level of nickel in developed countries in 2020.

7.10.2 Properties and applications

The density of nickel is 8.9 g/cm^3, its melting point is 1455°C, and its boiling point is 2732°C. Nickel is mainly used to make stainless steel and to make other alloys stronger and better able to withstand high temperatures and corrosive environments. More than 65% of nickel is used in stainless steels (Aalco, 2005), and nickel-containing grades make up 75% of stainless steel production (Nickel Institute, 2019). In super alloys, nickel is mainly used for the production of turbines, and is vital for power generation, aerospace, and military applications (British Geological Survey, 2008). Copper-nickel alloys (75%/25%) are used for coins, and computer hard disks are reliant on nickel plating, as are CD and DVD masters (British Geological Survey, 2008). A growing share of nickel is used in batteries for electric vehicles, with the current use of nickel in batteries estimated at 3%–4% of the global nickel consumption (British Geological Survey, 2018). Vehicle electrification is expected to accelerate demand for nickel in batteries (Roskill Information Services Ltd, 2017). Currently, the most important type of battery used in electric vehicles is the lithium nickel manganese cobalt oxide type (NMC). The cathode of these Li-ion batteries

[9] This section is based on a publication of Henckens, M.L.C.M., Worrell, E., 2020. Reviewing the availability of copper and nickel for future generations. The balance between production growth, sustainability and recycling. J. Cleaner Prod. 264. Available from: 2020, 121460.

consists of 30%–72% nickel (British Geological Survey 2018). In 2016, 39% of Li-ion batteries contained nickel, and this is expected to increase to around 58% in 2025 (Nickel Institute, 2018). Although it is known that the future share of electric vehicle use in transport will increase drastically, it is still speculative to take the current use of raw materials as a point of departure for the future. Research into higher performance batteries is intensive, and the chemistry of vehicle batteries may change. One forecast is that nickel use in batteries for electric vehicles will grow by 39% annually between 2017 and 2025 (Hamilton, 2018). With such growth rates, the nickel demand for electric vehicles will be 1.1 Mt in 2030 (British Geological Survey, 2018). This is more than 50% of the 2017 global nickel production (Glencore, 2018). Nickel end uses in the United States and globally are given in Table 7.10.1.

7.10.3 Production and resources

7.10.3.1 Production

The annual production of nickel has grown rapidly (see Fig. 7.10.1).

Table 7.10.1 End uses of nickel in the United States and worldwide in 2015 (US Geological Survey, 2018b).

	Nickel applications in the United States in 2015		Global primary nickel use
	Metric tons of contained nickel	**(%)**	**(%)**
Stainless steel	129,000	65%	66%
Super alloys	25,600	13%	10%
Other Ni alloys	13,000	7%	
Electroplating	7490	4%	9%
Other (cast iron, chemicals, electric magnet expansion alloys, Ni–Cu and Cu–Ni alloys, alloy steel, batteries, catalysts, ceramics, coinage, other alloys containing nickel)	23,897	12%	15%
Total	198,987	100%	100%

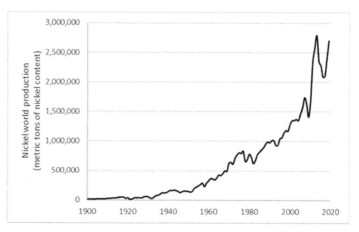

Figure 7.10.1 World nickel production between 1900 and 2019 (US Geological Survey, 2017, 2018, 2019, 2020).

It is observed that, although nickel consumption is stabilizing in developed countries, primary nickel production is increasing rapidly on a global scale. Between 1900 and 2015, global nickel production increased by 4.9% annually.

Global nickel production in 2019 was 2.7 Mt (US Geological Survey, 2018a), and the main nickel-producing countries were Indonesia, the Philippines, Russia, Canada, New Caledonia, Canada, Australia, and China in this order (see Fig. 7.10.2). Russia was previously the world's most important nickel producer, but that role has been taken over by Indonesia and the Philippines, with 30% and 16% respectively of the world nickel production in 2019.

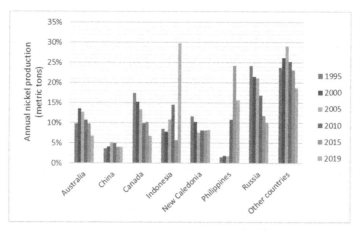

Figure 7.10.2 Main nickel-producing countries (US Geological Survey, 2020).

7.10.3.2 Resources

The average reported concentration of nickel in the Earth's crust is 68 ppm (see Table 2.3). Nickel is a carrier metal, which means that nickel is mainly mined for its own sake. The main by-products of nickel mining are silver, gold, and copper. The nickel concentration in sulfide deposits ranges mostly between 0.2% and 2% (Hoatson et al., 2006; European Commission, 2014), and between 1.0% and 1.6% in lateritic ores (European Commission, 2014). At present, sulfide deposits are the primary source of mined nickel (British Geological Survey, 2008), although 60% of nickel is found in laterites (British Geological Survey, 2008). Extensive nickel resources are also found in manganese crusts and nodules on the ocean floor (US Geological Survey, 2018b).

Mudd and Jowitt (2014) estimate total nickel resources at 296 Mt. If we use the approach that the extractable nickel resources are a maximum of about 0.01% of the total amount of nickel in the upper 3 km of the Earth's crust (UNEP, 2011; Erickson, 1973; Skinner, 1976), nickel resources are much higher at 8000 Mt (rounded). A third approach for estimating the ultimately extractable nickel resources is to extrapolate the results of assessments of the extractable resources of other elements. US Geological Survey (2000) compared the results of 19 assessments of the estimated total (identified plus undiscovered) deposits of gold, silver, copper, lead, and zinc in the United States in the upper 1 km of the continental crust of the United States (a) with the identified resources of the same elements in the United States that far (b). The ratios a/b were 4.82 for zinc, 3.88 for silver, 2.67 for lead, 2.2 for gold, and 2.12 for copper, with an average of 3.14. We applied this average ratio to the most recent US Geological Survey data regarding the identified global resources of nickel: which is 130 Mt (US Geological Survey, 2020). Extrapolating the result to the upper 3 km crust, this approach results in globally available nickel resources of $3 \times 3.14 \times 130 = 1000$ Mt (rounded). Hence, we have three—rather divergent—estimates for the ultimate nickel resources: 300 Mt (rounded), 1000 Mt, and 8000 Mt. We will take the whole range into consideration in the analysis of how to achieve sustainable nickel production.

7.10.4 Current stocks and flows

According to data from 2000 (Reck et al., 2008), the nickel recovery rate in the nickel production stage was 82%, the end-of-life recycling rate was

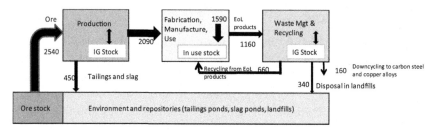

Figure 7.10.3 Simplified global anthropogenic nickel cycle for 2016, based on relative flows in 2000 (Reck et al., 2008). The quantities are in 1000 metric tons of contained nickel. The arrow widths are a rough indication of flow magnitudes. The point of departure is a nickel production of 2090 kt in 2016 (Nickel Institute, 2019). The recovery efficiency is 82%, end-of-life recycling rate 57%, build-up in in-use stock 59%, and downcycling is 14%. IG stock is industrial, commercial, and government stocks.

57%, and the increase in the nickel in-use stock was about 59% of the annual nickel consumption. There is relatively little dissipation of nickel to the environment through usage, due to the non-dissipative use of nickel in steel, alloys, and batteries. In 2000, downcycling of nickel was about 14% of end-of-life nickel (Reck et al., 2008). We have extrapolated these data to the 2016 nickel flows, resulting in the simplified global anthropogenic nickel cycle for 2016 depicted in Fig. 7.10.3.

7.10.5 Sustainable nickel flows

7.10.5.1 Introduction

We will investigate whether and under which conditions it is possible to combine a sustainable nickel extraction rate with increasing services of nickel for the average citizen of the world to the service level of nickel in developed countries in 2020. Nickel consumption per capita in the developed countries in 2020 was estimated at about 2 kg per person per year (see Table 5.3).

7.2.5.2 Increase in recovery efficiency

The current nickel recovery efficiency at the production stage is about 80% (Reck et al., 2008). With a further decreasing ore grade, it will be difficult to increase the nickel recovery efficiency substantially. We assume that the future recovery rate of nickel will remain the same as at present.

7.10.5.3 Substitution

Though substitution of nickel by other materials is possible in some cases (US Geological Survey, 2018a), nickel is so essential in its different applications that large-scale substitution hardly seems possible without a substantial loss of quality (European Commission, 2014; British Geological Survey, 2008).

7.10.5.4 Material efficiency improvement

Apart from recycling, a main tool for improving material efficiency is to increase the product lifetimes. The potential of "nickel-saving stainless steels" is also interesting in this respect (Oshima et al., 2007). It is assumed that, apart from recycling, a 10% material efficiency improvement is feasible in the future. In a general way, this estimate is underpinned in Chapter 7.1.

7.10.5.5 Increase in recycling

The nickel content of nickel-bearing scrap is currently about 8%−9%. The nickel scrap processing industry consists of only a few companies operating at an international level that collect nickel-containing scrap from all over the world (International Nickel Study Group, 2019). Nickel is recycled by blending steel scrap with different nickel concentrations. The purpose is to obtain a mix with a nickel content that can be used again for the production of stainless steels. The current nickel recycling rate from end-of-life products is 57%. If nickel is a minor constituent, for example in low-alloy steels and in plating, it is not economically attractive enough to include these products in the nickel cycle, which is the case for about 14% of the total nickel scrap (Reck et al., 2008). In these cases, nickel becomes a constituent of carbon steel or copper scrap cycles and is in fact *downcycled*. We can also say that this nickel is dissipated outside the nickel cycle, as it becomes unrecoverable for uses that take advantage of nickel's properties. The end-of-life recycling rate of nickel can be increased by increasing the fraction of separately collected nickel-containing end-of-life products and by increasing the nickel recycling efficiency.

For the future, we assume a maximum recyclability of 70% for nickel. This could be achieved by reducing both the downcycling and disposal of nickel by about one third, for instance.

Table 7.10.2 Different mixes of measures to achieve a sustainable nickel production rate, while realizing a global nickel service level that is similar to the nickel service level in developed countries in 2020. The boldfaced rows include the main (combinations of) measures.

		Sustainability ambition (years of sufficient availability)					
		1000		500		200	
Available nickel resources[a,b]	Gt	5.4	8.0	2.7	4.3	1.1	1.7
Sustainable nickel production[c]	Mt/year	5.4	8	5.4	8.6	5.35	8.7
Aspired future nickel service level	kg/cap/year	2.0	2.0	2.0	2.0	2.0	2.0
Assumed future substitution	%	0%	0%	0%	0%	0%	0%
Assumed increase in material efficiency	**%**	**10%**	**0%**	**10%**	**0%**	**10%**	**0%**
Aspired future nickel consumption level	g/cap/year	1.80	2.00	1.80	2.00	1.80	2.00
Corresponding annual nickel consumption	Mt/year	18.0	20.0	18.0	20.0	18.0	20.0
Assumed future dissipation	%	0%	0%	0%	0%	0%	0%
Accumulation in in-use-stock	%	0%	0%	0%	0%	0%	0%
End-of-life nickel	Mt	18.0	20.0	18.0	20.0	18.0	20.0
End-of-life recycling	Mt	12.6	12.0	12.6	11.4	12.7	11.3
Required end-of-life recycling rate[c]	**%**	**70%**	**60%**	**70%**	**57%**	**70%**	**57%**

The assumed maximum recovery efficiency in the future is 82%, and the assumed maximum substitutability of nickel in the future is 0%. The assumed material efficiency increase is 10%, the assumed minimum in-use dissipation rate of remaining products is 0%, and the assumed maximum end-of-life recyclability of nickel in the future is 70%.
[a]Low estimate: 0.3 Gt; high estimate: 8 Gt (see Section 7.10.3.2).
[b]Current recovery efficiency: 82% (see Section 7.10.4).
[c]Current end-of-life recycling rate: 57% (see Section 7.10.4).

7.10.6 Conclusions

Table 7.10.2 presents the nickel recycling rates that are required to achieve a sustainable nickel production rate while simultaneously realizing a nickel service level for the entire world population that is equal to the nickel service level in developed countries in 2020.

The conclusions are:

- A sustainability ambition of 1000 years can be achieved with ultimately available nickel resources of >5.4 Gt in combination with an end-of-life recycling rate of nickel of 70%. If the ultimately available nickel resources are >8 Gt, the end-of-life recycling rate needs to be increased only slightly.
- A sustainability ambition of 500 years can be achieved with ultimately available nickel resources of >2.7 Gt in combination with an end-of-life recycling rate of nickel of 70%. If the ultimately available nickel resources are >4.3 Gt, no measures are necessary.
- A sustainability ambition of 200 years can be achieved with ultimately available nickel resources of >1.1 Gt in combination with an end-of-life recycling of nickel of 70%. If the ultimately available nickel resources are >1.7 Gt, no measures are required.
- If the ultimately available nickel resources are <1.1 Gt, none of the three sustainability ambitions is achievable for nickel.

References

Aalco, 2005. Stainless Steels — Introduction to Grades, Properties and Applications, May 20 2005. ID=2873, sponsored by Aalco — Ferrous and non-ferrous metals stockist. https://www.amazon.com/article.aspx?Article.

British Geological Survey, 2008. Nickel. www.mineralsuk.com.

British Geological Survey, 2018. Battery Raw Materials, May 2018. www.MineralsUK.com.

Erickson, R.L., 1973. Crustal occurrence of elements, mineral reserves and resources. U. S. Geol. Surv. Prof. Pap. 820, 21—25.

European Commission, 2014. Report on Critical Raw Materials for the EU, Non-Critical Raw Materials Profiles. DG Enterprise and Industry.

Glencore, 2018. The EV revolution and its impact on raw materials. In: IEA Seminar on E-Mobility, March 2018.

Hamilton, C., 2018. Battery Raw Materials, the Fundamentals, Presentation at the International Energy Agency Workshop on Batteries for Electric Mobility. BMO Capital Markets.

Hoatson, D.M., Subhash, J., Jaqyes, A.L., 2006. Nickel sulphide deposits in Australia, Characteristics, resources and potential. Ore Geol. Rev. 29, 177—241. Cited by British Geological Survey, 2008.

International Nickel Study Group, 2019. Recycling of Nickel Containing Products. http://www.insg.org/recycling.aspx.

Mudd, G.M., Jowitt, S.M., 2014. A detailed assessment of global nickel resource trends and endowments. Econ. Geol. 109, 1813—1841.

Nickel Institute, 2018. Nickel energizing batteries. Nickel 33 (1).

Nickel Institute, 2019. Stainless Steel: The Role of Nickel. https://www.nickelinstitute.org/about-nickel/stainless-steel/.

Oshima, T., Habara, Y., Kuroda, K., 2007. Efforts to save nickel in austenitic stainless steels. ISJI 47 (3), 359—364.

Reck, B.K., Gordon, R.B., 2008. Nickel and Chromium Cycles: Stocks and Flows Project Part IV, pp. 55—59. www.tms.org/jom.html.

Reck, B.K., Müller, D.B., Rostowski, K., Graedel, T.E., 2008. Anthropogenic nickel cycle: insights into use, trade and recycling. Environ. Sci. Technol. 42, 3394—3400.

Roskill Information Services Ltd, 2017. Nickel Market Beware — Batteries Can No Longer Be Ignored. Roskill Information Services Ltd, London, UK. Press release, April 20, 2017, cited by US Geological Survey, 2018b.

Skinner, B.J., 1976. A second iron age ahead? Am. Sci. 64, 158—169.

UNEP, International Panel on Sustainable Resources management, 2011. Estimating Long-Run Geological Stocks of Metals, International Panel on Sustainable Resources Management, Working Group on Geological Stocks of Metals, Working Paper.

US Geological Survey, 2000. 1998 Assessment of undiscovered deposits of gold, silver, copper, lead and zinc in the United States. Circular 1178.

US Geological Survey, 2017. Nickel Statistics, January 2017.

US Geological Survey, 2018a. Nickel, Mineral Commodity Summaries, January 2018.

US Geological Survey, 2018b. Nickel, 2015 Minerals Yearbook, June 2018.

US Geological Survey, 2019. Nickel, Mineral Commodity Summaries, January 2019.

US Geological Survey, 2020. Mineral Commodity Summaries, Nickel, 1996-2020.

CHAPTER 7.11

Silver

7.11.1 Introduction

Silver is an important metal for modern society, not only in jewelry, but increasingly in electronic applications like solar panels. The annual production of silver has grown rapidly.

For different sustainability aspirations, we investigate under which conditions we can achieve a sustainable silver extraction rate while simultaneously realizing a global service level of silver for technological applications at the same level as the service level of silver for technological applications in developed countries in 2020.

7.11.2 Properties and applications

Silver has a density of 10.49 g/cm^3 at 20°C, a melting point of 962°C, and a boiling point of 2162°C. It is soft, malleable, and ductile. Silver is relatively inert to atmospheric oxygen and very corrosion-resistant. These properties of silver also determine its applications. Silver has been used for a long time in fashion, for personal adornment, and for decoration purposes in jewelry and silverware, and also in high-quality musical instruments. The surface tarnishing on silver objects is caused by the reaction of silver with acidic compounds in the air, such as SO_2 (US Geological Survey, 2018).

A major use of silver remains in jewelry (20% of all silver use in 2017) and silverware (6% of all silver use in 2017) (US Geological Survey, 1996—2018). For a long time, silver was used for coins, but this use ended after the 1960s because of the high and increasing demand for silver for other uses. Sterling silver, with 92.5% silver and 7.5% copper, has been the standard for silver flatware and silver hollowware since the 14th century. Less expensive tableware has a silver coating that is 20—30 microns thick (The Silver Institute, 2019e). The standard in silver jewelry is also Sterling silver.

During the 20th century, the application of silver in photography was very important. Halogen compounds of silver are light sensitive, which means that these compounds have been used since the 1820s in photography, including photography for medical purposes like X-rays. However,

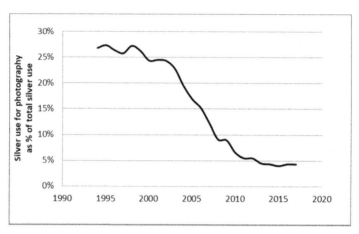

Figure 7.11.1 Silver use for photography as a percentage of total silver use (based on data from The Silver Institute, 2004, 2014, 2018).

silver use for photography has declined in favor of digital photography since the 1990s: from more than 25% of global silver use, to less than 5% within a period of only 15 years between 2000 and 2015 (see Fig. 7.11.1). In 2001, only 2.4% of the 1.6 billion cameras in the world were digital (The Silver Institute, 2002); however, this picture has changed drastically.

Silver is an excellent conductor of electricity and heat, having the highest conductivity of any element (Butterman and Hilliard, 2004). This property determines silver's increasing use in electrical contacts in all sorts of machines, cars, and electrical and electronic products. In electronic products, silver is used for the wiring on printed circuit boards, and nearly all of the electrical connections in a modern car are silver-coated (The Silver Institute, 2018). The on–off switches on most electrical and electronic equipment use a silver membrane, and silver is often used in high-performance spark plugs (The Silver Institute, 2019a). Furthermore, electrically heated silver coatings and conductive lines are used to remove frost and condensation from windshields and rear windows. A typical vehicle may contain as much as 16 g of silver (US Geological Survey, 2013). In 2017, automotive manufacturing used approximately 1500 tons of silver (The Silver Institute, 2018), which is 6% of the global silver production in 2017. Electrical vehicles therefore contain more silver than cars based on internal combustion. It is expected that domestic chargers for electric vehicles will represent a potential silver demand of around 300 tons by 2030 (CRU Consulting, 2018). Computers and televisions contain an average of 0.02–0.03 wt.% of silver. The silver concentration in printed circuit boards within computers, mobile phones, and televisions is about 10 times higher: between 0.1 and 0.33 wt.% of silver, with an average of 0.16% (US Geological Survey, 2013). The silver concentration in printed circuit boards is higher than the silver grade in ores currently mined, which is in the order of 0.1 wt.%. Furthermore, RFID (radio frequency identification) tags, which are used in many products, have silver-based inks.

Silver is used in the manufacturing of PV cells, which are the unit compounds of a PV solar panel. Silver-containing layers are pasted on the front and rear sides of a silicon solar cell (CRU Consulting, 2016) then, when light strikes the silicon wafer in the solar panel, electrons are set free and the silver coating carries the electricity for immediate use to the electrical grid, or for storage in a battery.

Silver is also a necessary component of newer solar cell technologies, such as dye-sensitized solar cells (DSSC) and organic photovoltaics (OPV) (Grandell and Thorenz, 2014).

Between 2008 and 2015, the amount of silver used in PV cells declined from 0.5 g per cell to approximately 0.1 g per cell (see Fig. 7.11.2). However, the production of PV cells increased by a factor of 10 in the same period, which led to an increase in global silver use for PV cells, with a dip between 2011 and 2014 due to a temporary stagnation in PV cell production (see Fig. 7.11.3).

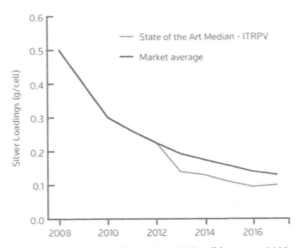

Figure 7.11.2 Amount of silver per photovoltaic (PV) cell between 2008 and 2017 (The Silver Institute 2018).

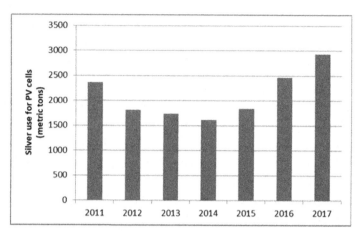

Figure 7.11.3 Global silver use in PV cells between 2011 and 2017 (based on data from The Silver Institute, 2018).

If wind and PV capacity need to keep the global temperature rise below 2°C, the silver needed to produce enough PV cells in 2040 will be twice the global silver production in 2017 (Van Exter et al., 2018), unless silver in solar panels is partly or wholly substituted by another material. However, this would have a negative impact on the performance of the PV cells.

Silver is used in high-quality, high energy-density silver oxide and silver-zinc button batteries in applications such as watches, cameras, toys, hearing aids, mobile phones, and laptops (Grandell and Thorenz, 2014). In November 2018, silver oxide batteries accounted for over 24% of all primary battery sales in Japan (Battery Association of Japan, 2019). The silver-zinc battery is a secondary rechargeable battery with one of the highest energy densities. Silver oxide batteries contain about 31 wt.% silver (US Geological Survey, 2013). The estimated amount of silver contained in silver oxide batteries becoming available as scrap every year is 500 metric tons (US Geological Survey, 2013).

Silver is an excellent reflector of light, which is why silver is used for plane and spherical mirrors and heat-reflecting glass. Silver-coated glass is also used in cars to reflect the rays of the sun. Because of its superior light reflectivity characteristics, silver is used in concentrated solar power systems, in which solar light is concentrated by mirrors into a small area, and the heat is converted into electricity.

Silver is also a catalyst for the production of ethylene oxide (EO) and formaldehyde, which are both important base chemicals. The main applications of EO are in PET, used for the packaging of liquids and for the manufacturing of polyester for textiles. EO is also used as a raw material for the production of glycol, which is used as antifreeze in car-cooling systems. About 25% of EO production is used as antifreeze in vehicles (The Silver Institute, 2019b).

Silver has antimicrobial properties and is used in biocides and for coating medical instruments such as needles, catheters, stethoscopes, breathing tubes, and surgical tools, and also for cooking utensils to prevent infections. Silver-impregnated activated carbon filters are used to purify drinking water at the point of use. These filters remove any remaining chlorine and the silver functions as an effective bactericide. Until very recently, silver sulfur diazine (SSD) was one of the most widely used salves to treat burns (The Silver Institute, 2019d). Moreover, silver-mercury amalgams have been used for a long time in tooth restoration.

Because of its low friction characteristics, silver is used for high-performance bearings at high temperatures, for instance in airplanes. A film of silver coating between moving steel parts enables lubrication of those steel parts (US Geological Survey, 2013).

In the nuclear power sector, especially in pressurized water reactors (PWR), silver is used in control rods in an alloy of 80% silver, 15% indium, and 5% cadmium. In other types of nuclear reactors, boron carbide is preferred. The average amount of silver in a PWR is 1700 kg (CRU Consulting, 2018). The silver total demand of the nuclear power sector is therefore relatively small; currently about 40 tons of silver annually.

Silver is also used for high-quality brazing. Brazing is the joining of two pieces of different metal compositions at temperatures above 600°C by means of a metal alloy that adheres to both at high temperatures (US Geological Survey, 2013). Silver brazing alloys consist of silver-copper, silver-copper-zinc, or silver-copper-zinc-cadmium (Butterman and Hilliard, 2004). Silver brazes and solders combine high strength, ductility, and thermal conductivity (The Silver Institute, 2019c).

The most important silver alloy is silver-copper. Sterling silver is a mixture of 92.5% silver with 7.5% copper, and coin silver consists of 90% silver and 10% copper. Yellow and green 10—18 karat gold are alloys of gold, copper, silver, and minor amounts of zinc in which silver accounts for 4—35 wt.% (Butterman and Hilliard, 2004).

Fig. 7.11.4 and Table 7.11.1 present global end uses of silver and their development over time.

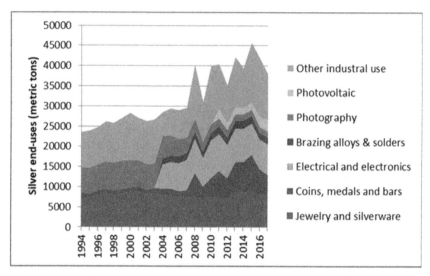

Figure 7.11.4 End uses of silver since 1994. "Other industrial use" includes end use as catalyst, plus "photovoltaic" prior to 2010, plus "electrical and electronics" and "brazing alloys and solders" prior to 2004 (based on data from The Silver Institute, 2004, 2014, 2018).

Table 7.11.1 Silver end uses in 2017 (based on data from The Silver Institute, 2018).

Use	Percentage
Jewelry	21%
Coins and bars	15%
Silverware	6%
Electrical and electronics	24%
Brazing alloys and solders	6%
Photography	4%
Photovoltaic	9%
Catalyst	1%
Other industrial use	15%
Total silver use in 2017	100%

7.11.3 Production and resources

7.11.3.1 Production

In 2017, 72% of the silver produced was obtained as a by-product from lead-zinc mines (36%), copper mines (23%), and gold mines (12%), and 28% as a principal product from dedicated silver mines (US Geological Survey, 2018; Butterman and Hilliard, 2004; Wiebe, 2018). When silver is produced as a by-product of gold, copper, lead, or zinc production, its production requires proportionately many steps, which means that the overall silver recovery efficiency at the mining and production stages is relatively low.

The development of global silver production development since 1900 is presented in Fig. 7.11.5.

Between 1980 and 2015, the production of silver increased by 2.47% annually (US Geological Survey, 2017), despite the decreasing use of silver in photography. The use of silver in electrical and electronic devices, and especially in solar cells, is increasing. Since the 1960s, annual silver production has tripled. In 2019, the main silver-producing countries were Mexico (23%), Peru (14%), and China (13%) (see Fig. 7.11.6 and Table 7.11.2).

7.11.3.2 Resources

The silver abundance in the Earth's crust is 0.069 parts per million (see Section 3.2). This is almost 30 times that of gold. The economically minable silver grade from mines with silver as the principal product in the United States was about 0.07% or 700 ppm in 2001, which is about 10,000 times higher than the average silver concentration in the Earth's crust (Butterman and Hilliard, 2004). The silver grade in copper, zinc, lead, and

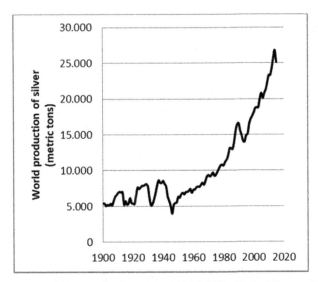

Figure 7.11.5 Silver production since 1900 (US Geological Survey, 2017).

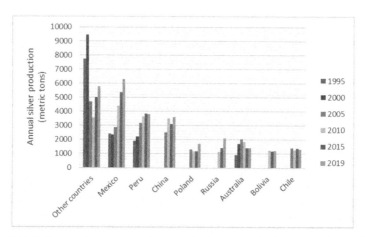

Figure 7.11.6 The most important silver-producing countries between 1995 and 2019 (US Geological Survey, 1996–2020).

Table 7.11.2 The most important silver-producing countries in 2019 (US Geological Survey, 2020).

Mexico	23%	Bolivia	4%
Peru	14%	Chile	5%
China	13%	Australia	5%
Poland	6%	Other countries	21%
Russia	8%	World total	100%

gold mines, from which silver is extracted as a by-product, is much lower (about 100 ppm). However, even these low silver grades are still 1000 times higher than the average silver concentration in the Earth's crust (see Fig. 7.11.7).

Based on the approach that the extractable silver resources are a maximum of about of 0.01% of the total amount of silver in the upper 3 km of the Earth's crust (UNEP, 2011; Erickson, 1973; Skinner, 1976), extractable silver resources are about 8 Mt (see Table 2.3). A thorough assessment of US silver resources in 1998 indicated 620,000 tons of silver in the upper 1 km of the continental crust in the United States (160,000 tons identified and 460,000 tons undiscovered) (US Geological Survey, 2000). This means that the ratio between total estimated silver resources and identified resources in the United States in 1998 was 620,000/160,000 = 3.88. For the estimation of the ultimate silver resources in the world, we apply this ratio of 3.88 to the most recent data regarding identified global silver resources. These are not provided by US Geological

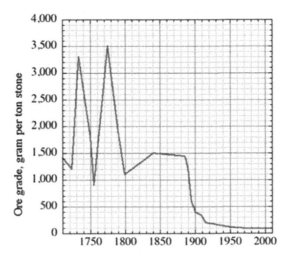

Figure 7.11.7 Ore grade of silver between 1700 and 2000 (gram per ton) (Sverdrup et al., 2014).

Survey, therefore we make an estimate by using data of other metals. The US Geological Survey provides a recent figure both on globally identified resources and on the reserve base for 14 metals (copper, iron, lead, lithium, magnesium, mercury, molybdenum, nickel, platinum group metals, rhenium, strontium, titanium, vanadium, and zinc) in 2009. We use this ratio, which is 2.8, to estimate a proxy for the identified global silver resources. This proxy is justified in our eyes, because the world outside the United States has been explored to a lesser extent than the United States. The global reserve base of silver in 2009 was 570,000 tons. We apply a factor $3.88 \times 2.8 = 10.9$ to obtain the estimate for the ultimately exploitable silver resources in the world. This results in nearly 20 million tons ($10.9 \times 0.57 \times 3$) of ultimately available silver in the upper 3 km of the Earth's continental crust. Hence, we have two estimates for the ultimately available silver resources (8 Mt and 20 Mt), both of which we will consider in the framework of the analysis of the conditions for sustainable silver production.

The US Geological Survey estimates total silver production from prehistory through 2001 at 1.26 million metric tons (Butterman and Hilliard, 2004). Between 2002 and 2017, global silver production was 0.36 Mt. Therefore, total silver production from prehistory through 2017 is 1.6 Mt (Sverdrup et al., 2014). Of the 1.6 Mt of silver mined since prehistory, 1.2 Mt can be traced, mainly as jewelry, silverware, coins, medals, and bars, privately owned or state owned, and about 0.4 Mt is missing. It is thought that about half of this amount has dissipated into the environment via loss, diffusion in water, and into other metals and via landfills (Sverdrup et al., 2014). Most of this (about two thirds) is thought to be lost as disposed of electronic and electronic scrap (Sverdrup et al., 2014). The other half of the missing silver is thought to be still above ground in government strategic stocks and private hidden holdings. This means that about 10%—15% of mined silver is irretrievably lost and 85%—90% is still retrievable.

7.11.4 Current stocks and flows

7.11.4.1 Recovery efficiency

The recovery efficiency of silver at production is estimated at 73% (Johnson et al., 2005). This is relatively low, because most silver is produced as a by-product of other metals (copper, zinc, gold, and lead).

7.11.4.2 Increase in in-use stocks

It is striking that the average amount of silver in society per world citizen has stayed remarkably stable for a long time: approximately between 100 g and 200 g per person (see Table 7.11.3). It should be realized that throughout history, silver has been practically only used in jewelry, silverware, and coins. The industrial use of silver at a significant level only started in the late 19th century, for photography. Nowadays, 60% of silver end use is for industrial purposes.

Because of the silver sequestered in jewelry, silverware, coins, and medals, there is a large reservoir of above-ground silver. The lifetime of

Table 7.11.3 Silver stock in society (Sverdrup et al., 2014).

Date	Estimated silver stock in society (×1000 tons)	Global population (millions)	Grams silver per person
300 BC	10	150	66
200	40	300	133
800	20	300	67
1490	40	300	134
1550	100	400	250
1720	124	600	206
1840	275	1400	196
2012	1271	7200	162

silver in other applications varies from a few years (electronic products), to 10 years (electrical products), and 30 years (e.g., in PV cells), and possibly centuries in jewelry, silverware, coins, medals, and bars. For 1997, it was estimated that the amount of end-of-life silver was 55% of the silver consumption globally (primary plus secondary silver) (Johnson et al., 2005). There are large differences between countries, depending on the proportion of silver used for jewelry and silverware compared to the rest of silver use (Johnson et al., 2005). This means that, in 1997, the in-use stock increased by 45% of the consumption in that year.

Due to the substantial increase in the global EO and formaldehyde production capacity, the amount of silver catalyst used in the production processes is also increasing. This is a part of the in-use stock. The expected increase is from about 4200 tons of silver in 2015 to 4800 metric tons of silver in 2020, resident in EO plants (The Silver Institute, 2019b).

Silver stocks are relatively important. In addition to the in-use stock, these include the stock at the mining and production stages and the stock at the waste management stage. These stocks are held because silver is a precious material. In times of economic uncertainty, people and institutions are inclined to buy precious metals like gold and silver, and in times of relatively low silver prices, silver is stocked, awaiting better times. The amounts of silver in the stock at the production stage and the waste management stage are not exactly known, but they can be estimated by mass balancing. Silver-containing scrap may also be stored until it is sufficiently rewarding to recycle the contained silver.

The application of silver in photography has decreased from 25% of total use in 1997 to 6% of total use in 2017. Other silver uses have increased. Silver in other applications has a longer lifetime than silver in photography. This means that the flow of silver into the in-use stock might have increased considerably since 1997. However, we did not find specific data on this subject. In the simplified silver flow scheme of Fig. 7.11.9, it has been assumed that the annual flow into the in-use silver stock has increased from 45% to 60% of the gross annual silver consumption.

7.11.4.3 In-use dissipation

The only in-use dissipation of silver is in applications such as bactericides and salves; however, this is a negligible amount.

7.11.4.4 Recycling

Recycled silver is a substantial part of the total silver supply (see Fig. 7.11.8).

Fig. 7.11.8 presents the share of recycled silver as a percentage of total silver use between 1994 and 2016. After a long period when this share was about 20%–25%, it declined to under 15% in just a few years between 2013 and 2015.

In the past, photographic materials made an important contribution to the production of secondary silver. Because of the decline in the use of silver in photography, other silver applications have become relatively more important for silver recycling.

Because the decline in recycled silver as a percentage of total silver use took only place from about 2013, this decline cannot have been caused by the decrease in silver use for photography. We can only explain this short-term decline in the relative amount of recycled silver as being caused by speculation; by silver scrap being kept in stock temporarily until the silver price goes up again. It is expected that the amount of recycled silver will return to the level of the 1990s (The Silver Institute, 2018).

In 2004, 80% of secondary silver still originated from the photographic industry and 6% from electrical and electronic scrap in the United States (Butterman and Hilliard, 2004). In 2011, at the global scale, this was about 15% from photography and about 50% from other industrial applications, mainly electrical and electronic scrap (Grandell and Thorenz, 2014).

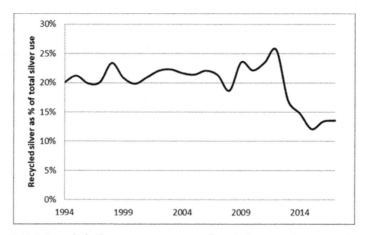

Figure 7.11.8 Recycled silver as a percentage of total silver use since 1994 (based on data from The Silver Institute, 2004, 2014, 2018).

The remainder of recycled silver (35%) came from jewelry, silverware, and coins. The recovery rate of silver from the photography sector is in the order of 50% (Grandell and Thorenz, 2014; US Geological Survey, 2013), mostly from spent development solutions, films, and plates. A large proportion of photographic paper is discarded as municipal solid waste or as paper waste.

From these data, it can be derived that about 20% of end-of-life silver is currently of nontechnological origin (with a recycling rate of 90%), and 80% is of technological origin (with a recycling rate of 40%). This leads to an overall current silver end-of-life recycling rate of 50%.

It is estimated that, globally, no more than 25% of electrical and electronic waste was separately collected and processed for recycling in 2017. Of this amount, only 20% is documented (Baldé et al., 2017). This means that a large part of WEEE is recycled by the informal sector, probably mainly in developing countries. The average recycling rate of WEEE in the EU was about 30% in 2016 (Eurostat, 2019). A substantial proportion of end-of-life electrical and electronic products is traded from developed countries to developing countries, where it starts a second life (Ongondo et al., 2011). In 2009, about 18% of silver in electronic scrap was collected and recycled in the United States, mostly from computers, printers, keyboards, computer displays, televisions, and mobile phones (US Geological Survey, 2013). Electronic scrap also includes other metals of interest such as copper, gold, indium, palladium, platinum, and some rare earth elements (US Geological Survey, 2013). The silver content in electronic scrap varies between 0.02% and 0.5% (Grandell and Thorenz, 2014).

Regarding the silver in silver oxide batteries, only a very small part of this is recycled. In Canada, the recycling rate of silver from batteries was only 2% in 2013 (US Geological Survey, 2013), with the rest disposed of in landfills. On an annual basis, silver disposal to landfills is estimated to be in the order of 500 metric tons of silver globally (US Geological Survey, 2013). The value of silver in silver oxide button batteries is about US$ 0.05—0.10 per battery, so it may be worthwhile considering a deposit money system to increase the recovery of silver oxide batteries.

Silver in spent catalyst that has been used for the production of EO and formaldehyde is reactivated by the same specialized companies that deliver the catalyst to the EO- and formaldehyde-producing chemical industry. A small part of the deactivated catalyst needs replacement and is recycled to the general silver recycling loop (CRU Consulting, 2016). The resulting overall silver recycling rate for the sector is more than 98% (Hilliard, 2003).

Data on current silver recycling from end-of-life solar panels are not available, but we estimate that this is very little or none at all, as the number of end-of-life solar panels remains limited. This will certainly change in the future, when much larger numbers of end-of-life solar panels become available. Much research is taking place into the best ways of recycling solar panels, including the different metals and silicon.

In 1997, the end-of-life recycling rate of silver was about 57% (Johnson et al., 2005). We have not found more recent figures on the end-of-life silver recycling rate but, due to the decline in silver use for photography and the relative increased user for electrical and electronic applications and in solar cells, which have lower end-of-life silver recycling rates, it may well be possible that the overall end-of-life silver recycling rate has decreased between 1997 and 2017. In the simplified silver flow scheme of Fig. 7.11.9, an end-of-life recycling rate of 50% has been assumed. This means that the same amount of silver in end-of-life products is discarded to the environment.

7.11.4.5 Disposal and downcycling

Dissipation of silver in end-of-life products to the environment, via landfill, bottom ash and fly ash from waste incineration plants, and other ways is estimated to be about 14% of silver use. This does not include silver in tailings and slag from the silver production stage, estimated at about 27% of silver contained in silver-containing ores. A small part of silver is down-cycled via secondary steel melting processes, for instance the silver contained in automotive vehicles.

7.11.4.6 Current silver flows summarized

The above data are summarized in the simplified global anthropogenic silver cycle scheme presented in Fig. 7.11.9.

7.11.5 Sustainable silver flows

7.11.5.1 Introduction

We define silver extraction as sustainable if the ultimately available silver resources in the Earth's crust are sufficient for a future world population of 10 billion people for a period of 1000, 500, or 200 years, depending on the

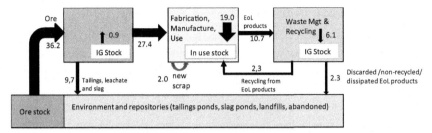

Figure 7.11.9 Simplified global silver flow scheme for the year 2017. Flows are in thousands of metric tons. Assumptions are an end-of-life recycling rate of 50%, and that tailings and slags are 27% of ore. The build-up in in-use stock is estimated at 60% of the gross silver consumption. The arrow widths are a rough indication of flow magnitudes. IG stock is industrial, commercial, and government stocks. The point of departure is the anthropogenic silver cycle in 1997 (Johnson et al., 2005). This is adapted to 2017 data regarding primary silver production, silver use, and the amount of recycled silver in 2017 (The Silver Institute, 2018). The flows to and from IG stocks are the result of mass balancing.

aspired service level of silver. In this section, we investigate whether and under which conditions a sustainable production rate of silver can be achieved while simultaneously bringing the service level of silver for technological applications to the same level as in developed countries in 2020. In developed countries, silver consumption is estimated at about 20 g/cap/year (see Table 5.3). About 60% of this, or 10 g/capita/year (rounded), is for technological applications of silver. We first analyze what the potentials are for saving on silver extraction by increasing the recovery efficiency, by substitution, by increasing material efficiency, by using in-stock silver, and by increasing the end-of-life recycling rate.

7.11.5.2 Increase in recovery efficiency at mining and production stages

The current silver recovery efficiency at the mining stage is 73% (Johnson et al., 2005). The costs of silver recovery will not stop increasing because of the ever lower silver grades in mined ores. Hence, even when assuming an increasing market price for silver and further improvements in technology, it is not certain that a substantial increase in the silver recovery rate may become profitable. We assume a future maximum silver recovery rate of 80%.

7.11.5.3 Increase in-use stock

We have to distinguish between silver that is used for jewelry, silverware, coins, and bars, and silver used for other purposes (industrial, medical). The in-use silver stock will continue to grow with increasing GDP, as people will buy more jewelry and silverware. The number of solar panels and other electrical and electronic equipment will also gradually grow in many countries, until it reaches developed country levels. Eventually, if we assume a stabilization in the world population, the in-use silver stock will stop increasing. At this point, the silver use in new products will be approximately the same as the silver output via end-of-life products. Like gold and PGMs, silver is special in this respect because it is a precious metal. A substantial part of the in-use stock of silver therefore consists of silver in jewelry, silverware, coins, metals, and bars. About 85%–90% of all the silver produced since prehistory is still available (see Section 7.11.3). In fact, silver in this form will always be available for other, possibly more essential, applications, and this part of the in-use silver stock is therefore a buffer for the future.

7.11.5.4 Substitutability

For some applications, there are possibilities to replace silver with other materials. The substitution of silver is possible in the following applications (Butterman and Hilliard, 2004):

- In photography: using digital photography
- In coins and medals: using other metals or by digital money, such as bitcoins
- In electrical and electronic uses: by aluminum, copper, gold, palladium, and platinum
- In solar cells: by copper (Grandell and Thorenz, 2014) and nickel phosphide (CRU Consulting, 2016)
- In jewelry: by other less scarce metals, precious stones, and special polymers
- In tableware: by stainless steel
- In silver batteries: by non-silver batteries
- In silver-containing brazes and solders: by brazes and solders containing no or less silver
- In dental fillings: by non-silver amalgams and certain polymers
- In surgical pins: by tantalum
- In mirrors: by aluminum and rhodium.

However, these substitutes generally provide a lower performance than silver, or they are as scarce as silver. In most industrial uses, silver is difficult to substitute while maintaining the same performance. Silver as a catalyst will also be difficult to replace. For the long term, we assume a maximum substitutability of 10% of silver in current applications on an overall basis.

7.11.5.5 Increase in material efficiency

A good example of material efficiency regarding silver use is the decreased use of silver in solar cells (see Fig. 7.11.2). After a striking reduction of 80% in just five years between 2008 and 2017, silver usage in solar cells is still slowly decreasing (CRU Consulting, 2016). Silver use in electrical and electronic applications may also have material efficiency potential, by reducing the amount of silver used per contact point. However, the possibilities are limited. We assume a maximum material efficiency improvement compared to the actual situation of 10% (see Chapter 7.1).

7.11.5.6 Increase in end-of-life recycling rate for technological applications

The current end-of-life recycling rate of silver from end-of-life technological products is estimated at 40%. To achieve a higher silver recycling rate from technological products, special measures are needed, such as identifying and labeling silver-containing products and making their design recycling-oriented. It must become easy to selectively dismantle and recycle the silver-containing parts of a product. Silver must be stopped from migrating to the steel cycle, as it will then move to the dust phase of electric arc furnaces (EAF) and will be very difficult to recycle because it alloys with other elements (Nakajima et al., 2011). Producers' responsibility for increasing silver recycling will be essential in this respect. Higher silver prices will probably lead to lower and more dispersed silver use in various applications. Therefore, future silver recycling from technological applications may become more difficult and costly (US Geological Survey, 2013). Silver from nanosilver applications may never be recovered in large amounts, because of the very small quantity of silver mixed with several other materials.

Solar panels have a lifetime of about 20–30 years. Hence, the stream of end-of-life solar panels is quickly increasing and will become significant from about 2030. There is much research ongoing into efficient and energy- and environment-friendly methods of solar panel recycling. At sufficient turnover, solar panel recycling can be profitable (Kang et al., 2015): the annual net profit of recycling 3 GW of solar panels annually could be as much as US\$ 70 million to US\$ 200 million. As the main products of solar panel recycling are silicon, silver, aluminum, copper, and glass, the profit is based on the market prices of these materials in 2015. The profit of avoiding landfill is also included in the business model. The contribution of silver in the total financial returns of the materials from the solar panel recycling process would be about 25% (Kang et al., 2015). Even with a further decrease in the silver content of solar panels, the business model remains profitable. For the promotion of efficient solar cell recycling, it will be important to impose producers' responsibility on the manufacturers of solar cells, to force producers to continue to develop better recyclable solar panels.

We assume a future maximum recycling rate of silver from end-of-life technological products of 65%.

7.11.6 Conclusions

Table 7.11.4 presents the necessary measures to achieve sustainable silver production for technological applications, while simultaneously realizing a global service level of silver that is equal to the service level of silver in developed countries in 2020. The share of silver used for technological applications has been about 60% during the last 20 years. Table 7.11.4 presents the required end-of-life silver recycling rates for achieving sustainable silver production.

The conclusions are as follows:
- A sustainability ambition of 1000 years is not achievable for silver.
- A sustainability ambition of 500 years is achievable with >13 Mt of ultimately available silver resources combined with 80% recovery efficiency, 10% substitution, a 10% material efficiency increase, and 65% end-of-life recycling. No silver will be available for nontechnological applications. With 20 Mt of ultimately available silver resources, silver-saving measures still need to be taken to achieve the sustainability goal.
- A sustainability ambition of 200 years is achievable with >8 Mt of ultimately available silver resources combined with 80% recovery efficiency, 10% substitution, a 10% material efficiency increase, and 48% end-of-life recycling. No silver will be available for nontechnological applications. From an amount of 12 Mt of ultimately available silver

Table 7.11.4 Different mixes of measures to achieve a sustainable silver production rate, while realizing a global silver service level for technological applications that is similar to the silver service level in developed countries in 2020. The boldfaced rows include the main (combinations of) measures.

		Sustainability ambition (years of sufficient availability)			
		500		200	
Ultimately available silver resources[a]	Mt	13	20	8	12
Assumed future recovery efficiency[b]	%	**80%**	**80%**	**80%**	**73%**
Future availability of silver resources	Mt	14	22	8	12
Sustainable silver production[e]	kt/year	28	44	42	60
Aspired future silver service level for technological applications	g/capita/year	10	10	10	10
Assumed future substitution	%	**10%**	**10%**	**10%**	**0%**
Assumed increase in material efficiency	%	10%	10%	10%	0%
Aspired future silver consumption level for technological applications	g/capita/year	8.1	8.1	8.1	10
Corresponding annual silver consumption	kt/year	81	81	81	100
Assumed in-use dissipation of remaining products[c]	%	0%	0%	0%	0%
Silver in end-of-life technological applications	kt/year	81	81	81	100
Required silver recycling of remaining products	kt/year	53	37	39	40
Required silver recycling rate from technological applications[d]	%	**65%**	**46%**	**48%**	**40%**

The assumed future maximum recovery efficiency is 80%, and the assumed future maximum substitutability of silver is 10%. The assumed material efficiency increase is 10%, the assumed minimum in-use dissipation rate of remaining products is 0%, and the assumed future maximum recyclability of silver from end-of-life technological products is 65%.

[a]Low estimate: 8 Mt; high estimate: 20 Mt (see Section 7.11.3.2).
[b]Current recovery efficiency: 73% (see Section 7.11.4).
[c]Current in-use dissipation rate: 0% (see Section 7.11.4).
[d]Current end-of-life recycling rate from technological applications: 40% (see Section 7.11.4).
[e]It is assumed that the world population stabilizes at 10 billion people.

resources, no measures need to be taken compared to the current situation and silver will be available for nontechnological applications depending on the extent of the measures taken.

References

Baldé, C.P., Forti, V., Gray, V., Kuehr, R., Stegmann, P., 2017. The Global E-Waste Monitor 2017. United Nations University, International Telecommunication Union, International Solid Waste Association. ISBN 978-92-808-9054-9.

Battery Association of Japan, 2019. Monthly Battery Sales Statistics. http://www.baj.or.jp/e/statistics/02.php.

Butterman, W.C., Hilliard, H.E., 2004. Silver, Mineral Commodity Profiles, US Geological Survey, Open-File- Report 2004-1251.

CRU Consulting, 2016. Prospects for Silver Demand in Ethylene Oxide and Photovoltaics. The Silver Institute.

CRU Consulting, 2018. The Role of Silver in the Green Revolution. Prepared for The Silver Institute.

Erickson, R.L., 1973. Crustal occurrence of elements, mineral reserves and resources. U. S. Geol. Surv. Prof. Pap. 820, 21–25.

Eurostat, 2019. Waste Statistics – Electrical and Electronic Equipment. Downloaded from: https://ec.europa.eu/eurostat/statisticsexplained.

Grandell, L., Thorenz, A., 2014. Silver supply risk analysis for the solar sector. Renewable Energy 2014, 157–165.

Hilliard, H.E., 2003. Silver Recycling in the United States in 2000, US Geological Survey, Circular1196-N, 2003.

Johnson, J., Jirikowic, J., Bertram, M., Van Beers, D., Gordon, R.B., Henderson, K., Klee, R.J., Lanzano, T., Lifset, R., Oetjen, L., Graedel, T.E., 2005. Contemporary anthropogenic silver cycle: a multilevel analysis. Environ. Sci. Technol. 39, 4655–4665.

Kang, D., White, T., Thomson, A., 2015. PV Module Recycling: Mining Australian Rooftops, Presentation at 2015 Asia-Pacific Solar Research Conference.

Nakajima, T., Takeda, O., Takahiro, M., Kazuyo, M., Nagasaka, T., 2011. Thermodynamic analysis for the controllability of elements in the recycling process of metals. Environ. Sci. Technol. 45, 4929–4936.

Ongondo, F.O., Williams, I.D., Cherrett, T.J., 2011. How are WEEE doing? A global review of the management of electrical and electronic wastes. Waste Manag. 31, 714–730.

Skinner, B.J., 1976. A second iron age ahead? Am. Sci. 64, 158–169.

Sverdrup, H., Koca, D., Ragnarsdottir, K.V., 2014. Investigating the sustainability of the global silver supply, reserves, stocks in society and market price using different approaches. Resour. Conserv. Recycl. 83, 121–140.

The Silver Institute, 2002. World Silver Survey 2002.

The Silver Institute, 2004. World Silver Survey, 2004.

The Silver Institute, 2014. World Silver Survey, 2014.

The Silver Institute, 2018. World Silver Survey 2018.

The Silver Institute, 2019a. Silver and Your Automobile, 2019. https://www.silverinstitute.org.

The Silver Institute, 2019b. Silver Catalysts, 2019. https://www.silverinstitute.org.

The Silver Institute, 2019c. Silver in Brazing and Soldering, 2019. https://www.silverinstitute.org.

The Silver Institute, 2019d. Silver in Medicine — Past, Present and Future. https://www. silverinstitute.org.

The Silver Institute, 2019e. Silver Tableware, 2019. https://www.silverinstitute.org.

UNEP, International Panel on Sustainable Resources management, 2011. Estimating Long-Run Geological Stocks of Metals, International Panel on Sustainable Resources Management, Working Group on Geological Stocks of Metals, Working Paper, April 2011.

US Geological Survey, 1996—2020. Mineral Commodity Summaries, Silver, January 1996-January 2020.

US Geological Survey, 2000. 1998 assessment of undiscovered deposits of gold, silver, copper, lead and zinc in the United States. Circular 1178, 2000.

US Geological Survey, 2013. The Life Cycle of Silver in the United States in 2009, Scientific Investigations Report 2013-5178.

US Geological Survey, 2017. Silver Statistics, January 2017.

US Geological Survey, 2018. Mineral Commodity Summaries, Silver, January 2018.

US Geological Survey, 2020. Mineral Commodity Summaries, Silver, January 2020.

Van Exter, P., Bosch, S., Schipper, B., Sprecher, B., Kleijn, R., 2018. Metal Demand for Renewable Electricity Generation in The Netherlands, Report for Ministry of Infrastructure and Water Management.

Wiebe, J., April 2018. The Silver Market in 2017. The Silver Institute.

CHAPTER 7.12

Tin

7.12.1 Introduction

Tin is a material that has been used by humanity for a very long time. Because of its ability to harden copper and make it more malleable, people have combined tin with copper to make bronze objects since about 3500 BC. Bronze is a copper-tin alloy with between 10% and 30% tin. Currently, the most important application of tin is in solder, which is mainly used in electronic equipment. Other uses include tin chemicals, tinplate, lead-acid batteries, and miscellaneous alloys.

The purpose of this section is to investigate the conditions for the sustainable extraction of primary tin. The sustainable extraction of tin is defined as a extraction that can be maintained for an agreed period of time, while simultaneously realizing a tin service level for the world population at a service level that is equal to the tin service level of developed countries in 2020, at an affordable price.

7.12.2 Properties and applications

An important property of tin is its low melting point ($231.93°C$) combined with a rather high boiling point ($2602°C$). This determines some of tin's important applications, such as in solder and the production of float glass. The density of tin at room temperature is about 7.3 g/cm^3.

The applications of tin are presented in Table 7.12.1 and Fig. 7.12.1.

Almost half of tin consumption is in *solders*. Solders are used to connect metallic parts, especially in electrical and electronic equipment (>80%). Solders used to be a mixture of lead and tin, but because of the toxic properties of lead, lead-containing solders are prohibited in many applications. The European Directive on the Restriction of Hazardous Substances has prohibited the use of lead-containing solder in most electronic products in Europe since 1/7/2006. Most countries have also prohibited the use of lead-containing solder for fitting copper pipes in drinking water systems. In 2018, more than 74% of solders were lead-free (ITA, 2020b), and the ambition was to increase this to 95% by 2023. However, some high-reliability sectors such as the military and aerospace still have concerns

Table 7.12.1 Global tin applications in 2018.

Tin application	
Solder	47%
Chemicals	18%
Tinplate	13%
Lead-acid batteries	6%
Copper alloys	6%
Other applications	10%
Total	100%

Based on data from ITA, 2020a. Global Resources & Reserves, Security of Long-Term Tin Supply, 2020 Update.

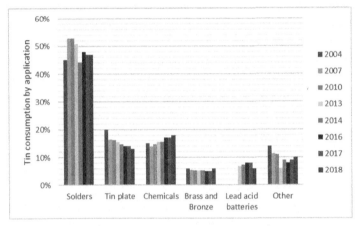

Figure 7.12.1 Global tin consumption by application. *(Based on data from ITRI (2014, 2015a, 2015b, 2017a), from ITA (2018a), from personal communication with ITA (2020b) and from Yang et al. (2017).)*

that it may be more difficult to reach the same quality with lead-free solders. Due to the miniaturization of electronics, lead-free tin-based solders are reaching their limits for components with very small termination sizes and pitches (ITRI, 2015a).

Apart from applications in electronic and electrical equipment, tin-lead and lead-free solders are used in industrial applications such as soldering copper water pipes and joining solar cells. According to data from ITA (cited by Yang et al., 2017), this was 17% of all solders in 2007 and 2010, and is still estimated at close to 20%.

Despite the growth in the electronics market, the contribution of solder consumption to total tin consumption has been fairly stable during the last

15 years (see Fig. 7.12.1), due to miniaturization. Once this effect mitigates in the next five years, the solder market growth could resume.

Tinplate is tin-coated steel, which is mainly used for cans and containers for food, beverages, oil, and chemicals. Steel cans are stronger than cartons or plastic and less fragile than glass. Aluminum is less costly than tin–plated steel, is more malleable, and has the same resistance to corrosion, but cannot be used for food. Hence, aluminum cans are a serious competitor of tin-plated cans for beverage and aerosol applications, but tin retains a major share in food packaging. The market is mature and tinplate production has also been stable over the last decade. Tin use has been in some decline due to economization, although this may now have reached a minimum.

Tin has a wide variety of applications in *chemicals*. The largest use of tin in this sector is as a PVC stabilizer. A polymer stabilizer retards degradation of the polymer, including oxidation, UV damage, thermal degradation, ozonolysis, and photo-oxidation. Organo-tin compounds used as a stabilizer have been under the scrutiny of health authorities, especially in Europe, but most higher risk compounds have now been phased out. Other applications of tin in chemicals are as a catalyst, for example in the production of silicones and polyurethane; in adhesives; in glass coatings for several purposes such as for increasing scratch resistance, for decoration, for conductivity, for de-icing, electroluminescent displays, radio frequency shielding and reduction of energy losses; in chemicals for tinning objects through electroplating; in glazes as an opacifier; in pigments, in cement additives, in brake pads, in flame-retardants, in biocides, and in anticancer agents (ITRI, 2015b).

Tin is also used for the production of float glass. Molten glass is spread out on a liquid tin bath to provide a flat surface. Although the amount of liquid tin used for the production of float glass is substantial, the annual tin losses from these baths are quite limited (40 g per ton of glass produced, or less than 0.5 per mill; ITA, 2020b). Nevertheless, in 2016, the use of tin for the production of float glass resulted in a tin use of 2% of the total tin consumption (ITA, 2020c).

Pewter has been used for centuries for flatware, tankards, and decorative and commemorative items. Pewter is an alloy consisting of 85%—99% tin with the remainder copper, antimony, or lead (Hull, 1992).

Bottle capsules for premium wines and spirits can be made of tin alloy foil.

Though not a real application, it is worthwhile mentioning that tin is a necessary micronutrient in the human diet. The daily intake of tin by humans is about 10 mg (Kamilli et al., 2017). Hence, the annual tin intake of the current world population is between 2 and 3 kt.

7.12.2.1 Tin in copper alloys (*brass and bronze*)

Bronze contains 70%—90% copper, and between 10% and 30% tin, often with the addition of small amounts of other metals such as aluminum, manganese, nickel, or zinc. Brass contains 55%—70% copper and 30%—45% zinc, often with the addition of small amounts of other metals such as lead and tin. Currently, bronze is used for applications such as sculptures, ship propellers, many types of bearings, clips, bells, and electrical connectors. Tin use in brass is much less. Tin-containing brass is used for corrosion-resistant applications, gun metal, and naval brass.

Tin in *lead-acid batteries* is part of both the positive lead-calcium battery grids (up to 1.6%) and the negative grid of lead-calcium batteries (up to 0.4%). Tin is added to the grids to improve casting and cycling performance. Furthermore, 2% tin is contained in lead-tin alloy posts and straps connecting the grids. Some tin is used as solder and tin sulfate can be used to counter corrosion (ITA, 2018b). The sectors in which tin-containing lead-acid batteries are used include automotive (starting, lighting, ignition), transport (for example, forklift trucks), and infrastructure (for telecoms, power supply backup) (ITA, 2018b).

7.12.2.2 Future developments

The International Tin Association expects the use of tin in lead-acid batteries to gradually decrease, possibly with an increase in the use of tin in other types of batteries (ITRI, 2017a; ITA, 2020d). In connection with the electrification of the transport sector, intensive research is ongoing to develop new high-performance batteries, including Li-ion batteries. Tin can be used as an electrode material not only in Li-ion batteries, but, possibly also in sodium, magnesium, and other new types of batteries.

Indium tin oxides are used in opto-electronic applications such as in liquid crystal displays and thin-film photovoltaics, because they are transparent and electrically conductive (Kim et al., 1999; Goe and Gaustad, 2014).

The use of organo-tin compounds as PVC stabilizers and polymer catalysts is expected to continue to increase in line with the growth in construction markets in particular.

Solder markets are likely to benefit strongly from 5G-based electronics technologies, electric vehicles, smart factories, and other new markets related to the "fourth industrial revolution" in the 2025—30 timescale.

Other new applications of tin are likely in relation to climate change technologies, including thermoelectric materials for converting heat into electricity, for hydrogen production, in fuel cells, and as carbon-capture catalysts (ITA, 2020d).

7.12.3 Production and resources

7.12.3.1 Production

Tin mining is focused on the extraction of cassiterite (tin dioxide, SnO_2). Cassiterite is mined by dredging, hydraulic mining, or open-pit mining (Sutphin et al., 1992). Hydraulic mining utilizes a powerful jet of water to dislodge minerals from weathered deposits, while dredging extracts cassiterite from placer deposits by filtering out the heavier tin mineral from the gangue minerals.

Tin production tripled in the 20th century, but—in contrast to a number of other raw materials—tin production growth was linear rather than exponential (see Fig. 7.12.2). World tin production in 2019 was 310,000 tons.

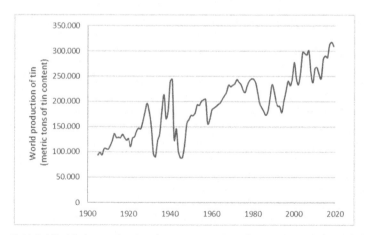

Figure 7.12.2 World tin production between 1905 and 2019 (US Geological Survey, 2017, 2020).

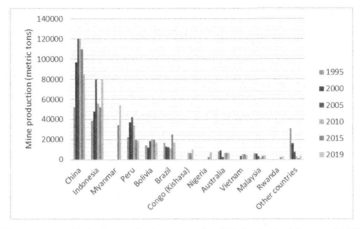

Figure 7.12.3 Main tin-producing countries (US Geological Survey, 2020).

In 2019, three countries were responsible for 70% of global tin production: China (27%), Indonesia (27%), and Myanmar (15%) (Fig. 7.12.3).

7.12.3.2 Resources

The average reported concentration of tin in the Earth's crust is about 2 ppm (see Table 2.3). According to Kamilli et al. (2017), about 20% of the world's identified tin resources occur as hard-rock veins. However, tin is mainly recovered from cassiterite-containing placer deposits, which form as a result of weathering when more resistant minerals are removed from their original location and deposited in stream beds. The tin grades in existing deposits are mainly between 0.1% and 1%, and the ore tonnage is between 10 and 100 kt (see Fig. 7.12.4).

According to US Geological Survey, 2020, tin reserves are 4.7 Mt. The US Geological Survey does not provide data on identified tin resources. In 2009, the US Geological Survey estimated the reserve base of tin at 11 Mt (US Geological Survey, 2009). However, according to the International Tin Association, world tin resources in known deposits or occurrences are 15.4 Mt (ITA, 2020a). The ITA stresses that this number only refers to known resources, "*which can be considered relatively few compared to the vast potential for new discoveries*" (ITA, 2020a, p. 14). However, these deposits may not all be economically viable, and cannot be considered reserves. The ITA estimates that tin reserves total 5.5 Mt (ITA, 2020a).

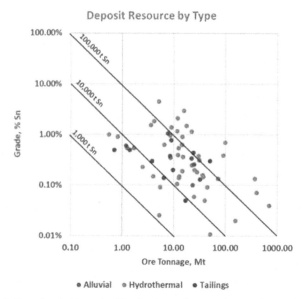

Figure 7.12.4 Singular tin deposits. The horizontal axis indicates the total ore tonnage, the vertical axis the tin grade, and the diagonal lines the tin contained in the deposit (ITA, 2020a).

A generic approach to estimate the ultimately available tin resources is that about 0.01% of the total amount of tin in the upper 3 km of the Earth's crust could be extracted (UNEP, 2011; Erickson, 1973; Skinner, 1976). According to this approach, tin resources could be as much as 300 Mt (rounded).

Another generic approach to estimate the ultimately available tin resources is to extrapolate the results of assessments of the extractable resources of other elements. The US Geological Survey (2000) compared the results of 19 assessments of the estimated total (identified plus undiscovered) deposits of gold, silver, copper, lead, and zinc in the United States in the upper 1 km of the continental crust of the United States (a) with the identified resources of the same elements in the United States thus far (b). The ratios a/b were 4.82 for zinc, 3.88 for silver, 2.67 for lead, 2.2 for gold, and 2.12 for copper, with an average of 3.14. As the US Geological Survey does not provide data regarding the identified tin resources, these were estimated by applying the average ratio between the identified resources and the reserve base of 14 raw materials, which is 2.8 (Table 2.5). Extrapolating the result to the upper 3 km crust, this approach results in globally available tin resources of $3 \times 3.14 \times 2.8 \times 11 = 300$ Mt (rounded). Strikingly, this is the same result as that according to the generic approach based on the upper crustal abundance of tin.

The two generic approaches for estimating the ultimately available tin resources result in an amount that is 20 times higher than the presently known tin deposits.

7.12.4 Current stocks and flows

7.12.4.1 Recovery efficiency

The recovery efficiency of tin from tin ore is higher than that of many other raw materials. According to Izard et al. (2010), the total amount of tin-containing slag and tailings generated by tin mining and beneficiation between 1927 and 2005 was 1860 kt from a total amount of 17,000 kt of tin ore. This means that the recovery efficiency is about 89%. In its 2020 survey of global tin smelters, the ITA estimated that global smelter recoveries averaged 96.8% (ITA, 2020b).

7.12.4.2 Dissipative use

There is practically no in-use dissipation of tin. Tin used in chemicals (as a stabilizer, catalyst, adhesive, glass coating, etc.) is not dissipated during use, but is disposed of in landfills or by incineration when the product in which it is contained reaches its end-of-life phase.

7.12.4.3 Increase in in-use stock

Some tin products, such as bronze, have a long lifetime, but that covers only a relatively small fraction of the total tin use. The average lifetime of tin-containing products is short in comparison to metals such as copper and iron, and is estimated at about 15 years (see Table 7.12.2).

Given the linear increase in tin production, the relatively short average lifetime of tin products leads to a limited increase in the global in-use stock. We estimate the annual increase in the in-use stock of tin to be about 20% of the annual tin consumption. This is based on the approach as described in Chapter 7.1.

7.12.4.4 Recycling

Table 7.12.3 shows that the end-of-life recycling rate of non-exported end-of-life tin products in the United States was 24% in 2005.

Table 7.12.2 Lifetimes of tin applications (own estimate based on UNEP (2010) and Gerst and Graedel (2008)).

	Estimated lifetime (years)	Fraction, global 2018	Contribution to average lifetime (years)
Solders	15	47%	7.05
Tinplate	0.5	13%	0.07
Chemicals	10	18%	1.8
Brass and bronze	40	6%	2.4
Batteries	8	6%	0.48
Other	15	10%	1.5
Total		100%	13.3
Average lifetime (rounded years)			15

Table 7.12.3 End-of-life tin in the United States in 2005. The figures are in kt.

	Recycling	Downcycling	Disposal (landfill or incineration)	Export	Total	Fraction of non-exported end-of-life tin
Tin in cans and containers		4.6	2.6		7.2	24%
Tin in WEEE			2.5	1.5	4	8%
Miscellaneous products with tin alloys	4.4		5.3		9.7	32%
Tin in lead-acid batteries	3				3	10%
Tin in chemicals			7.9		7.9	26%
Total	7.4	4.6	18.3	1.5	31.8	100%
Fraction of non-exported end-of-life tin	24%	15%	60%		100%	

Data are from Izard, C.F., Müller, D.B., 2010. Tracking the devil's metal: historical global and contemporary U.S. tin cycles. Resour. Conserv. Recycl. 54, 1436–1441.

Table 7.12.3 shows that tin recycling in the United States in 2005 was limited to the recycling of tin alloys and lead-acid batteries. According to data from 2005, the recycled old scrap tin alloys consist of about 40% lead-tin alloys and 60% copper-tin alloys (bronze/brass) (Izard and Müller, 2010). Bronzes and brasses are recycled within the copper-tin alloy cycle.

Tin contained in lead–tin alloys and tin in end-of-life batteries is kept in the lead–tin cycle, and most is reused in batteries or used to produce new lead–tin alloys.

A substantial part of tin-plated can and container waste is downcycled to the steel cycle. Tin is only a minor part of cans and containers and so prior detinning of end-of-life cans is barely economical currently, except in some specific cases.

For solders, most recycling takes place during the construction of the final product. Waste and dross from the manufacturing process are often remelted in-house, creating an internal loop. This is also known as new scrap, which is discussed below. However, as with tinplate, solders in electronic and electrical equipment waste are not yet recycled due to economics. There is very little tin recycling from solder currently.

Tin in chemicals is not recycled because it is dispersed in products in very low concentrations.

Overall, the resulting end-of-life recycling rate of postconsumer tin is quite low in comparison to many other metals. Based on its survey data, the International Tin Association estimated that the tin end-of-life recycling rate was 21% in 2018.

Apart from the above postconsumer scrap, there is also the flow of new scrap, which is generated during the fabrication and manufacturing stages of the different tin products. Generally, the amount of new scrap is somewhat larger than the amount of recycled postconsumer scrap. In 2018, gross tin consumption including old scrap and new scrap was 456,000 tons, and the net consumption of tin in consumer products was 372,000 tons. In 2019, the recycling input rate (RIR), which is the use of total recycled tin (old scrap plus new scrap) as a proportion of the gross tin consumption (primary tin plus old scrap plus new scrap), was about 32% (ITA, 2019). Of this, 14% was old scrap and 18% was new scrap.

7.12.4.5 Downcycling

According to Izard and Müller (2010), about 65% of the tin in plated cans and containers in the United States was downcycled to the steel cycle in 2005, and the rest (35%) was disposed of in landfills and incinerators. According to the same authors, there is no other downcycling of tin. The overall picture is that about 80% of nonrecycled tin was disposed of and 20% was downcycled in 2005 in the United States.

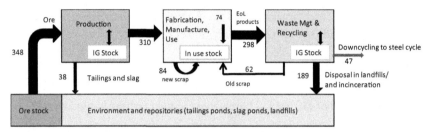

Figure 7.12.5 Simplified global tin cycle for 2018. The amounts are in kt. The points of departure are: tin recovery efficiency is 89%, annual inflow into in-use stock is 20% of net consumption, and the end-of life recycling rate is 21%. The produced amount of tin and the new scrap flow are based on data from the International Tin Association (ITA, 2020b). The ratio between downcycling and disposal is based on data from Izard and Müller (2010). IG stock is industry and government stocks. The arrow widths are a rough indication of the volume of the respective flows.

7.12.4.6 Disposal

Disposed end-of-life tin is composed of tin-plated cans and containers (about 15%), waste from electronic and electric equipment with solder (about 15%), miscellaneous products with tin alloys (about 30%), and products with tin-based chemicals such as PVC, glass, pigments, cement, adhesives, and medical products with tin (about 40%). The percentages between brackets indicate the approximate relative amounts in the United States in 2005 (Izard and Müller, 2010). It is striking that about 60% of end-of-life tin is disposed of. Disposal may be higher in some regions (such as Asia) and lower in others (such as Europe); however, data are not yet available at this level.

7.12.4.7 Current tin flows summarized

Fig. 7.12.5 extrapolates the above flow data to the situation in 2018. The conclusion is that tin consumption is still far from circular.

7.12.5 Sustainable tin flows
7.12.5.1 Introduction

In this section, we investigate under which conditions it is possible to achieve sustainable tin extraction, while simultaneously realizing a tin service level globally that is equal to the tin service level in developed countries in 2020.

In 2018, the average global consumption of tin per capita was about 50 g tin per year (derived from data of the International Tin Association, ITA 2020b), whereas in developed countries tin consumption was estimated at about 200 g/capita/year (see Table 5.3).

We first analyze what the possibilities are to economize on the use of tin.

7.12.5.2 Increase in recovery efficiency

The reported recovery efficiency of tin is already very high (89%). Hence, it is assumed that this will not increase any further.

7.12.5.3 Substitution

In some of its applications, tin can be replaced by other products without tin. Tin in cans and containers can be substituted by aluminum, glass, paper, plastic, or tin-free steel in some use sectors, although tin has a robust position in its major application for food packaging. In solder, tin can be substituted by epoxy resins in some smaller sectors such as displays and by alternative interconnection technologies in high-end miniaturized products, but its mainstream uses are technically and commercially unlikely to be substituted in the foreseeable future. Bronze can be replaced in some applications by aluminum alloys, other copper-based alloys, and plastics. Plastics can replace some tin-containing bearing metals (US Geological Survey, 2020). Lead-acid batteries can and are being replaced by other types of batteries, notably lithium-ion batteries, in sectors where there can be a cost and performance parity. Developments are taking place in the field of batteries because of the ongoing transition to fossil-free transport.

The conclusion is that the substitution of tin may reduce the use of primary tin to some extent, but that there are significant technical and commercial barriers. We therefore assume a maximum substitutability of 40%.

7.12.5.4 Increase in material efficiency

We assume a future increase in the material efficiency of 10% (see Chapter 7.1).

7.12.5.5 Increase in recycling

Tin can technically be recycled from tin-plated cans by detinning end-of-life cans before recycling them to the steel cycle. This has a historical precedent, but at the moment this process is not profitable. Moreover, it seems to make more sense to replace tin-plated cans with aluminum cans or plastic cans wherever possible, instead of trying to recycle tin from tin-plated cans.

At the moment, the global recovery and recycling of WEEE is limited to something between 15% and 20% (Yang et al., 2017), and tin is barely recovered from recycled WEEE scrap. With regard to the recovery of materials from WEEE, the focus is on more precious materials such as gold, silver, and copper, and on the more voluminous materials such as steel and aluminum. Currently, recycling of tin from WEEE is not economic. However, this does not mean that tin would not be recyclable from WEEE scrap from a technical point of view. Yang et al. (2017) present an overview of the results of the scientific literature in this field. The conclusion is that it is possible to recover 85% to almost 100% of tin from WEEE using thermal or hydrometallurgical methods.

At the moment, almost 50% of the tin in consumer products is used as solder in electronic and electrical products, but this is likely to grow in the longer term as electronic miniaturization tails off and new market opportunities for electrical and electronic interconnections open up. Solder is likely to remain tin's most important application. This means that, in the framework of the reduction of primary tin use, it is important to increase both midstream recycling from electronics production and the end-of-life recycling rate of tin in WEEE, which is very limited at the moment. For the time being, tin recycling from WEEE remains an additional cost factor.

It can be derived from the 2005 data provided by Izard and Müller (2010) that tin in end-of-life brass and bronze applications and in end-of-life batteries is almost completely recycled.

Recycling of tin from its chemical applications seems barely possible, because of the low tin concentrations and high dispersion in the different end products.

A considerable part of the non-brass/bronze alloys of tin is disposed of in landfills and by incineration. This was about 17% of all non-exported end-of-life tin and almost 30% of all disposed of tin in 2005 in the United States. This affects a large variety of applications. The large-scale collection and recovery of tin from this type of end-of-life tin is deemed not to be easy. Nevertheless, if a substantial increase in tin recycling is necessary, this option needs to be further explored.

Table 7.12.4 Different mixes of measures to achieve a sustainable tin production rate, while realizing a global tin service level that is similar to the tin service level in developed countries in 2020. The boldfaced rows include the main (combinations of) measures.

		Sustainability ambition (years of sufficient availability)		
		460	200	
Ultimately available tin resources	Mt	300	300	300
Assumed future recovery efficiency[a]	%	89%	89%	89%
Future availability of tin resources	Mt	300	300	300
Sustainable tin production[d]	kt/year	652	1500	1500
Aspired future tin service level	g/capita/year	200	200	200
Assumed future substitution	%	**40%**	**5%**	**0%**
Assumed increase in material efficiency	%	10%	0%	0%
Aspired future tin consumption level	g/capita/year	108	190	200
Corresponding annual tin consumption	kt/year	1080	1900	2000
Assumed in-use dissipation of remaining products[b]	%	0%	0%	0%
Tin in end-of-life products	kt/year	1080	1900	2000
Required tin recycling of remaining products	kt/year	428	400	500
Required tin recycling rate[c]	%	**40%**	**21%**	**25%**

The assumed future maximum recovery efficiency remains 89%, and the assumed future maximum substitutability of tin is 40%. The assumed material efficiency increase is 10%, the assumed minimum in-use dissipation rate remains 0%, and the assumed future maximum recyclability of end-of-life tin is 40%.
[a]Current recovery efficiency: 89% (see Section 7.12.4).
[b]Current in-use dissipation rate: 0% (see Section 7.12.4).
[c]Current end-of-life recycling rate: 21% (see Section 7.12.4).
[d]It is assumed that the world population stabilizes at 10 billion people.

The conclusion is that the most feasible option for further increasing tin recycling from end-of-life products other than from brass, bronze, bearings, and lead-acid batteries, from which tin is already recycled, seems to be to increase the recycling of tin from solders in WEEE and possibly from tinplate and cans. It will be more difficult to recycle tin from tin chemicals and miscellaneous products with tin alloys. If we assume that a maximum of

30% of tin from solders, 60% of tin from tinplate and lead-acid batteries, 85% from copper alloys, and 90% of tin from other applications can ultimately be recycled, then the overall end-of-life tin recycling rate can be increased to a maximum of 40% from the current 21%. This will be quite a challenging task, however.

7.12.6 Conclusions

Table 7.12.4 presents different measures needed to achieve a sustainable tin production rate.

The conclusions are as follows:

- A sustainability ambition of 1000 and 500 years is not achievable for tin. The longest period of time that tin can be available for a world population of 10 billion people at a service level that is the same as the tin service level in developed countries in 2020, is 460 years. This can only be achieved with the most far-reaching measures: 40% substitution, a 10% material efficiency improvement, and 40% end-of-life recycling.
- A sustainability period of 200 years is achievable. Maintaining the current end-of-life recycling rate requires substitution of 5% of tin, without material efficiency improvement. Without substitution and material efficiency improvement, the end-of-life recycling rate must be increased from the current 21% to 25% in the future.

References

Erickson, R.L., 1973. Crustal occurrence of elements, mineral reserves and resources. U.S. Geol. Surv. Prof. Pap. 820, 21—25.

Gerst, M.D., Graedel, T.E., 2008. In-use stocks of metals: status and implications. Environ. Sci. Technol. 42 (19), 7038—7045.

Goe, M., Gaustad, G., 2014. Identifying critical materials for photovoltaics in the US: a multi-metric approach. Appl. Energy 123, 387—396.

Hull, C., 1992. Pewter. Osprey Publishing, pp. 1—5. ISBN 978-0-7478-0152-8.

ITA, 2018a. ITA Survey Shows Weaker Tin Use Growth in 2018.

ITA, 2018b. Tin for the Future, an Introduction to the Tin Market and the International Tin Association.

ITA, 2019. Tin Users See Weaker Markets. Press release October 16, 2019.

ITA, 2020a. Global Resources & Reserves, Security of Long-Term Tin Supply, 2020 Update.

ITA, 2020b. Personal Communication with Dr Jeremy Pearce of the International Tin Association, 14-4-2020.

ITA, 2020c. Insight on Tin Use in Glass Production. Derived from: https://www. internationaltin.org/insight-on-tin-use-in-glass-production/on 13-4-2020.

ITA, 2020d. Tracking New Technologies. Derived from: https://www.internationaltin. org/new-technologies/on 13-4-2020.

ITRI, 2014. Tin Use and Recycling 2013.

ITRI, 2015a. Solders Technology Roadmap 2015, Tomorrow's Solders, February 2015.

ITRI, 2015b. Tin Chemicals, Roadmap 2015, Challenging but Growing, May 14, 2015.

ITRI, 2017a. ITRI Survey Shows Robust Growth in Tin Use. Press Release September 27, 2017.

Izard, C.F., Müller, D.B., 2010. Tracking the devil's metal: historical global and contemporary U.S. tin cycles, Resources. Conserv. Recycl. 54, 1436−1441.

Kamilli, R., Kimball, B.E., Carlin, J.F., 2017. Tin, Chapter S of Critical Mineral Resources of the United States − Economic and Environmental Geology and Prospects for Future Supply, Professional Paper 1802S, US Department of the Interior, US Geological Survey, 2017.

Kim, H., Gilmore, C., Piquw, A., Horwitz, J., Mattoussi, H., Murata, H., Kafafi, Z., Chrisey, D., 1999. Electrical, optical and structural properties of indium tin oxide thin films for organic light-emitting devices. J. Appl. Phys. 86 (11), 6451.

Skinner, B.J., 1976. A second iron age ahead? Am. Sci. 64, 158−169.

Sutphin, D.M., Sabin, A.E., Reed, B.L., 1992. Tin − International Strategic Minerals Inventory Summary Report, p. 9. ISBN 978-0-941375-62-7.

UNEP, 2010. International Resource Panel, Metal Stocks in Society, Scientific Synthesis, 2010.

UNEP, 2011. International Panel on Sustainable Resources Management, Estimating Long-Run Geological Stocks of Metals, Working Group on Geological Stocks of Metals. Working Paper, April 2011.

US Geological Survey, 2000. 1998 Assessment of undiscovered deposits of gold, silver, copper, lead and zincin the United States. Circular 1178, 2000.

US Geological Survey, 2009. Mineral Commodity Summaries, Tin, January 2009.

US Geological Survey, 2017. Tin Statistics, 2017.

US Geological Survey, 2020. Mineral Commodity Summaries, Tin, 1996-2020.

Yang, C., Tan, Q., Liu, L., Dong, Q., Li, J., 2017. Recycling tin from electronic waste: a problem that needs more attention. ACS Sustain. Chem. Eng. 5, 9586−9598.

CHAPTER 7.13

Tungsten

7.13.1 Introduction

Tungsten is an essential material for high-quality industrial tools that are used for cutting, drilling, sawing, and abrading. Tungsten is also widely known for its application as filament in conventional household light bulbs, although this application is disappearing for energy-saving reasons. Tungsten has been identified as an important material for the European economy, as it is difficult to substitute tungsten in most of its applications. Additionally, tungsten production is mainly concentrated in just one country, outside of the European Union. Therefore, the European Commission considers tungsten to be a critical raw material.

The purpose of this section is to investigate how to achieve the sustainable extraction of tungsten, while simultaneously realizing a global service level of tungsten that is equal to the service level of tungsten in developed countries in 2020.

7.13.2 Properties and applications

Tungsten is an exceptional material. It has the highest melting point (3422°C) and the highest boiling point (5930°C) of all metals. It has about the same density (19.3 g/cm^3) as gold and a relatively high conductivity for heat and electricity. Additionally, it has a low thermal expansion coefficient. Its hardness exceeds that of many steels, and it also has a high wear resistance and a high tensile strength.

This combination of properties determines tungsten's uses. Table 7.13.1 presents available public data on the applications of tungsten in intermediate products.

The different tungsten intermediate products have a wide array of applications.

7.13.2.1 Tungsten carbide

Tungsten carbide is "cemented" to form cemented carbides in a binder matrix of cobalt or nickel alloy. Cemented carbides combine a high

Table 7.13.1 Tungsten applications.

	2010 (Leal-Ayala et al., 2015)	2010 (European Commission, 2013)	2012 (MSP REFRAM, 2016)	2016 (International Tungsten Industry Association, 2018)
Tungsten carbide	54%	60%	55%	65%
Steels and alloys	27%	23%	25%	17%
Tungsten metal	13%		15%	10%
Chemicals	6%		5%	8%
Fabricated products		17%		
Total	100%	100%	100%	100%

hardness and strength, toughness, and plasticity (International Tungsten Industry Association, 2020). Cemented carbides are used in cutting tools, drills and drill heads, knives, saws, turning tools, milling insets, and wear parts. Tungsten carbide products are widely used in the mining, construction, and energy sectors (e.g., for asphalt reclamation from roads, tunnel boring, drilling), and the automobile and aerospace industries. Tungsten carbide is also used in jewelry and as the ball in ballpoint pens.

7.13.2.2 Steels and alloys

Tungsten is alloyed with steels and in super alloys. In high-speed steels, tungsten allows a high productivity level in industrial metal cutting. The tungsten content of high-speed steels can vary from 1.5% to 20% (International Tungsten Industry Association, 2020). High-performance skating blades, for example, are made of high-speed steel. Stainless Super Duplex Steel contains 1% tungsten (International Tungsten Industry Association, 2020). Stellites are cobalt-chromium-tungsten alloys and are used in applications where a tough wear-resistant material and corrosion resistance are required, such as bearings, valve seats and pistons, and mill inliners (International Tungsten Industry Association, 2020). In super alloys, tungsten is used in aerospace, industrial, and marine turbines. Super alloys are nickel, cobalt, or iron-based alloys with high contents of tungsten, molybdenum, tantalum, and rhenium (International Tungsten Industry Association, 2020). Combinations with highly conductive metals, such as copper or silver, make tungsten useful in arc welding applications. Furthermore, tungsten is the main component (90%–98%) in tungsten heavy metal alloys. Nickel with iron, cobalt, or copper serve as a binder. Tungsten heavy metal alloys are used as counterweights in airplanes, helicopter blades, X-ray and gamma ray radiation shielding, darts, weights in golf club heads, and in penetrating projectiles (International Tungsten Industry Association, 2020).

7.13.2.3 Tungsten metal

Because of its very high melting point, tungsten metal is produced as powder that is pressed into parts, sintered, and worked (rolled, forged, swaged, or wire drawn) to the desired form (e.g., wires, sheets, and rods) (International Tungsten Industry Association, 2020). Therefore, tungsten metal parts are formed by powder metallurgy, and not through melting and casting. Tungsten wires are used in light bulbs, X-ray tubes, vacuum tubes, and heating elements, and tungsten sheets are used as radiation shielding. Tungsten metal is also used as a gold substitute in jewelry. Fishing weights are made of tungsten-weighted plastics to replace conventional lead weights (International Tungsten Industry Association, 2020), and tungsten shot and ammunition have replaced lead.

7.13.2.4 Chemicals

Tungsten (IV) sulfide and tungsten (II) sulfide are high-temperature lubricants (Mang and Dresel, 2007), and a component of catalysts for hydrodesulfurization (Delmon et al., 1999). Tungsten oxides are used as a pigment in ceramic glazes, and tungsten trioxide is part of the selective catalytic reduction catalysts in coal-fired power plants for converting nitrogen oxides to nitrogen. The tungsten oxide contributes to the physical strength of the catalyst and extends the catalyst life (Spivey, 2002).

7.13.3 Production and resources

7.13.3.1 Production

Global tungsten production has tripled since the 1960s, reaching 85,000 tons in 2019 (see Fig. 7.13.1). China has a dominant share of the global tungsten production, and has produced between 70% and 90% of the world's tungsten for many decades (see Figs. 7.13.2 and 7.13.3). In 2019, the next most important tungsten producers after China were Vietnam (6% of global production) and Russia and Mongolia (each with 2% of global tungsten production) (see Table 7.13.2).

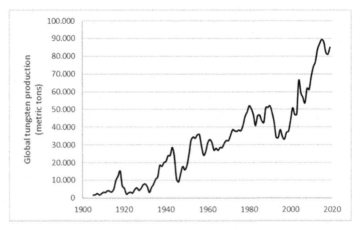

Figure 7.13.1 World tungsten production between 1905 and 2019 (US Geological Survey, 2017, 1996−2020).

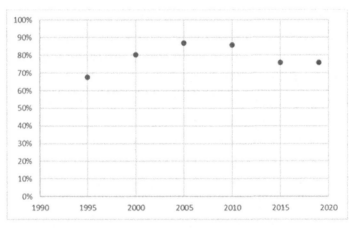

Figure 7.13.2 China's share of global tungsten production (US Geological Survey, 1996−2020).

7.13.3.2 Resources

The average reported concentration of tungsten in the Earth's crust is about 1.5 ppm (see Table 2.3). The ores from which tungsten is extracted are wolframite and scheelite, with 76.5% and 80.6% tungsten trioxide (WO_3), respectively. Scheelite is calcium tungstate ($CaWO_4$), and wolframite is a mixture of ferrous tungstate ($FeWO_4$) and manganous tungstate ($MnWO_4$). The ore grades of exploited tungsten depots are between 0.06% and 1.2%

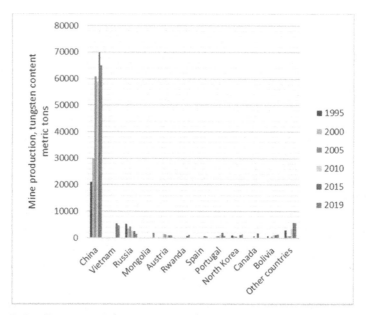

Figure 7.13.3 Tungsten-producing countries between 1995 and 2019 (US Geological Survey, 1996–2020).

Table 7.13.2 Contribution of each of the tungsten-producing countries in 2019 (US Geological Survey, 1996-2020).

China	76%	Spain	1%
Vietnam	6%	Portugal	1%
Russia	2%	North Korea	2%
Mongolia	2%	Canada	0%
Austria	1%	Bolivia	2%
Rwanda	1%	Other countries	7%

tungsten (Leal-Ayala et al., 2015). Using traditional beneficiation techniques, tungsten in the extracted ore is concentrated to market-grade concentrates with WO_3 contents between 15% and 75% WO_3. This corresponds to a 12%–60% tungsten content (Leal-Ayala et al., 2015). This product can be used directly as an alloying element in steel, but most is hydrometallurgically transformed into ammonium paratungstate (APT). APT is a powder with 89.5% WO_3 or 71.6% tungsten (Leal-Ayala et al., 2015). APT can be transformed in tungsten metal powder for the production of tungsten carbide, alloys, chemicals, or tungsten wires, rods, and sheets.

According to US Geological Survey (2020), tungsten reserves are 3.2 Mt. The US Geological Survey does not provide data on identified tungsten resources. In 2009, the US Geological Survey estimated the reserve base of tungsten at 6.3 Mt (US Geological Survey, 1996-2020). Apart from China, Canada, Kazakhstan, Russia, and the United States also have significant tungsten resources (US Geological Survey, 2020). According to Hinde (2008), tungsten resources are approximately 7 Mt.

To estimate the ultimately available tungsten resources, we can use the approach that the extractable tungsten resources are a maximum of about 0.01% of the total amount of tungsten in the upper 3 km of the Earth's crust (derived from UNEP, 2011; Erickson, 1973; Skinner, 1976). According to this approach, tungsten resources could be as high as 200 Mt (rounded).

Another approach for estimating the ultimately extractable tungsten resources is to extrapolate the results of assessments of the extractable resources of other elements. US Geological Survey (2000) compared the results of 19 assessments of the estimated total (identified plus undiscovered) deposits of gold, silver, copper, lead, and zinc in the United States in the upper 1 km of the continental crust of the United States (a) with the identified resources of the same elements in the United States that far (b). The ratios a/b were 4.82 for zinc, 3.88 for silver, 2.67 for lead, 2.2 for gold, and 2.12 for copper, with an average of 3.14. As the US Geological Survey does not provide data regarding the identified tungsten resources, we estimated these by applying the average ratio between the identified resources and the reserve base of 14 raw materials, which is 2.8 (see Table 2.5). Extrapolating the result to the upper 3 km crust, this approach results in globally available tungsten resources of $3 \times 3.14 \times 2.8 \times 6.3 = 200$ Mt (rounded). Strikingly, this is the same result as according to the generic approach based on the upper crustal abundance of tungsten.

In the evaluation of the feasibility of a sustainable use of tungsten, we assume 200 Mt of ultimately available tungsten resources.

7.13.4 Current stocks and flows

7.13.4.1 Recovery efficiency

According to Leal-Ayala et al. (2015), the recovery rate of tungsten in the beneficiation stage varies between 60% and 90%. In line with their publication, we assume an average recovery rate of 75%.

7.13.4.2 Losses during fabrication and manufacturing

According to Leal-Ayala et al. (2015), tungsten losses during fabrication and manufacturing were 5% of the total processed tungsten (primary plus secondary tungsten) in 2010. According to the International Tungsten Industry Association (2018), this was 2% in 2016. We assume 5% losses.

7.13.4.3 New scrap

According to the International Tungsten Industry Association (2018), the amount of new (or prompt) scrap in 2016 was about 8% of the total amount of tungsten processed.

7.13.4.4 Dissipative use

An important part of tungsten is used in abrasive applications like cutting and drilling. In these processes, a small amount of tungsten is abraded from the working tools. Additionally, some in-use dissipation of tungsten takes place through the erosion of tungsten electrodes and tungsten wires in lighting. Tungsten used in chemicals is not dissipated during use, but will be disposed of in landfills or by incineration when the product in which it is contained reaches its end-of-life phase.

We did not find quantitative data on the in-use dissipation of tungsten, but estimate total in-use dissipation of tungsten at about 5% of the tungsten carbide use, which is about 3% of total tungsten use.

7.13.4.5 Increase in in-use stock

Many tungsten-containing products have a rather short lifetime, especially the abrasive tungsten carbide applications. The overall average lifetime of tungsten-containing products is estimated at four years (see Table 7.13.3).

Based on this average lifetime and the increase in tungsten consumption, we estimate the current annual increase in the in-use stock of tungsten to be about 10% of the annual tungsten consumption. This is based on the approach described in Chapter 7.1.

7.13.4.6 Recycling

According to the International Tungsten Industry Association (2018), new scrap recycling from fabrication and manufacturing was about 8% in 2016.

The recycling of tungsten from end-of-life products is about 30% (Leal-Ayala, 2015; International Tungsten Industry Association, 2018). This mainly concerns scrap from tungsten carbide products, heavy metal alloys, super alloys with a relatively high tungsten content, and tungsten sheets. According to the International Tungsten Industry Association, an average

Table 7.13.3 Estimate of the lifetimes of tungsten applications.

		Estimated product lifetime (Harper and Graedel, 2008; Gerst and Graedel, 2008; and own estimates)	Estimated average	Fraction of applications (Leal-Ayala et al., 2015)	Contribution to average lifetime
		Years	Years	%	Years
Chemicals		5	5	6%	0.3
Tungsten metal	Wire	5	8	13%	1.04
	Sheets	10			
	Rods	10			
Steels and alloys	Heavy metal alloys	10	9	27%	2.43
	High–speed steels	1			
	Cast steel	10			
	Super alloys	10			
	Tool steels	10			
	Other	10			
Tungsten carbide	All	0.8	0.8	54%	0.432
Total				100% Rounded	4.202 4 years

of 55% of cutting tools is recycled and about 40% of the other tungsten carbide products (new scrap and old scrap). This leads to an overall recycling rate of tungsten carbide of 46% (International Tungsten Industry Association, 2018). With a share of tungsten carbide of 65% of all tungsten applications in 2016, this results in a tungsten recycling rate of 30% due to the recycling of tungsten carbide products only. The bulk of this is old scrap.

The recycling of tungsten from super alloys and steels contributes about 2.5% to overall tungsten recycling. The bulk of this is new scrap (International Tungsten Industry Association, 2018).

The contribution of tungsten recycling from tungsten metals (wires, sheets, rods) is about 2%, with the bulk being new scrap.

Tungsten recycling from chemicals is about 0.5% of total recycling, which is mostly old scrap (Industrial Tungsten Industry Association, 2018). This does not include the recycling of tungsten-containing catalysts, which is considered to be part of the catalyst usage and is periodically carried out by the catalyst provider. In 2016, this resulted in a recycled tungsten content of 35%, of which about 75% was old scrap and about 25% new scrap (International Tungsten Industry Association, 2018).

7.13.4.7 Downcycling

Steels with a relatively low tungsten content are recycled in the steel cycle. The tungsten contained is lost for specific tungsten applications. This affected about 20% of tungsten applications in 2010 (Leal-Ayala et al., 2015).

7.13.4.8 Disposal

The remainder of end-of-life products is disposed of. Currently, this affects about 50% of the tungsten in end-of-life products.

7.13.4.9 Current tungsten flows summarized

The flow data in the previous section are summarized in Table 7.13.4. On the basis of the flow data in Table 7.13.4, Fig. 7.13.4 presents the global tungsten flows in 2019.

7.13.5 Sustainable tungsten flows

7.13.5.1 Introduction

In this section, we analyze under which conditions a sustainable extraction rate of tungsten is feasible, while simultaneously realizing a global service level of tungsten that is equal to the service level of tungsten in developed countries in 2020. The tungsten consumption in developed countries in 2020 is estimated at about 60 g/capita/year (see Table 5.3).

Below, we explore the possibilities for economizing on tungsten extraction.

7.13.5.2 Increase in recovery efficiency

The current recovery efficiency varies between 60% and 90% (see Section 7.13.4). We assume that it will be possible to increase the average recovery efficiency in the future to 80%.

Table 7.13.4 Summary of tungsten flow data.

Source	Leal-Ayala et al. (2015)	International Tungsten Industry Association (2018)	Own estimate	Points of departure for global tungsten flow scheme 2019 (Fig. 7.13.4)
Year	2010	2016		
Recovery rate	75%			75%
Process losses	5%	2%		5%
Prompt scrap	8%			8%
In-use dissipation			3%	3%
To in-use stocks		2%	10%	10%
End-of-life recycling rate	31%	30%		30%
Downcycling			20%	20%

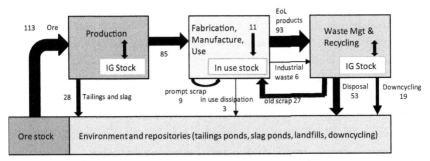

Figure 7.13.4 Simplified global tungsten cycle for 2019. The amounts are in kt. Based on: recovery efficiency 75%; process losses 5%; prompt scrap 8%; annual inflow into in-use stock 10% of consumer products; in-use dissipation 3% of consumer products; downcycling 20% of end-of-life products; and end-of life recycling rate 30%. The produced amount of tungsten in 2019 is based on data from US Geological Survey (2020). For the other assumptions, refer to the text. IG stock is industry and government stocks. The arrow widths are a rough indication of the volume of the respective flows.

7.13.5.3 Substitution

Potential alternatives for tungsten carbide are often molybdenum-based. However, molybdenum is geologically scarcer than tungsten. Hence, molybdenum is not a suitable substitution candidate for tungsten.

Tungsten carbide can be substituted with niobium carbide, titanium carbide, ceramics, and ceramic-metallic composites. However, the quality is lower (US Geological Survey, 2020). In lighting, tungsten can be replaced by light-emitting diodes. For radiation shielding, tungsten can be substituted with lead or depleted uranium (US Geological Survey, 2020), though such a replacement is not very attractive from the viewpoint of health and the environment.

The general conclusion is that substitution of tungsten is difficult in most of its applications without a loss of quality, as illustrated in Table 7.13.5.

On the basis of these data, we assume a maximum substitutability of tungsten of 30%.

7.13.5.4 Increase in material efficiency

We assume that it will be possible in general to increase the material efficiency by about 10% (see Chapter 7.1).

7.13.5.5 Decrease in in-use dissipation

The main in-use dissipation of tungsten is caused by the use of tungsten carbide products in abrasive applications. We assume that it will be possible to decrease this to the extent that the overall in-use dissipation of tungsten can be decreased from 3% to 2%.

7.13.5.6 Increase in end-of-life recycling rate

Products with a high tungsten content, such as tungsten carbide products and some tungsten-based alloys, are the easiest to collect and recycle. For other tungsten products with a relatively low tungsten content, a business case for recycling will be more difficult. This concerns chemicals in

Table 7.13.5 Substitutability of tungsten (European Commission, 2013).

Cemented carbides	0.7
Tungsten alloys	0.7
Super alloys	1.0
Alloy steels	0.7
Fabricated products	1.0

The meaning of the scores is as follows: 1, not substitutable; 0.7, substitutable at high cost and/or loss of performance.

particular, but also tungsten in consumer products with small amounts of tungsten metal, such as lamps, ballpoints, and mobile phones. The growing complexity of tools also makes tungsten recycling more difficult. The recycling of tungsten from steels with a relatively low tungsten content is also financially unattractive.

However, based on the current mix of tungsten products, an overall end-of-life tungsten recycling rate of 50%–60% seem feasible in the future. We therefore assume a maximum future end-of-life recyclability of tungsten of 60%. This is, for instance, possible through a combination of increasing the end-of-life recycling rate from 30% to 60%, decreasing downcycling from 20% to 10%, and decreasing disposal from 50% to 40%.

7.13.6 Conclusions

Table 7.13.6 presents mixes of conditions that are required to achieve sustainable tungsten production, while simultaneously realizing a global service level of tungsten that is equal to the service level of tungsten in developed countries in 2020. It is assumed that the world population will stabilize at 10 billion people.

The conclusions are as follows:

- A sustainability period of 1000 years can be achieved with ultimately available tungsten resources of >145 Mt combined with a tungsten recovery efficiency of 80%, tungsten substitution of 30%, in-use dissipation of 2%, and an end-of-life recycling rate of 60%. If the ultimately available tungsten resources are 200 Mt, the required measures can be less severe.

- A sustainability period of 500 years can be achieved with ultimately available tungsten resources >73 Mt combined with a tungsten recovery efficiency of 80%, tungsten substitution of 30%, in-use dissipation of 2%, and an end-of-life recycling rate of 60%. If the ultimately available tungsten resources are 200 Mt, hardly any extra measures are required to achieve the sustainability ambition of 500 years compared to the current situation.

- A sustainability period of 200 years can be achieved with ultimately available tungsten resources of >85 Mt without any further measures compared to the current situation.

Table 7.13.6 Different combinations of measures for achieving a sustainable tungsten production rate, with a tungsten service level at the global scale that is equal to the level in developed countries in 2020. The boldfaced rows include the main (combinations of) measures.

		Sustainability ambition (years of sufficient availability)				
		1000		500		200
Ultimately available tungsten resources	Mt	145	200	73	200	85
Assumed future recovery efficiency[a]	%	80%	80%	80%	75%	75%
Future availability of tungsten resources	Mt	155	213	78	200	85
Sustainable tungsten production[d]	kt/year	155	213	156	400	425
Aspired future tungsten service level	g/ capita/ year	60	60	60	60	60
Assumed future substitution	%	**30%**	**30%**	**30%**	**0%**	**0%**
Assumed increase in material efficiency	%	10%	10%	10%	0%	0%
Aspired future tungsten consumption level	g/ capita/ year	38	38	38	60	60
Corresponding annual tungsten consumption	kt/year	378	378	378	600	600
Assumed in-use dissipation of remaining products[b]	%	2%	2%	2%	3%	3%
Tungsten in end-of-life products and industrial waste	kt/year	370	370	370	582	582
Required tungsten recycling of remaining end-of-life products and industrial waste	kt/year	223	165	222	200	175
Required tungsten recycling rate from end-of-life products and industrial waste[c]	%	**60%**	**44%**	**60%**	**34%**	**30%**

Assumes ultimately available tungsten resources of 200 Mt and a world population of 10 billion people. The assumed future maximum recovery efficiency is 80%, the assumed future maximum substitution is 30%, and the assumed future maximum material efficiency improvement is 10%. The assumed future minimum in-use dissipation is 2% and the assumed future maximum end-of-life recycling rate is 60%.
[a]Current recovery efficiency: 75% (see Section 7.13.4).
[b]Current in-use dissipation rate: 3% (see Section 7.13.4).
[c]Current end-of-life recycling rate: 30% (see Section 7.13.4).
[d]It is assumed that the world population stabilizes at 10 billion people.

References

Delmon, B., Froment, G.F., 1999. Hydrotreatment and hydrocracking of oil fractions. In: Proceedings of the 2nd International Symposium, 7th European Workshop, Antwerpen, Belgium, November 14−17, 1999.

Erickson, R.L., 1973. Crustal occurrence of elements, mineral reserves and resources. U. S. Geol. Surv. Prof. Pap. 820, 21−25.

European Commission, 2013. Report on Critical Raw Materials for the EU, Critical Raw Materials Profiles. DG Enterprise and Industry.

Gerst, M.D., Graedel, T.E., 2008. In-use stocks of metals: status and implications. Environ. Sci. Technol. 42 (19), 7038−7045.

Harper, E.M., Graedel, T.E., 2008. Illuminating Tungsten's life cycle in the United States: 1975-2000. Environ. Sci. Technol. 42, 3835−3842.

Hinde, C., 2008. Tungsten. Mining Journal Special Publication.

International Tungsten Industry Association, 2018. Recycling of Tungsten, Current Share, Economic Limitations and Future Potential. Newsletter.

International Tungsten Industry Association, 2020. Information on Tungsten: Sources, Properties and Uses. Retrieved on 2-5-2020. https://www.itia.info/tool-steels.html.

Leal-Ayala, D., Allwood, J.M., Petavratzi, E., Brown, T.J., Gunn, G., 2015. Mapping the global flow of tungsten to identify key material efficiency and supply security opportunities. Resour. Conserv. Recycl. 103, 19−28.

Mang, T., Dresel, W., 2007. Lubricants and Lubrication. John Wiley & Sons.

MSP-REFRAM project, 2016. Match between Supply and Demand of Refractory Metals in the EU, MSP-REFRAM − D1.3, Funded by the European Union's Horizon 2020 Research and Innovation Programme.

Skinner, B.J., 1976. A second iron age ahead? Am. Sci. 64, 158−169.

Spivey, J.J., 2002. Catalysts, Royal Society of Chemistry, 2002.

UNEP, 2011. International Panel on Sustainable Resources Management, Estimating Long-Run Geological Stocks of Metals, Working Group on Geological Stocks of Metals. Working Paper, April 2011.

US Geological Survey, 2000. 1998 Assessment of undiscovered deposits of gold, silver, copper, lead and zincin the United States. Circular 1178, 2000.

US Geological Survey, 2017. Tungsten Statistics, January 2017.

US Geological Survey, 1996−2020. Mineral Commodity Summaries, Tungsten.

US Geological Survey, 2020. Mineral Commodity Summaries, Tungsten.

CHAPTER 7.14

Zinc

7.14.1 Introduction

The main application of zinc is the protection of steel from corrosion. The future scarcity of zinc might therefore have a substantial impact on society.

The purpose of this section is to investigate whether and how the extraction of primary zinc can be reduced to and maintained at a sustainable rate, while simultaneously raising the global level of zinc's services to the zinc service level in developed countries in 2020.

7.14.2 Properties and applications

For a metal, zinc's melting point and boiling point are relatively low, at 419.5 and 907°C, respectively. The surface of the pure metal tarnishes quickly, forming a protective layer by reacting with carbon dioxide in the air. This property determines the main use of zinc: to protect iron and steel from corrosion. The major applications of zinc are presented in Fig. 7.14.1.

Galvanizing is by far the most important application of zinc. Galvanizing provides a zinc coating on another metal, generally steel, to protect the metal from corrosion. Zinc is more reactive than iron and will be oxidized (corroded) first, until it completely corrodes away. It is therefore not zinc itself but the resulting surface layer of zinc oxide and zinc carbonate that is responsible for the protection. It is relatively easy to apply zinc to many different surfaces and zinc remains a relatively cheap material. Galvanized steel can also be easily coated with paint. The zinc applications in expressway guardrails, lamp posts, car bodies, and so on, are well known. The protective layer of zinc oxide is a physical barrier in itself; however, even when this layer is damaged, zinc can serve as a sacrificial anode and combines (before steel) with oxygen or other reactive chemicals in the air (or water). In this way, zinc is an important element for cathodic protection, for instance to protect buried pipelines or a ship's steel rudders, keel, and propellers, especially in relatively corrosive seawater (International Zinc Association, 2013a).

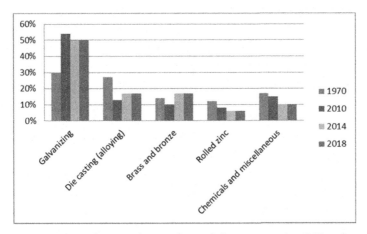

Figure 7.14.1 Main end uses of zinc. *(Derived from International Zinc Association, 2014a. http://wwwzincorg/basics/zinc_statistics and Statista, 2019. Distribution of Zinc Consumption in 2018, Worldwide in 2018, by End Use, Statistacom/statistics; Available from: www.statist.com/statistics/449828/world-wide-distribution-of-boron-end-use-by-application. (Accessed 15 October 2019).)*

Die casting is characterized by forcing molten metal into a mold cavity under high pressure. Most die castings are made from nonferrous metals, for example, zinc, copper, aluminum, magnesium, lead, pewter, and tin-based alloys (International Zinc Association, 2013a). Zinc die casts are used for example in bathroom fixtures, door and window hardware, tools, electronic components, and automotive components.

Brass and bronze are both alloys with copper as the main component and mixtures of other metals. When zinc is the main other metal, the alloy is called *brass*. Brass is a copper-zinc alloy containing 3%—45% zinc, depending on the type of brass. Brass is more ductile and stronger than copper and has better corrosion resistance. Therefore, brass is used for example in communication equipment, hardware, musical instruments, and water valves. Bronze is a copper-tin alloy to which zinc may be added (International Zinc Association, 2013a).

Rolled zinc has an important application as roofing, gutters, and downpipes (International Zinc Association, 2013a).

Chemicals Zinc is applied in chemicals as pigment (zinc oxide, zinc sulfate, and zinc sulfide), flame-retardant (zinc chloride), and biocides (zinc dithiocarbamate, zinc naphtenate, zinc pyrithione). Another application of zinc oxide is as a vulcanizing accelerator in rubber tires. Furthermore, zinc compounds are used as an additive to animal (pig) fodder and lubricants (International Zinc Association, 2013a).

Miscellaneous Zinc may be included in various materials such as solder and casting aid. Furthermore, zinc is used as an anode material in batteries, for example in lithium batteries, alkaline batteries, and zinc-air batteries. Zinc products are used in blasting grit (International Zinc Association, 2013a).

7.14.3 Production and resources

7.14.3.1 Production

Zinc ores are extracted in more than 50 countries, and the total global zinc production was about 13 Mt in 2018.

The development of the annual world zinc production over time is presented in Fig. 7.14.2.

China (33%), Peru (12%), Australia (7%), and the United States (6%) were the four countries with the largest zinc production in 2018 (see Fig. 7.14.3). Data are derived from US Geological Survey (2019).

7.14.3.2 Resources

Identified zinc resources in the world total 1.9 Gt (US Geological Survey, 2019).

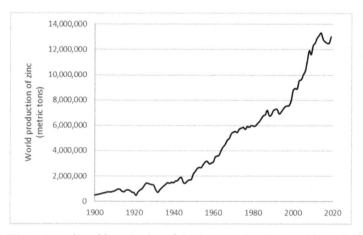

Figure 7.14.2 Annual world production of zinc between 1900 and 2018 (US Geological Survey, 2017a, 2017b, 2018, 2019, 2020).

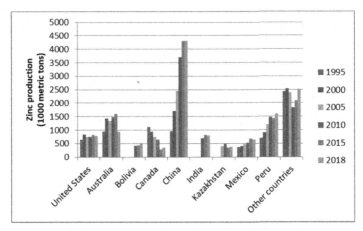

Figure 7.14.3 Main zinc-producing countries (1000 metric tons) (US Geological Survey, 2019).

A thorough assessment of US zinc resources in 1998 indicated 265,000 kt of zinc in the upper 1 km of the continental crust in the USA (55,000 kt identified and 210,000 kt undiscovered) (US Geological Survey, 2000). This means that the ratio between the total estimated zinc resources and identified zinc resources in the United States was 265,000/55,000 = 4.82 in 1998. To estimate the ultimate zinc resources in the world, we apply this ratio of 4.82 to the most recent data on identified global zinc resources (1.9 Gt). This results in about 30 Gt (rounded) of ultimately available zinc resources in the upper 3 km of the Earth's continental crust ($3 \times 4.82 \times 1.9$).

Another approach is that the economically extractable amount of a mineral is a maximum of about 0.01% of the total amount of that mineral in the upper 1 km of the Earth's continental crust (UNEP, 2011a; Erickson, 1973; Skinner, 1976). Based on this approach, and with an upper crustal abundance of 72 ppm[10] the extractable amount of zinc in the upper 1 km of the Earth's crust is about 3 Gt (rounded). Assuming that future technological development makes zinc mining feasible to a depth of 3 km, the ultimately available zinc resources according to this approach could reach 9 Gt (rounded).

[10] The scientific literature provides different values for the crustal abundance of elements. We have taken the average value of seven sources: McLennan (upper crustal abundance), 2001; Darling, 2007; Barbalace, 2007, Webelements, 2007; Jefferson Lab, 2007, Wedepohl, 1995, Rudnick and Fountain, 1995.

Hence, we have two estimates for the ultimately available zinc resources: 9 Gt and 30 Gt. We consider both in the framework of our considerations regarding a sustainable zinc production.

7.14.4 Current stocks and flows

Much work on analyzing the anthropogenic zinc cycle has been done in the framework of the Yale's Stocks and Flows project. This project aims to quantify global, regional, and national amounts of mined, refined, fabricated, and manufactured metals (Graedel et al., 2005; Graedel and Cao, 2010).

7.14.4.1 Recovery efficiency at the mining and processing stages

The purpose of the first processing steps of metal ores is enrichment (or beneficiation). These steps typically consist of milling, grinding, and flotation. If the ore consists of metal sulfides, as is the case with zinc, the next step is roasting, in which the metal sulfide is oxidized to metal oxide. In the literature, the data on losses of zinc during primary zinc production and processing vary. According to James et al. (2000), referred to by Gordon et al. (2003), the average zinc losses are 5% (tailings) in the enrichment steps, and another 7% (slag) during the roasting process, giving a total of 12% (Gordon et al., 2003). According to Guo (2010), zinc losses in tailings and smelting slag in China were about 15% in 2006. According to Plachy (2001), the processing losses of zinc during domestic zinc production in the United States were about 10% in 1998. Meylan and Reck (2016) provide a more comprehensive global picture: compared to the quantity of primary zinc extracted from the Earth's crust, the quantities lost during the various steps from mining to manufacturing are 19% in total (tailings: 13%; slags: 5%; losses during fabrication and manufacturing: 1%). We use the figures of Meylan and Reck (2016) in our analysis.

7.14.4.2 Dissipation

Compared to other metals, a relatively large part of zinc use is dissipative by the nature of its use. This is especially the case for zinc in chemicals (i.e., pigments, animal feed, rubber) and zinc used for galvanizing.

The zinc runoff from die casts and brass and bronze objects, is very little, because die casts are usually not exposed to wet situations, and because zinc in brass and bronze objects is not concentrated on the metal surface, but distributed through the metal object as a whole.

The zinc emission from rolled zinc (i.e., used on roofs and in gutters), expressed in grams per square meter, is comparable to the zinc emission from exposed galvanized products. However, the thickness of zinc in rolled zinc is much higher (10—50 times higher) than the thickness of zinc on galvanized products. Therefore, zinc dissipation from rolled zinc is negligible compared to the dissipation from galvanized products.

We therefore focus on dissipation from zinc-containing chemicals and galvanized products.

7.14.4.2.1 Zinc dissipation from products with zinc-containing chemicals.

The most important applications of zinc-containing chemicals are presented in Table 7.14.1.

The total amount of zinc in these applications was about 15% of total zinc use in 2010 (International Zinc Association, 2014a). This proportion differs per country, as well as the various sub-applications. For example, zinc in pig fodder will be completely dissipated in the environment via pig manure, while rubber tires, which contain an average of 1.6 wt.% of zinc, will partly emit zinc through wear of the tires. Based on data from Vos and Janssen (2008), it can be calculated that the average zinc emissions from rubber tires due to wear are about 8% of the total zinc content in tires.

Based on these data, we estimate the total dissipation of zinc from zinc-containing chemicals during usage to be in the order of 25% (the applications in biocides, lubricating oil, tires, and pig fodder).

7.14.4.2.2 Dissipation from galvanized products

Estimates of the zinc runoff rate from exposed zinc surfaces vary. For example, Japanese researchers estimated the yearly runoff rate at about 2% on the basis of the total weight of the zinc layer (Tabayashi et al., 2009). On the basis of the research of Wallinder et al. (1998), the Netherlands Center for Water Management (2008) provides the following formula for zinc runoff from horizontal surfaces that are exposed to rain, depending on the average SO_2 concentration in the air: zinc runoff rate $(g/m^2/year) = 1.36 + 0.164 \times [SO_2](\mu g/m^3)$. This zinc runoff rate needs to be corrected for the spatial orientation of the objects. Measurements by Mourik et al. (2003) show that the runoff rate from horizontal zinc surfaces is almost four times higher than from vertical surfaces. The average correction factor in The Netherlands is 0.84. Assuming an average SO_2 concentration of $2\ \mu g/m^3$ at present (National Institute for Public Health and Environment

Table 7.14.1 Applications of zinc-containing chemicals and the most important routes of zinc at the end-of-life stage. The estimates are from the author, unless indicated otherwise.

	Percentage of zinc in zinc-containing chemicals (%) (derived from Annema and Ros, 1994)	Dissipation through use	Disposal of end-of-life products (landfill or incineration)	Recycling
Pigment		0%	100%	0%
Flame-retardant		0%	100%	0%
Biocide, fungicide for potatoes	10%	100%		0%
Oil additive in lubricating oil		100%		0%
Blasting grit		Little	Most	0%
Vulcanizing accelerator for rubber tires	6%	8%	92%	0%
Additive in pig fodder	14%	100%		0%
Zinc anodes		Little	Most	
Other applications (solder, casting aid, batteries)		Little	Most	

(2012), this leads to an average zinc runoff rate of 1.7 g/m^2/year. If we assume an average zinc amount on galvanized products of 200 g/m^2 (based on data from the International Zinc Association, 2014b), this leads to the conclusion that zinc dissipation from exposed galvanized products is about 1 wt.% per year, which is half of the estimation of Tabayashi et al. (2009). Note that not all galvanized products are exposed, and those that might be exposed are not continuously exposed.

7.14.4.2.3 Dissipation summarized

According to Meylan and Reck (2016), zinc dissipation in 2010 was about 8 wt.% of the total amount of zinc entering the usage phase in that year.

Assuming a dissipation of 25% of zinc in chemicals and other miscellaneous uses, and 0% zinc dissipation from die casts, brass, bronze, and rolled

Table 7.14.2 Zinc dissipation from various zinc applications. The usage proportions are from the International Zinc Association (2014a). Total end use in 2010 is 12.9 million tons (International Zinc Association, 2013b). The dissipation figures are explained in the text.

	2010 usage of zinc	Zinc dissipation through use	Remaining zinc after dissipation
	(%)	(%)	(%)
Galvanizing	54	8	50
Die casts	13	0	13
Brass and bronze	10	0	10
Rolled zinc	8	0	8
Chemicals and miscellaneous	15	25	11
Total	100	8	92

zinc, this leads to the conclusion that zinc dissipation from galvanized products is about 8 wt.% (see Table 7.14.2). This is a plausible result in view of the above considerations on zinc dissipation from galvanized steel. If we assume an average lifetime of galvanized products of 20 years, 25% exposure of the galvanized products, and a yearly dissipation of 1.5 wt.%, then the total yearly loss from galvanized products is 7.5 wt.%.

7.14.4.3 Increase in in-use stocks

According to Yokota et al. (2003), the average lifetimes of zinc-containing products are 34.5 years in construction, 28.9 years in buildings, and 12.1 years in machines. In emerging economies like China and India, the economy is growing so fast that the input of zinc to the anthropogenic stock is much larger than the output as end-of-life products. In developed countries like Japan, the accumulation of zinc in the in-use stocks has almost stabilized (Tabayashi et al., 2009). At a global scale, the in-use stock of zinc is still growing. In 2010, this build-up was estimated at almost 40% of the quantity of zinc entering the usage phase (Meylan and Reck, 2016). In the further future, it can be expected that the global in-use stocks will gradually stabilize and an equilibrium will be reached between the zinc flow into the usage phase and the flow out of the usage phase (by dissipation and in end-of-life products).

7.14.4.4 Recycling

Until recently, zinc recycling concentrated on die casts and rolled zinc, which are almost completely recycled. We assume a 90% end-of-life zinc recycling rate from die casts and rolled zinc, based on data from the International Zinc Association (2013b) and Van Beers et al. (2007). Until relatively recently, zinc was hardly recycled from galvanized products. However, zinc recycling from galvanized products is now increasing, due to improved dezincing technologies for steel scrap and hydrometallurgical leaching of zinc from electric arc furnace dust. Zinc recycling from galvanized steel will depend on the collection rate of end-of-life galvanized steel products and the zinc recovery from these products.

Brass and bronze are almost completely recycled due to their valuable copper content. If not remelted to brass or bronze, zinc moves to the dust phase. However, dusts from the copper industry are not directly suitable for application in the primary zinc industry, because of their halogen content (Antrekowitsch et al., 2014; Tabayashi, 2009). Recycling of zinc from zinc-containing chemicals is difficult, because of the overall low zinc concentrations in the end-of-life product or because of the inherent dissipative use (in pig fodder, fungicide). Theoretically, zinc could be recovered from applications in anodes, tires, batteries, and solder. However, we shall assume that zinc recycling from chemicals is 0%.

Overall, zinc has a fairly high recyclability, although actual zinc recycling data vary. According to data on the website of the International Zinc Association (2013b), the overall global end-of-life recycling rate of zinc is 60%. Graedel et al. (2005) quantified recycled zinc flows from end-of-life products at 41% in 1994.

In the later article of Graedel et al. (2011), the end-of-life zinc recycling rate is estimated to be >50%. This is similar to the figure provided by UNEP (2011b). However, Meylan and Reck (2016) estimate a global end-of-life zinc recycling rate of just 33% in 2010. The recycling rates vary regionally: Asia 24%, Europe 46%, Latin America 35%, North America 42%, and China 21% (Meylan and Reck, 2016). Spatari et al. (2003) provide an end-of-life recycling rate of 34%. For the current situation, we follow the data of Meylan and Reck (2016) (a global end-of-life zinc recycling rate of 33%).

Assuming an overall end-of-life zinc recycling rate of 33%, a 90% zinc recycling rate from die casts and rolled zinc, and 0% recycling of zinc from chemicals, we conclude that the current zinc recycling from galvanized products must be about 30% (see Table 7.14.3).

Table 7.14.3 Current zinc recycling rates of end-of-life zinc products and the resulting quantity of secondary zinc.

	Zinc in end-of-life products after dissipation (see Table 7.14.2	Zinc to in-use stocks (40%, see Section 7.14.4.3)	Remaining zinc in end-of-life products	Current zinc recycling rate from end-of-life products (see Section 7.14.4.4)
	(%)	(%)	(%)	(%)
Galvanized products	50	23	26	30
Die casts	13	6	7	90
Brass and bronze	10	5	5	0
Rolled zinc	8	4	4	90
Chemicals and miscellaneous	11	0	11	0
Total	92	38	54	33

7.14.4.5 Current zinc flows summarized

The figures provided in the previous subsections result in the zinc flow diagram presented in Fig. 7.14.4.

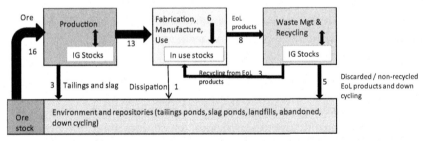

Figure 7.14.4 Estimated global zinc flows in 2018. The indicated amounts are expressed in Mt. The figures may not add up exactly because of rounding. Assumptions are: zinc recovery efficiency: 81% of ore; in-use dissipation: 8% of consumption; to in-use-stocks: 40% of consumption; and end-of-life recycling rate 33%. The arrow widths are indicative of the relative volume of the flows. IG stock is industry and government stocks.

7.14.5 Sustainable zinc flows

7.14.5.1 Introduction

In this section, we explore under which conditions it is possible to realize a sustainable primary zinc production rate while simultaneously increasing the global service level of zinc to the service level of zinc in developed countries in 2020. This is about 8 kg/capita/year (see Table 5.3).

First, we explore the possibilities and limitations of increasing the zinc recovery efficiency at the mining stage, the substitution of zinc, and the material efficiency, decreasing in-use dissipation of zinc, and increasing the end-of-life recycling rate of zinc.

7.14.5.2 Increase in recovery efficiency at the mining and processing stages

Zinc losses during primary production and processing are currently about 19%. Technically, these losses could be reduced (e.g., by improving the flotation process, and by the chemical recovery of zinc from tailings and slags). However, the gains of such measures need to be economically balanced with their costs. The higher the price of primary zinc, the more attractive it will be to further reduce zinc losses during primary production and processing. Improving the potential recovery efficiency of the primary enrichment steps, and the exploitation of ever lower grade zinc ores, will depend on technological developments and the energy and raw material costs of mining. Because of ever lower grades of exploited zinc ores in combination with the increasing remoteness and depth of zinc mines, we assume no further increase in the zinc recovery efficiency.

7.14.5.3 Substitutability of zinc

7.14.5.3.1 General

There is an abundance of literature on zinc recycling and its potential, but systematic and quantitative research into the substitutability of zinc in its various applications is limited.

Substitutability is used as one of the indicators for measuring the vulnerability of society to the scarcity of a mineral. According to Oakdene (2013) in a report in the framework of the EU Raw Materials Initiative, the substitutability score of zinc is as follows:

-   Galvanizing: 0.7 (substitutable at high cost and/or loss of performance)
-   Brass and bronze: 0.5 (substitutable at medium cost and/or some loss or performance)

- Zinc-based alloys: 0.7 (substitutable at high cost and/or loss of performance)
- Zinc in chemicals: 0.1 (not substitutable)

According to US Geological Survey, 2020, galvanized sheet can be substituted for "... *aluminum and plastics ... ... in automobiles.*" Furthermore, zinc can be substituted for "... *aluminum alloys, cadmium, paint and plastic coatings in other applications. Aluminum- and magnesium-base alloys are major competitors for zinc-base die casting alloys. Many elements are substitutes for zinc in chemical, electronic, and pigment uses.*"

Zinc anodes are 100% substitutable by aluminum anodes and/or magnesium anodes (derived from data from Roovaart and van Duijnhoven, 2007). On the other hand, the application of zinc in batteries is strongly growing and zinc as a fodder additive cannot be replaced.

7.14.5.3.2 Zinc substitution scenarios

It can be concluded that, in quite a number of applications, zinc is substitutable to a certain extent. However, it is not possible to provide a precise figure. We shall consider three scenarios: (1) a low zinc substitution scenario (20% substitution), (2) a medium zinc substitution scenario (50% substitution), and (3) a high zinc substitution scenario (70% substitution). Table 7.14.4 shows how these three different substitution outcomes could possibly be reached. In line with the data provided by Oakdene (2013), we have assumed that the substitution of brass and bronze is more easily achieved than the substitution of galvanizing and die casts. Following Annema and Ros (1994), we have assumed that rolled zinc can be easily substituted.

Table 7.14.4 Three zinc substitution scenarios.

	Zinc applications in 2010	Low substitution scenario	Medium substitution scenario	High substitution scenario
Galvanized products	54%	15%	50%	70%
Die casts	13%	15%	50%	70%
Brass and bronze	10%	30%	60%	90%
Rolled zinc	8%	90%	90%	90%
Chemicals and miscellaneous	15%	0%	25%	45%
Total/average	100%	20%	50%	70%

7.14.5.4 Material efficiency improvement

We have, prudently, assumed an overall material efficiency potential of 10% on top of the dissipation reduction and recycling measures (see Section 7.1.3).

7.14.5.5 Dissipation reduction

Zinc dissipation from galvanized products can be reduced by covering galvanized products with an organic coating or, depending on the requirements and circumstances, by replacing zinc with another type of corrosion protection layer. According to Sullivan and Worsley (2002), organic coatings are very effective (>90% reduction of zinc dissipation to less than 0.25 g/m^2 over 16 months) in reducing zinc dissipation from galvanized steel. Research by Gouman (2004) also concludes that a zinc dissipation reduction of >90% is possible by covering galvanized surfaces with an impermeable plastic coating. Hence, the adequate protection of galvanized products with a plastic coating can reduce the total zinc runoff from galvanized products from 7.5% to less than 1%. To be on the safe side, we assume that the in-use dissipation of zinc from galvanized products can be decreased to 2%.

The zinc runoff from rolled zinc, die casts, brass, and bronze is negligible.

As for zinc-containing chemicals, we assume that zinc dissipation cannot be further reduced, except by substituting zinc. This means an in-use dissipation rate of zinc-containing chemicals of 25%.

This means that the minimum in-use dissipation from the remaining products after substitution in the three different substitution scenarios will be:
- 5% in the no substitution scenario
- 6% in the low substitution scenario
- 3% in the medium substitution scenario
- 8% in the high substitution scenario

7.14.5.6 Increase in end-of life recycling

Since the current recycling rate of die casts and rolled zinc is already high, we focus on the recycling of zinc from galvanized steel and from brass and bronze. A small contribution to zinc recycling could be provided by recycling zinc from the ashes of incinerated rubber tires (Martin et al., 2001). However, we do not take this into account in our analysis.

The current recycling rate of zinc from galvanized steel is estimated at 30% (see Section 7.14.4.4). Galvanized steel scrap is usually recycled to electrical arc furnaces (EAF), in which new steel is produced. Zinc contained in steel scrap is mainly "gassed off" and concentrated in the EAF dust. A relatively small part of zinc is contained in the slags from steel production. EAF dust contains an average of about 20% zinc according to US Geological Survey (1998), Jha et al. (2001), Nakajima et al. (2003), Antrekowitsch et al. (2014), Spatari et al. (2003), and Ma et al. (2011).

Dust from copper smelters may contain more than 40% zinc (Antrekowitsch, 2014). Both zinc in EAF dust and zinc from brass and bronze recycling can be completely recycled. EAF dust can be recycled in the Waeltz process, resulting in Waeltz oxide with more than 40% zinc that can be used for zinc oxide production or to produce zinc metal or zinc alloys (Spatari 2003). Zinc from recycled brass and bronze contains harmful contaminants and is not suitable for direct application in the primary zinc industry. Its recycling requires a specific approach including hydrometallurgical or pyrometallurgical processes (Antrekowitsch et al., 2014). Hence, zinc in recycled brass and bronze remains in the copper cycle and is not recycled to the zinc cycle.

The conclusion is that, technically, zinc in galvanized steel and brass and bronze is recyclable to a high extent. However, due to economic reasons, low landfill costs, and low costs of primary zinc, zinc recycling from galvanized steel is still limited at the global scale.

Though the recycling of zinc from waste incineration residues is technically possible, it is not economically viable (Fellner et al., 2015; Shen and Forssberg, 2003).

By using specific hydrometallurgical and pyrometallurgical processes, zinc can practically always be recovered to a high extent from any industrial waste or end-of-life product (Rabah and El-Sayed, 1995; Jha et al., 2001; Basir and Rabah, 1999). However, its economic viability thus far is questionable.

We assume that the recyclability of zinc from galvanized steel and from brass and bronze can be increased to 80% in the future.

Table 7.14.5 summarizes the assumed future recyclability of different zinc applications compared with current recycling rates.

After substitution, the maximum overall end-of-life recycling rate of zinc from the remaining products is as follows:
-   No substitution: end-of-life recycling rate: 70% of zinc in remaining products
-   Low substitution scenario: end-of-life recycling rate: 70% of zinc in remaining products

Table 7.14.5 Assumed future zinc end-of-life recycling rates.

	Current end-of-life zinc recycling rate (see Table 7.14.3)	Assumed maximum future zinc end-of-life recycling rate
Galvanized products	30%	80%
Die casts	90%	90%
Brass and bronze	0%	80%
Rolled zinc	90%	90%
Chemicals and miscellaneous	0%	0%

- Medium substitution scenario: end-of-life recycling rate: 78% of zinc in remaining products
- High substitution scenario: end-of-life recycling rate: 64% of zinc in remaining products

7.14.6 Conclusions

Table 7.14.6 presents the conditions under which sustainable zinc production can be achieved for different sustainability ambitions. This assumes that the world population stabilizes at 10 billion people and that the global service level of zinc is 8 kg/capita/person. The ultimately available zinc resources are assumed to be between 9 Gt and 30 Gt.

The conclusions are as follows:

- A sustainability ambition of 1000 years is achievable with ultimately available zinc resources of 9 Gt, if combined with 70% substitution, 10% material efficiency improvement, and 63% recycling from the remaining zinc-containing end-of-life products. With 30 Gt of ultimately available zinc resources, the measures can be focused on increasing the end-of-life recycling from the current 33% to 66%.
- A sustainability ambition of 500 years is achievable with ultimately available zinc resources of 9 Gt combined with 50% substitution, 10% material efficiency improvement, and 52% recycling from the remaining zinc-containing end-of-life products. With ultimately available zinc resources of >28 Gt, no measures are necessary compared with the current situation.
- A sustainability ambition of 200 years is achievable without any measures, with ultimately available zinc resources >11.2 Gt. The minimum amount of ultimately available zinc resources for realizing a sustainability ambition of 200 years is 1.8 Gt.

Table 7.14.6 Different mixes of measures to achieve a sustainable zinc production rate, while realizing a global zinc service level that is similar to the zinc service level in developed countries in 2020. The boldfaced rows include the main (combinations of) measures.

		Sustainability ambition (years of sufficient availability)						
		1000			**500**		**200**	
Ultimately available zinc resources[a]	Gt	9	9	30	9	28	1.8	11.2
Sustainable zinc production[e]	Mt/year	9	9	30	18	56	9	55
Aspired future zinc service level	kg/capita/year	8	8	8	8	8	8	8
Assumed future substitution	%	**70%**	**50%**	**0%**	**50%**	**0%**	**70%**	**0%**
Assumed increase in material efficiency[b]	%	10%	10%	0%	10%	0%	10%	0%
Derived future zinc consumption level	kg/capita/year	2.2	3.6	8	3.6	8	2.2	8
Corresponding annual zinc consumption	Mt/year	22	36	80	36	80	22	80
Assumed in-use dissipation of remaining products[c]	%	8%	3%	5%	3%	8%	8%	8%
Zinc in end-of-life products	Mt/year	20	35	76	35	74	20	76
Required zinc recycling of remaining products	Mt/year	13	27	50	18	24	13	25
Required zinc end-of-life recycling rate[d]	%	**63%**	**77%**	**66%**	**52%**	**33%**	**64%**	**33%**

The assumed future maximum recovery efficiency remains 81%, and the assumed future maximum substitutability of zinc depends on the scenario: 20%, 50%, or 70%. The assumed future maximum material efficiency increase is 10%, the assumed minimum in-use dissipation rate of remaining products depends on the substitution scenario, and the assumed future maximum recyclability of remaining products also depends on the substitution scenario. See text for more details.

[a]Low estimate: 9 Gt; high estimate: 30 Gt (see Section 7.14.3.2).
[b]Current recovery efficiency: 81% (see Section 7.14.4.1).
[c]Current in-use dissipation rate: 8% (see Section 7.14.4.2).
[d]Current end-of-life recycling rate: 33% (see Section 7.14.4.4).
[e]It is assumed that the world population stabilizes at 10 billion people.

References

Annema, J.A., Ros, J., 1994. Management of the zinc chain. Resour. Conserv. Recycl. 13, 209—219.

Antrekowitsch, J., Steinlechner, S., Unger, A., Rösler, G., Pichler, C., Rumpold, R., 2014. Zinc and Residue Recycling, Cht 9 in Handbook of Recycling, Edited by Ernst Worrell and Markus A Reuter.

Barbalace, K., 2007. Periodic Table of Elements. Retrieved from: 14-4-2007. http://environmentalchemistry.com/yogi/periodic.

Basir, S.M.A., Rabah, M.A., 1999. Hydrometallurgical recovery of metal values from brass melting slag. Hydrometallurgy 53, 31—44.

Darling, D., 2007. Elements, Terrestrial Abundance. http://www.daviddarling.info/encyclopedia/E/elterr.html.

Erickson, R.L., 1973. Crustal occurrence of elements, mineral reserves and resources. U. S. Geol. Surv. Prof. Pap. 820, 21—25.

Fellner, J., Ledere, J., Purgar, A., Winterstetter, A., Rechberger, H., Winter, F., Laner, D., 2015. Evaluation of resource recovery from waste incineration residues — the case of zinc. Waste Manag. 37, 95—103.

Gordon, R.B., Graedel, T.E., Bertram, M., Fuse, K., Lifset, R., Rechberger, H., Spatari, S., 2003. The characterization of technological zinc cycles. Resour. Conserv. Recycl. 39, 107—135.

Gouman, E., 2004. Reduction of zinc emissions from buildings; the Policy of Amsterdam, Environmental Services Department of the City of Amsterdam. Water Sci. Technol. 49 (3), 189—196.

Graedel, T.E., Allwood, J., Birat, J.-P., Buchert, M., Hagelüken, C., Reck, B.K., Sibley, S.F., Sonnemann, G., 2011. What do we know about metal recycling? J. Ind. Ecol. 15 (3), 355—366.

Graedel, T.E., Cao, J., 2010. Metal spectra as indicators of development. Proc. Natl. Acad. Sci. U.S.A. 107, 20905—20910.

Graedel, T.E., van Beers, D., Bertram, M., Fuse, K., Gordon, R.B., Gritsinen, A., Harper, E.M., Kapur, A., Klee, R.J., Lifset, R., Memon, L., Spatari, S., 2005. The multilevel cycle of anthropogenic zinc. J. Ind. Ecol. 9 (3), 67—90.

Guo, X., Zhong, J., Tian, Q., 2010. Substance flow analysis of zinc in China, Resources. Conserv. Recycl. 54, 171—177.

International Zinc Association, 2013a. wwwzincorg.

International Zinc Association, 2013. Zinc Recycling, Closing the Loop, wwwzincorg.

International Zinc Association, 2014a. http://wwwzincorg/basics/zinc_statistics.

International Zinc Association, 2014b. Zinc Coatings Protecting Steel, wwwzincworld.

James, S.E., 2000. Zinc production — a survey of existing smelters and refineries. In: Dutrizac et al editors Lead zinc 2000 Warrendale: TMS, pp. 205—225.

Jefferson Lab, 2007. It's Elemental, the Periodic Table of Elements. http//education.jlab.org/itselemental/index.html.

Jha, M.K., Kumar, V., Singh, R.J., 2001. Review of hydrometallurgical recovery of zinc from industrial wastes. Resour. Conserv. Recycl. 33, 1—22.

Ma, H., Matsubae, K., Nakajima, K., Tsai, M., Sha, K., Chen, P., Lee, C., Nagasaka, T., 2011. Substance flow analysis of zinc cycle and current status of electric arc furnace dust management for zinc recovery in Taiwan. Resour. Conserv. Recycl. 56, 134—140.

Martin, D., Garcia, M.A., Diaz, G., 2001. Zinc recovery from tire ashes, Recycling and reuse of used tires. In: Proceedings of the International Symposium, University of Dundee, 19—20 March, 2001.

Meylan, G., Reck, B.K., 2016. The Anthropogenic Cycle of Zinc: Status Quo and Perspectives, Resources, Conservation and Recycling. http://dxdoiorg/101016/jresconrec201601006.

Mourik, W van, van der Mijle Meijer, H.J., van Tilborg, W.J.M., Teunissen, R.J.M., 2003. Emissions of Building Materials: Determining Leaching Rates Based on Experimental Measurements, National Institute of Water Management and Wastewater Treatment, RIZA Report 2003, 027. ISBN 9036956412.

Nakajima, K., Matsubae-Yokoyama, K., Nakamura, S., Itoh, S., Nagasaka, T., 2003. Substance flow analysis of zinc associated with iron and steel cycle in Japan, and Environmental Impact Assessment of EAF dust Recycling Process. Resour. Conserv. Recycl. 39, 137–160.

National Institute for Public Health and the Environment, 2012. Large-scale concentration and deposition maps for The Netherlands, Reporting 2012, Rijksinstituut voor Volksgezondheid en Milieu, Grootschalige concentratie- en depositiekaarten Nederland, Rapportage 2012.

Netherlands Center for Water Management, 2008. Atmospheric Corrosion of Galvanized Steel and Sheet Zinc.

Oakdene, H.F., 2013. Study on Critical Raw Materials at EU Level, A Report for DG Enterprise and Industry of the European Commission.

Plachy, J., 2001. Zinc Recycling in the United States in 1998, US Geological Survey Circular 1196-D.

Rabah, M.A., El-Sayed, A., 1995. Recovery of zinc and some of its valuable salts from secondary resources and wastes. Hydrometallurgy 37, 23–32.

Roovaart, J.C., van Duijnhoven, N., 2007. Zinkanodes Binnenscheepvaart, Emissieschattingen Diffuse Bronnen, Werkdocument Nr 201088X, vol. 7. RIZA, p. 2007.

Rudnick, R.L., Fountain, D.M., 1995. Nature and composition of the continental crust, a lower crustal perspective. Rev. Geophys. 267–309.

Shen, H., Forssberg, E., 2003. An overview of recovery of metals from slags. Waste Manag. 23, 933–949.

Skinner, B.J., 1976. A second iron age ahead? Am. Sci. 64, 158–169.

Spatari, S., Bertram, M., Fuse, K., Graedel, T.E., Shelov, E., 2003. The contemporary European zinc cycle: 1-year stocks and flows. Resour. Conserv. Recycl. 39, 137–160.

Statista, 2019. Distribution of Zinc Consumption in 2018, Worldwide in 2018, by End Use, Statistacom/statistics; Derived from the Website on 16-12-2019.

Sullivan, J.H., Worsley, D.A., 2002. Zinc runoff from galvanized steel materials exposed in industrial/marine environment. Br. Corrosion J. 37 (4), 282–288.

Tabayashi, H., Daigo, I., Matsuno, Y., Adach, Y., 2009. Development of a dynamic substance flow model of zinc in Japan. ISIJ Int. 49 (8), 1265–1271.

UNEP, 2011a. International Panel on Sustainable Resource Management, Working Group on Geological Stocks of Metals. Working Paper, April 6, 2011.

UNEP, 2011b. International Resource Panel, Recycling Rates of Metals, a Status Report, 2011.

US Geological Survey, 2000. 1998 Assessment of undiscovered deposits of gold, silver, copper, lead and zinc in the United States. Circular 1178, 2000.

US Geological Survey, 2017a. Zinc Statistics, Modification January 2017.

US Geological Survey, 2017b. Mineral Commodity Summaries, Zinc, January 2017.

US Geological Survey, 2018. Mineral Commodity Summaries, Zinc, January 2018.

US Geological Survey, 2019. Mineral Commodity Summaries, Zinc, February 2019.

US Geological Survey, 2020. Mineral Commodity Summaries, Zinc, January 2020.

Van Beers, D., Kapur, A., Graedel, T.E., 2007. Copper and zinc recycling in Australia: potential quantities and policy options. J. Clean. Prod. 15, 862–877.

Vos, J.H., Janssen, M.P.M., 2008. EU-wide Control Measures to Reduce Pollution from WFD Relevant Substances, Copper and Zinc in the Netherlands. RIVM. Report 607633002/2008.

Wallinder, O., Verbiest, P., He, W., Leygraf, C., 1998. The influence of patina and pollutant levels on the runoff rate of zinc from roofing materials. Corrosion Sci. 40 (11), 1977−1982.

Webelements, 2007. Abundance in Earth's Crust. http://www.webelements.com/webelements/properties/tekst/image-flah/abund-crust.html.

Wedepohl, K.H., 1995. The composition of the continental crust. Geochim. Cosmochim. 59 (7), 1217−1232.

Yokota, K., Matsuno, Y., Yamashita, M., Adachi, Y., 2003. Int JLCA 8, 129.

CHAPTER 8

Mineral resources governance[1]

8.1 Introduction

The results of Chapters 2 and 3 indicate that the extractable amounts of the following eight materials may be exhausted within a century: antimony, bismuth, boron, copper, gold, indium, molybdenum, and silver.

By exhaustion of a mineral resource, we mean a lack of profitability or feasibility of further extraction of that resource due to (a combination of) financial, environment, energy, climate change, waste, water use, and social factors.

Exhaustion of a mineral resource has serious economic consequences for future generations, including a permanent, very high cost increase of the mineral resource in question. Exhaustion is the irreversible effect of careless and unrestricted use of scarce mineral resources. Exhaustion does not mean that there is nothing of the material left in the earth's crust anymore. Exhaustion means that extraction of the material is no longer profitable due to an accumulation of economic, environmental, climatic, and social reasons. The point of exhaustion is reached when further extraction will cost more wealth than it will produce. This includes the external costs of extraction. Like climate change and loss of biodiversity, the impact of exhaustion of mineral resources will be felt in the longer term, not immediately. Therefore, confronted with other pressing global problems, humanity may not realize the urgent need to act now to address the future exhaustion of certain mineral resources in an adequate way.

There are warnings about the future exhaustion of some mineral resources (e.g., Rankin, 2011; UNEP, 2011; Erickson, 1973; Nickless, 2017; Ragnarsdottir et al., 2012; Sverdrup et al., 2017). However, until

---

[1] This chapter is based on the publication Henckens MLCM, Biermann FHB, Driessen PPJ, Mineral resources governance: A call for the establishment of an International Competence Center on Mineral Resources Management, Resources, Conservation and Recycling, 141,2019, 255−263.

Governance of the world's mineral resources
ISBN 978-0-12-823886-8
https://doi.org/10.1016/B978-0-12-823886-8.00027-0

now, technological development in combination with expansion of brown-fields and better knowledge on ore-forming systems has neutralized the increasing costs of extracting lower-grade ores, deeper mines, and remoter mine locations fully or for a substantial part. The Escondida copper mine in Chile and the Grasberg copper mine in Indonesia are examples of large brownfield extensions. Consequently, market prices of almost all minerals have remained remarkably stable in real terms (US Geological Survey, 2017a). In Chapter 6 we demonstrated that, so far, the long-term geological scarcity of a resource has not significantly influenced its price trend. The market price of a mineral resource might only react structurally when that resource is almost exhausted. The free market price mechanism reacts to developments of today and tomorrow but does not necessarily take account of the interests of future generations, which are still at least decades ahead.

To overcome this limitation of the market and to anticipate the future exhaustion of scarce mineral resources, the governance of mineral resources and mineral resources policy at a global scale need to be improved and implemented more effectively very soon. We define mineral resources governance as the set of norms, values, and rules through which mineral resources are managed. Despite general intentions[2], at global level, inter-national mineral resources governance and international mineral resources policy are practically nonexistent, although they exist on a country-by-country basis like in Australia. Several countries and also the European Union (EU) have developed policies regarding raw materials that are critical for their economies (Chapter 4). We define mineral resources policy as a set of concrete goals for achieving a more sustainable use of mineral resources in the interests of future generations.

The main technical measures for improving the sustainable use of a mineral resource are the improvement of the recovery efficiency at the beneficiation stage, substitution of the resource by a less scarce material, a higher recycling rate from end-of-life products, improvements in material efficiency, and measures to limit dissipation of the resource during its use. Technological developments and governmental limitations may change the

---

[2] Already in 1972, the world's nations stated that *"the non-renewable resources of the earth must be employed in such a way as to guard against the danger of their future exhaustion"* (Prin-ciple 5 in the Declaration of the United Nations Conference on the Human Environ-ment in Stockholm, 1972). More recently, the report on the *"Implementation of Agenda 21 on Sustainable Development Goals"* declares as target 12.2: *"By 2030 achieve the sustain-able management and efficient use of natural resources"* (United Nations, 2015).

demand, for example, the banning of the use of lead in gasoline, the decrease of the (future) use of Platinum Group Metals in catalytic converters due to the transition to the use of electric cars, and the lesser use of cadmium in certain products due to its toxic characteristics. However, none of these will happen spontaneously. To be widely used, substitutes must be cheaper than the original. Normally, recycling must be cheaper than using the virgin resource from the mine.[3] Higher material efficiency must be rewarded. Thus, to ensure mineral resources policy results in the available technical reduction options being implemented, dedicated instruments are essential.

In response to impending resource exhaustion or scarcity, various authors (e.g., Sverdrup et al., 2017; Valero and Valero; 2010; Calvo et al., 2016; Ali et al., 2017; Christmann, 2017; Tilton et al., 2018) have advocated developing international resources policies. The effectiveness of policy instruments will depend on the specific characteristics of the scarce material, for example, substitutability, recyclability, and its applications.

In this Chapter, we analyze and assess 11 different policy instruments for achieving a more sustainable use of eight scarce materials: antimony, bismuth, boron, copper, gold, indium, molybdenum, and silver. In Section 8.2, we analyze the impact of market price, substitutability, and recyclability on the applicability of technical measures for reducing the consumption of scarce mineral resources. In Section 8.3, we analyze and assess 11 policy instruments for achieving a more sustainable extraction rate of scarce mineral resources. Our conclusions and a discussion on the results are presented in Section 8.4.

8.2 Market price, substitutability, and recyclability

Three factors have a significant impact on the applicability of technical measures for reducing the use of scarce mineral resources and on the effectiveness of different policy instruments: the raw material's market price, its substitutability, and its recyclability. This is underpinned in the below subsections.

---

[3] Recycling can also be triggered by the imbalance between offer and demand. For example, currently the demand for dysprosium and neodymium, two rare-earth elements (REE), is high compared to the demand for other REEs. Adjusting the total REE production to the highest demand of any REE would be an expensive solution and would necessitate stockpiling of REEs in lower demand. This makes recycling of dysprosium and neodymium from end-of-life consumer goods interesting (Binnemans et al., 2013).

8.2.1 Market price

The higher the market price of a resource, the greater the market pressure to substitute the mineral, and to recycle it, and the less need there is for governments to act to reduce its global extraction. The market price of gold is much higher than that of indium, silver, bismuth, molybdenum, antimony, molybdenum, copper, and boron (Table 8.1). A high market price incentivizes substitution and recycling of even small amounts of pricy metals, implying that they will need specific policies to reduce their use to a lesser extent than cheaper materials. We expect that the free market price mechanism of pricy materials such as gold will automatically lead to their thrifty use and to maximizing their recovery from end-of-life products. Of course, actual recycling and substitution will also depend on the feasibility of recycling and substitution. From certain applications, recycling will be difficult (meaning relatively expensive), such as, for instance, gold and silver from their electronic applications, molybdenum from its steel alloys, and indium from consumer products, mainly due to the low amounts per kg product and the complicated material compositions, in which they are often embedded. For these cases, an increasing market price will have relatively little effect on the application of these materials in products. But this circumstance will not affect the general theorem that a higher market price of a material leads to a more economical use of that material.

8.2.2 Substitutability

To the extent that it is easier to substitute one material by another in a certain application without losing services or incurring unpalatable extra costs, it will be more acceptable and easier to agree to ban that material from use in that application or to impose a resource levy on the material.

Table 8.2 summarizes the substitutability of the eight materials and shows that antimony and boron can be substituted to a substantial extent with existing technology: antimony in flame retardants, in batteries, and as PET catalyst; boron in its use in glass wool and in detergents. By contrast, substitution of gold, molybdenum, copper, and silver is very limited, given current knowledge and technology.

8.2.3 Recyclability

The ease of recycling a raw material affects the effectiveness of certain policy instruments such as the promotion/obligation of recycling-oriented design and subsidizing recycled raw material. Table 8.3 presents the present

Table 8.1 Market prices of eight scarce mineral resources (1998 US$ per kg in the United States) (US Geological Survey, 2017a).

Element	Year					
	1990	1995	2000	2005	2010	2015
Gold	15,500.00	13,300.00	85,300.00	11,900.00	29,500.00	25,700.00
Indium	287.00	415.00	178.00	790.00	413.00	358.00
Silver	193.30	177.10	152.00	197.00	481.00	347.00
Bismuth	9.79	9.08	7.72	7.19	14.40	11.50
Molybdenum	7.84	18.60	5.33	58.50	26.00	9.97
Antimony	2.25	5.38	1.36	2.95	6.61	4.95
Copper	3.38	3.26	1.84	3.20	5.74	3.89
Boron	0.90	0.80	0.89	0.78	0.51	0.35

Table 8.2 Ultimate substitutability.

Material	Ultimate substitutability	Reference
Antimony	90%	Section 7.2
Boron	60%	Section 7.4
Indium	50%	Section 7.8
Bismuth	30%	Section 7.3
Molybdenum	10%	Section 7.9
Copper	10%	Section 7.6
Silver	10%	Section 7.11
Gold	0%	Section 7.7

Table 8.3 Present recycling rate and recyclability of eight scarce materials, based on their current applications.

Material	Present recycling rate	Ultimate recyclability of end-of-life products (after substitution)	Reference
Gold	85%	>90%[a]	Section 7.7
Silver	50%	75%[b]	Section 7.11
Copper	45%	70%	Section 7.6
Antimony	20%	50%	Section 7.2
Molybdenum	20%	50%	Section 7.9
Bismuth	0%	50%	Section 7.3
Indium	0%[c]	50%[c]	Section 7.8
Boron	0%	20%	Section 7.4

[a]Overall gold recycling rate, including technological and non-technological products.
[b]Overall silver recycling rate, including technological and non-technological products.
[c]Recyclability of industrial waste of fabrication of semi-final products and manufacturing of consumer products.

recycling rates and the ultimate recyclability of the eight materials considered. For background, details, and references, it is referred to the respective sections in this book.

The conclusion is that gold is already recycled to a high extent, which can be only slightly increased. The recycling rates of the other seven materials considered can be increased substantially. Boron's recyclability is limited, but boron can relatively easily be substituted in some applications.

8.3 Eleven policy instruments analyzed and assessed

We now discuss and assess 11 policy instruments that might be used to promote substitution and recycling, thereby decreasing the necessity of extraction of virgin raw materials. Policy instruments are non-technical means to foster the achievement of policy goals. They are presented in order of increasing comprehensiveness and legally binding status. The criterion used in the assessment is *the relative effectiveness* of the different policy instruments for each of the eight materials investigated, that is the goal achievement expected from applying a certain instrument as compared to the goal achievement expected from applying other instruments for the same mineral resource.[4] The effectiveness assessments of 11 policy instruments for reducing the use of eight scarce raw materials are based upon our own expert judgment. It needs to be emphasized that they are relative: the deemed effectiveness of one instrument compared to the deemed effectiveness of other instruments. The assessments are ex ante, and they are our best guess based on limited data on the effectiveness of the same instruments in other frameworks. The effectiveness of these instruments cannot yet be tested in reality. Further research will be needed to underpin the results of this ex ante assessment. We conclude by summarizing the relative effectiveness of each of the instruments for achieving a more sustainable use of the eight materials considered.

8.3.1 No dedicated policies

Not having dedicated policies assumes that the price mechanism of the free market will react on time and will automatically solve problems of geological scarcity of mineral resources. Because, as a resource becomes scarcer, its price will rise, making substitution and recycling more attractive. This approach is advocated by resource optimists such as Barnett and Morse (1963), Simon (1980 and 1981), Maurice and Smithson (1984), and more recently by Lomborg (2001) and Diamandis and Kotler (2012).

As explained in Section 8.2.1, the price mechanism will work well for precious resources such as gold but much less or hardly for lower-priced commodities such as the other materials considered. It is uncertain

---

[4] Other criteria, for assessing a policy instrument's suitability, which are not directly aimed at goal achievement, are as follows:

- impact on national sovereignty, for example, over natural resources and in matters of taxation and enforcement.
- political acceptability, for example, fairness and equity, compensation rules, non-discrimination, polluter pays principle, and positive and negative economic effects
- ease of implementation, including criteria such as simplicity, enforceability, credibility, flexibility, resilience, transparency, accountability, and efficiency. See for example, Konidari and Mavrakis, 2007; Mees et al., 2014)

whether the market prices of these materials will rise at short notice in response to their geological scarcity. See Section 6 of this book. We conclude that this option is very effective for precious materials such as gold, but cannot be relied on for the other seven materials. We conclude that the other 10 policy instruments are relatively less effective for gold than for the other seven resources.

8.3.2 Guidelines, recommendations, and codes of conduct

Existing guidelines for multinational companies stimulating the economical use of natural resources include the *OECD Guidelines for Multinational Enterprises, the Sustainability Framework* of the International Finance Corporation of the World Bank, and the *Business Charter for Sustainable Development, Coalition for Environmentally Responsible Economies*—*the 10 environmental principles* of the International Chamber of Commerce. Other relevant guidelines are guidelines 14040 to 14049 of the International Standardization Organization (ISO) on life cycle assessment (LCA). LCA includes abiotic resource depletion.

Like Pattberg (2006), we consider this a relatively ineffective way of reducing the use of mineral resources. The main weakness of guidelines, recommendations, and codes of conduct is that they are voluntary, necessarily general, and often lack specific performance standards.

8.3.3 Eco-labeling

If an eco-label has continued government support and sufficient budget for a clear and consistent information campaign, it can influence consumer choice (Van Amstel et al., 2008). None of the existing eco-label systems includes the impact on exhaustion of mineral resources, so an option could be to include scarcity of mineral resources within existing eco-labels or to create a new dedicated label indicating the amount of scarce mineral resources a product contains. Given the plethora of eco-labels, the former is preferable. World Trade Organization (WTO) rules permit eco-labels related to conserving exhaustible natural resources, providing they do not discriminate between foreign and domestic products.

We assess eco-labeling as only moderately effective because it is facultative and its success depends on consumers' motivation and finances. The main factor determining the decision to buy is product price (Brécard et al., 2009; Teisl et al., 2002; Banerjee and Solomon, 2003).

8.3.4 Sustainable purchasing

Sustainable purchasing has a positive effect on the results of ISO 14001 environmental management implementation (Beamon, 1999; Chen, 2004). A condition for the effectiveness of sustainable purchasing of products by private business and government is the availability of clear information on the impact of products during their life cycle on the environment, climate change, and exhaustion of natural resources. Thus sustainable purchasing will be facilitated by eco-labels. Incorporating mineral resources scarcity into existing national eco-labels will make it easier for governments and business to use scarcity as a criterion in their purchase policy.

We assess that this instrument is only moderately effective because it is not obligatory.

8.3.5 Promoting recycling-oriented design

Recycling-oriented design increases the potential for product repair, the potential for reusing components containing scarce mineral resources, and the recyclability of the scarce mineral resources contained in a product and can help decrease the in-use dissipation of a scarce material (Ciacci et al., 2015). Unless made mandatory, recycling-oriented design will primarily remain dependent on the initiative of private producers. We found no scientific literature on the effectiveness of recycling-oriented design in general. However, we expect a similarity with the effectiveness of eco-labels. If consumers are informed about the recycling-oriented design of a product, some may be motivated to buy it even if the costs are higher (as in the case of Fairphone). However, as with eco-labels, we expect that product price will remain the main decisive factor for most consumers. To overcome the facultative character of recycling-oriented design, governments could require selected scarce materials only to be used in products from which they can be easily retrieved. Recycling-oriented design is ineffective for boron and gold: for boron, because of its low recyclability; for gold, because its recycling rate is already very high. This policy instrument could be more or less effective for recovery of the other materials considered. This depends on the difference between the current recycling rate and the maximum recyclability of the material. The higher the difference, the more effective the policy instrument. Another factor determining the effectiveness of this instrument is the extent that the instrument is more or less mandatory.

8.3.6 Subsidizing secondary materials

This would encourage the use of secondary (recycled) resources instead of primary (virgin) resources. Making the recycled materials cheaper than their virgin equivalents will encourage the market and the cheapest recycling technologies. Government subsidies for specific recycling technologies seem inadvisable because this might inhibit innovation of recycling technologies (Chen, 2005; Söderholm and Tilton, 2012). By subsidizing the use of secondary materials instead of primary resources, any recycling technology would be supported and not particular technologies.

Combining subsidies for recycled materials with scarce resource levies on the related primary mineral resource would put the financial burden of this instrument on the users of primary scarce materials instead of on ordinary tax payers. For boron and gold, this strategy is ineffective because of the low recyclability of boron and because gold is already recycled to a high extent. The instrument could be effective for the other materials considered because, nowadays, the market price of the primary material is mostly lower than the cost of an increased recycling. However, subsidizing a recycled material could discourage substitution, if the secondary material is kept cheap artificially. Subsidizing recycled materials needs a customized (i.e., national) approach. In Europe, it should be organized at the EU level to avoid waste streams flowing to countries with the highest subsidies.

The effectiveness of this instrument depends on the difference between the current recycling rate and the maximum recyclability of the material. The higher the difference, the more effective the policy instrument.

8.3.7 Prohibiting and/or taxing disposal of scarce materials

The subsidizing of recycled secondary materials may be complemented by prohibiting disposal (via landfill or incinerators) of products containing scarce mineral resources or by imposing levies on their disposal. This will encourage recycling of the products concerned (Calcott and Walls, 2000; European Commission, 2012). Like for the two previous instruments, the effectiveness of this instrument depends on the difference between the current recycling rate and the maximum recyclability of the material. The higher the difference, the more effective the policy instrument.

8.3.8 Scarce resource taxation

Taxing scarce resources makes products containing the taxed materials more expensive and less competitive (Gerlagh and van der Zwaan, 2006; Goulder and Schein, 2013), causing demand for these products to decrease.

Four schemes are possible (European Environment Agency, 2012): global resource taxation; taxation of extraction by resource countries; material input taxation at first use of the raw material; and taxation of consumption. Of these, taxation on consumption is the only scheme that looks feasible in the short- and midterm. Unlike the other three types of taxation, it can be organized and imposed nationally. The main challenge is to design a system that is not too complex yet reflects the use of scarce resources.

Countries have a limited scope regarding scarce resource taxation because their taxation policy cannot differ greatly from the related policy of neighboring countries. The obvious approach is coordination between neighboring countries regarding taxation of the sale and use of scarce mineral resources and products containing them, combined with border tax adjustments, if needed.

We assess that scarce resource taxation is more effective to the extent that a material is lower priced, to the extent that a material can be easier substituted for another material, and/or to the extent that the end-of-life recycling rate can be further increased. Hence, we assess that this policy instrument will be relatively more effective for antimony, boron, and indium and less for the other materials considered.

8.3.9 Banning

A well-known example is the successful worldwide phasing out of ozone-depleting substances in the Montreal *Protocol on Substances that Deplete the Ozone Layer* (1987). Polychlorinated biphenyls were banned in 2001 by the Stockholm *Convention on Persistent Organic Pollutants*. Via REACH (the EU Directive on the *Registration, Evaluation, and Authorization of Chemicals*, 2006), the EU has banned cadmium from certain applications. The EU *Restriction of Hazardous Substances* Directive (2009) limits the use of certain substances in electronical and electronic equipment. Similarly, geologically scarce mineral resources could be banned from use in certain applications. Priority should be given to applications in which scarce raw materials can easily be substituted by less scarce substances without loss of services and with negligible (or no) extra costs. A product ban should focus not on minor applications but on low-hanging fruits, that is, most applications of antimony (e.g., in flame retardants, lead−acid batteries, and as PET catalyst) and some applications of boron (e.g., the applications in glass wool and a number of detergents). Banning of applications does not offer an effective solution to reduce molybdenum, copper, silver, and gold because of their relatively low substitutability.

8.3.10 Extraction quotas established by resource countries

In theory, international quotas could very effectively reduce the global extraction rates to controllably sustainable rates without further government actions. This is especially the case, if the reserves of a scarce raw material are concentrated in just a few countries. A fast way of achieving this is that these countries agree on extraction quotas. Table 8.4 shows that the majority of the world reserves of geologically scarce resources are concentrated in just a few countries. Only copper and gold reserves are more dispersed than the reserves of the other six materials. Because of its very high price, we esteem that extraction quotas are not effective for the sustainable use of gold anyway.

The advantages of limiting the extraction of certain scarce mineral resources to a sustainable rate via an agreement between resource countries rather than via a global agreement under UN auspices are as follows:

- Resource countries keep sovereignty over the natural resources on their territory.
- Arranging an agreement between a few countries with similar interests will probably be faster than doing so for many countries with diverse interests.
- The transaction costs of an agreement between a few countries are much lower.

The most important disadvantages are as follows:

- Free riders (non-participating countries) may frustrate the agreement.
- Participating countries may fail to comply without adequate sanctions.
- User countries will depend more on the policy of a few resource countries and may look for their own resources of these elements.
- Poor user countries will find it more difficult to negotiate on a special position than in a UN framework.
- Single-resource countries could be more susceptible to geopolitical pressure.

An agreement between resource countries seems possible only if it is in their interest (financially or otherwise) in the short and long terms; at the least, the agreement must not be disadvantageous. This means that the resource countries should be allowed by the other countries to establish proportionally higher tonnage prices for the reduced amount of extracted mineral resources in such a way that they do not lose revenue. A current example of resource countries coordinating resources policies is the Organization of Petroleum Exporting Countries (OPEC). It is estimated that 70%–75% of the world's proven reserves are in OPEC countries (Lai, 2008).

Table 8.4 Percentage of world reserves (Derived from data of (US Geological Survey, 2008), (US Geological Survey, 2017b) and (US Geological Survey, 2020)).

	Antimony	Bismuth	Boron	Copper	Gold	Indium	Molybdenum	Silver
Australia	9%			10%	20%			16%
Bolivia	20%	3%						4%
Chile	31%	65%	3%	23%	4%		8%	5%
China			2%	3%		75%	47%	7%
Indonesia				3%	5%			
Mexico	1%	3%		6%	3%		1%	7%
Peru			0%	10%	4%	3%	16%	21%
Poland								18%
Russia	23%		3%	7%	11%	1%	6%	8%
South Africa					6%			
Turkey	7%		88%				4%	
United States	4%		3%	6%	6%	3%	15%	4%
Vietnam		14%						
Other countries	5%	15%		32%	41%	18%	4%	10%
Total	100%	100%	100%	100%	100%	100%	100%	100%
Number of countries with >70% of the reserves	3	2	1	>8	>8	1	3	5

OPEC's goal is to stabilize oil prices by fixing the total oil production of member countries and allocating production among the member countries by a quota system. The allocation is the result of negotiations between the member countries.

Although OPEC plays a significant role, its effectiveness fluctuates because of conflicts between OPEC countries, differences in their interests and wealth, and the interference of non-OPEC countries such as Russia. The OPEC experience shows the difficulty of achieving long-lasting, harmonious agreements between resource countries, especially when countries have conflicting interests and different cultures. However, resource countries could usefully draw on OPEC's experiences when discussing the possibilities of limiting the extraction of geologically scarce mineral resources with a view to using these more sustainably and conserving them for future generations.

An agreement between resource countries on limiting the extraction and export of scarce raw materials would not contravene the General Agreement on Tariffs and Trade (GATT). Notwithstanding the general rule prohibiting export restrictions, in Article XX(g), GATT allows export restrictions if these are related *"to the conservation of exhaustible natural resources"* and *"if such measures are made effective in conjunction with restrictions on domestic production or consumption."* Resource countries may themselves agree to reduce extraction, but to preclude potential free riders and non-compliant resource countries, it would be preferable to secure the approbation of the world community and the protection of the agreement by the United Nations and the WTO. These organizations could conditionally agree to ban free riders and non-compliant resource countries from the international market. The conditions could, for example, concern an international agreement on a phasing down scheme and trends in the resource price. If resource countries establish extraction quotas, they will not be obliged to compensate developing countries for the increased costs of importing the scarce mineral in question.

8.3.11 Establishing extraction quotas under the auspices of the United Nations

The purpose of a global agreement is to make existing UN Policy on the sustainable use of natural resources operational. The approach is elaborated in more detail in Section 9. An international arrangement under the auspices of the United Nations is complex because many different interests are involved (e.g., Dimitrov, 2013; Zelli and Pattberg, 2016; Abbott and Snidal, 2000; 2009; Biermann et al., 2009; Biermann, 2011; Widerberg, 2016). The principle of permanent sovereignty of nations over their natural resources will be particularly difficult to overcome in a global agreement because nations do not want to lose control over their national territory, even in the interests of addressing a *common concern of mankind*,[5] which might be the case for the conservation of scarce mineral resources for future generations.

In the past, the United Nations has tried to regulate commodity markets through International Commodity Agreements such as the International Sugar Agreement (1954–1983), the International Cocoa Agreement (1972–1988), the International Coffee Agreement (1962–1989), the International Natural Rubber Agreement (1980–1999), and the International Tin Agreement (1954–1985). Since 1976, these Agreements have been embedded in United Nations Conference on Trade and Development's Integrated Program for Commodities. The main purpose of the agreements was price stabilization. All have failed. According to Gilbert (1996), the explanation for these failures is *"public intervention in commodity markets is not easily rationalized within a climate in which competitive markets are encouraged and state interventions are seen as requiring clear justification in terms of market failure."*

There are no strong global institutions that could enforce global regulation on sustainable use of scarce resources. Even if an agreement were formulated and ratified at the level of the United Nations, individual countries would need to implement the necessary measures.

An advantage of an international arrangement under the auspices of the United Nations is that fairness and equity for poor user countries can be better guaranteed in principle. The United Nations could establish a system of financial compensation of poor user countries, which would need to be financed by the main resources users: the industrialized countries.

The idea of establishment of international extraction quotas in connection with depleting resources is not new. In 2002, the Oil Depletion Protocol was drafted (Heinberg, 2006).

---

[5] The *common concern of humankind* principle is included in, for example, the preambles of the 1992 United Nations Framework Convention on Climate Change and of the 1992 Convention on Biological Diversity. According to several scholars (Brunnée, 2007; Schrijver, 2008; Perrez, 2000; International Law Association, 2014), this principle implies that permanent sovereignty should be exercised for *the benefit of mankind*, which consists of present and future generations. Thus far, depletion of scarce mineral resources has not specifically been identified as a *common concern of mankind*.

8.4 Conclusions and discussion

Table 8.5 summarizes our preliminary conclusions regarding the relative effectiveness of the 11 different policy instruments for reducing the use of antimony, bismuth, boron, copper, gold, indium, molybdenum, and silver resources.

The conclusion is that there is no single recipe for achieving more sustainable use of mineral resources. Which instruments are most effective depends on the specific characteristics of the mineral resource. Determinative variables are a material's market price, substitutability, and recyclability.

We also conclude that not all scarce mineral resources need dedicated policy instruments to achieve their prudent use. Such instruments are unnecessary for highly precious materials such as gold. For the other materials considered, the most effective approach is an international agreement on the establishment of sustainable extraction quotas. This could be arranged by the resource countries or under the auspices of the United Nations.

Banning is very effective for applications that are easily substitutable. This concerns especially antimony and boron. Antimony can be relatively easily substituted in flame retardants, lead—acid batteries, as a catalyst for PET production, as heat stabilizer, and in ceramics. These applications account for over 85% of antimony's use. Boron could be banned from its application in glass wool because glass wool can be substituted by rock wool. This concerns about 40% of boron's use.

Scarce resource taxation is effective for antimony, boron, and indium.

Promoting or prescribing sustainable design, subsidizing recycled materials, and prohibiting (or taxing) disposal could be moderately effective for antimony, bismuth, copper, indium, molybdenum, and silver, but not for boron because boron is hardly recyclable from its applications and neither for gold because of its high market price.

Eco-labeling and sustainable purchasing are only moderately effective because they rely on consumer conscience. Guidelines, recommendations, and codes of conduct are not considered very effective for any of the considered materials.

Credible global agreements require strong global implementing institutions and global enforcement and judicial bodies. Though UN agreements are legally binding, there is no independent body to enforce them. Nevertheless, empirical research has revealed that international regimes make a difference (Andresen 2013; Breitmeier et al., 2006). On the one hand, worldwide commitment to the sustainable use of scarce mineral

Table 8.5 Authors' ex ante assessment of the relative effectiveness of 11 policy instruments for reducing the use of eight scarce raw materials. The scores are explained in Sections 8.3.1–8.3.11.

Policy instrument	Relative effectiveness[a]							
	Antimony	Bismuth	Boron	Copper	Gold	Indium	Molybdenum	Silver
No dedicated policies	–	–	–	–	++	–	–	–
Guidelines, recommendations, and codes of conduct	–	–	–	–	–	–	–	–
Eco-labeling	0	0	0	0	–	0	0	0
Sustainable purchasing	0	0	0	0	–	0	0	0
Promoting/prescribing recycling-oriented design	–/0	0/+	–	–/0	–	0/+	–/0	–/0
Subsidizing recycled scarce materials	–/0	0/+	–	–/0	–	0/+	–/0	–/0
Prohibiting or taxing disposal of products containing geologically scarce mineral resources	–/0	0/+	–	–/0	–	0/+	–/0	–/0
Resource taxation	+	0	++	–	–	+	–	–
Banning the sale and use of certain applications	++	0	++	–	–	++	–	–
Establishment of extraction quotas by resource countries	++	++	++	+	–	++	++	++
Establishment of extraction quotas by the United Nations	++	++	++	+	–	++	++	++

[a] ––, ineffective; –, slightly effective; 0, moderately effective; +, effective; ++, very effective; all compared to the other instruments.

resources will be achieved faster if broad international acceptance of other measures for achieving such use already exists (Falkner, 2013). On the other hand, an agreement on extraction quotas will make the other policy instruments redundant. In the next Chapter, we will elaborate the setup on an international agreement on the conservation and sustainable use of geologically scarce mineral resources.

References

Abbott, K.W., Snidal, D., 2000. Hard and soft law in international governance. Int. Organ. 54 (3), 421–456.

Abbott, K.W., Snidal, D., 2009. Strengthening international regulation through transnational new governance: overcoming the orchestration deficit. Vanderbilt J. Transnatl. Law 42, 501–578.

Ali, S.H., Giurco, D., Arndt, N., Nickless, E., Brown, G., Demetriades, A., Durrheim, R., Enriquez, M.A., Kinnaird, J., Littleboy, A., Meinert, L.D., Oberhänsli, R., Salem, J., Schodde, R., Schneider, G., Vidal, O., Yakovleva, N., 2017. Mineral supply for sustainable development requires resource governance. Nature 543, 367–372.

Andresen, S., 2013. International regimes effectiveness. In: Falkner, R. (Ed.), The Handbook of Global Climate and Environment Policy, Section 18, first ed. John Wiley & Sons Ltd.

Banerjee, A., Solomon, B.D., 2003. Eco-labeling for energy efficiency and sustainability: a meta-evaluation of US programs. Energy Pol. 31, 109–123.

Barnett, H., Morse, C., 1963. Scarcity and Growth, the Economics of Natural Resource Availability. John Hopkins University Press for resources for the Future, Baltimore, MD.

Beamon, B., 1999. Designing the green supply chain. Logist. Inf. Manag. 12 (4), 332–342.

Biermann, F., 2011. International Organizations and Global Environmental Governance: Toward Structural Reform, from Handbook of Global Environmental Politics, second ed.

Biermann, F., Davies, O., Van der Grijp, N., 2009. Environmental policy integration and the architecture of global environmental governance. Int. Environ. Agreements Polit. Law Econ. 1–25.

Binnemans, K., Jones, P.T., Van Acker, K., Blanpain, B., Mishra, B., Apelian, D., 2013. Rare earth economics: the balance problem. J. Occup. Med. 65 (7), 846–848.

Brécard, D., Hlaimi, B., Lucas, S., Perraudeau, Y., Salladarrée, F., 2009. Determinants of Demand for Green Products: an Application to Eco-Label Demand for Fish in Europe.

Breitmeier, H., Young, O.R., Zürn, M., 2006. Analyzing International Environmental Regimes, from Case Study to Database. MIT Press.

Brunnée, J., 2007. Common areas, common heritage and common concern. In: Bodansky, D., Brunnée, J., Hey, E. (Eds.), Oxford Handbook of International Environmental Law. Oxford University Press, Oxford.

Calcott, P., Walls, M., 2000. Can downstream waste disposal policies encourage upstream "design for environment". Econ. Waste 90 (2), 233–237.

Calvo, G., Mudd, G., Valero, A., Valera, A., 2016. Decreasing ore grades in global metallic mining: a theoretical issue or global reality? Resource 5, 36. https://doi.org/10.3390/resources5040036.

Chen, C.-C., 2004. Incorporating green purchasing into the frame of ISO 14000. J. Clean. Prod. 13, 927–933.

Chen, C., 2005. An evaluation of optimal application of government subsidies on recycling of recyclable waste. Pol. J. Environ. Stud. 14 (2), 137−144.

Christmann, P., 2017. Towards a more equitable use of mineral resources. Nat. Resour. Res. https://doi.org/10.1007/s1053-017-9343-6.

Ciacci, L., Reck, B.K., Nassar, N.T., Graedel, T.E., 2015. Lost by design. Environ. Sci. Technol. 49, 9443−9451.

Diamandis, P.H., Kotler, S., 2012. The Future Is Better than You Think. Free Press, New York.

Dimitrov, R.S., 2013. International negotiations. In: Falkner, R. (Ed.), Handbook of Global Climate and Environment Policy. John Wiley & Sons, Ltd.

Erickson, R.L., 1973. Crustal Occurrence of Elements, Mineral Reserves and Resources. US Geological Survey, pp. 21−25. Professional Paper 820,1973.

European Commission, 2012. Use of Economic Instruments and Waste Management Performances. Final report of BioIntelligence Service for the European Commission (DG ENV, Unit G.4, 10 April 2012.

European Environmental Agency, 2012. Resource Taxation and Resource Efficiency along the Value Chain of Mineral Resources. ETC/SCP Working Paper 3/2012.

Falkner, R., 2013. The nation-state, international society and the global environment. Section 15. In: The Handbook of Global Climate and Environment Policy, first ed. John Wiley & Sons Ltd.

Gerlagh, R., Van Der Zwaan, B., 2006. Options and Instruments for a Deep Cut in CO2 Emissions: Carbon Dioxide Capture or Renewables, Taxes or Subsidies.

Gilbert, C.L., 1996. International commodity agreements: an obituary notice. World Dev. 24 (1), 1−19.

Goulder, L.H., Schein, A., 2013. Carbon Taxes vs. Cap and Trade: A Critical Review. National Bureau of Economic Research. Working paper 19338.

Heinberg, R., 2006. The Oil Depletion Protocol. A Plan to Avert Oil Wars, Terrorism, and Economic Collapse, Clairview.

International Law Association, 2014. Legal Principles Relating to Climate Change. Washington Conference.

Konidari, P., Mavrakis, D., 2007. A multi-criteria evaluation method for climate change mitigation policy instruments. Energy Pol. 35, 6235−6257.

Lai, S.C.S., 2008. Oil Prices and the OPEC: Is There a Basis for International Action? Erasmus University, Rotterdam.

Lomborg, B., 2001. The Skeptical Environmentalist. Cambridge university Press, United Kingdom.

Maurice, C., Smithson, C.W., 1984. The Doomsday Myth, 10,000 Years of Economic Crisis. Hoover Institution Press, Stanford University.

Mees, H.L.P., Van Soest, D., Driessen, P.P.J., Van Rijswick, M.H.F.M., Runhaar, H., 2014. A method for the deliberate selection of policy instrument mixes for climate change adaptation. Ecol. Soc. 19 (2), 58.

Nickless, E., 2017. Resourcing future generations: a contribution by the earth science community. Nat. Resour. Res. https://doi.org/10.1007/s11053-017-9331-x.

Pattberg, P., 2006. The influence of global business regulation: beyond good corporate conduct. Bus. Soc. Rev. 111 (3), 241−268.

Perrez, F.X., 2000. Cooperative Sovereignty: From Independence to Interdependence in the Structure of International Environmental Law. Kluwer Law International, The Hague.

Ragnarsdottir, K.V., Sverdrup, H.U., Koca, D., 2012. Assessing Long Term Sustainability of Global Supply of Natural Resources and Materials, Sustainable Development − Energy, Engineering and Technologies − Manufacturing and Environment. Section 5, pp. 83−116.

Rankin, W.J., 2011. Minerals, Metals and Sustainability. CSIRO.

Schrijver, N., 2008. Sovereignty over Natural Resources: Balancing Rights and Duties, 1997, reprinted in 2008. Cambridge University Press, Cambridge.

Simon, J.L., 1980. Resources, population, environment: an oversupply of false bad news. Science 208, 1431–1438.

Simon, J.L., 1981. The Ultimate Resource. Princeton University Press, Princeton, NJ.

Söderholm, P., Tilton, J.E., 2012. Material Efficiency, an economic perspective. Resour. Conserv. Recycl. 61, 75–82.

Sverdrup, H.U., Ragnarsdottir, K.V., Koca, D., 2017. An assessment of metal supply sustainability as an input to policy: security of supply extraction rates, stocks-in-use, recycling, and risk of scarcity. J. Clean. Prod. 140, 359–372.

Teisl, M.F., Roe, B., Hicks, R.L., 2002. Can eco-labels tune a market? Evidence from Dolphin-safe labeling. J. Environ. Econ. Manag. 43, 339–359.

Tilton, J.E., Crowson, P.C.F., DeYoung Jr., J.H., Eggert, R.G., Ericsson, M., Guzmán, J.I., Hunphreys, D., Lagos, G., Maxwell, P., Radetzki, M., Singer, D.A., Wellmer, F.W., 2018. Public policy and future mineral supplies. Resour. Pol. https://doi.org/10.1016/j.resourpol.2018.01.006.

UNEP International Panel on Sustainable Resource Management, April 2011. Estimating Long-Run Geological Stocks of Metals. Working Group on Geological Stocks of metals, Working paper.

United Nations, 2015. Declaration of the high level political forum on sustainable development. In: Sustainable Development Goal no 12.2, New York, September 25–27, 2015.

US Geological Survey, 2008. Mineral Commodity Summaries, Indium.

US Geological Survey, 2017a. Historical Statistics for Mineral and Material Commodities in the United States. https://minerals.usgs.gov/minerals/pubs/historical-statistics, 17-11-2017.

US Geological Survey, 2017b. Mineral Commodity Summaries, Bismuth.

US Geological Survey, 2020. Mineral Commodity Summarie.

Valero, A., Valero, A., 2010. Physical geonomics: combining the exergy and Hubbert peak analysis for predicting mineral resources depletion. Resour. Conserv. Recycl. 54, 1074–1083.

Van Amstel, M., Driessen, P., Glasbergen, P., 2008. Eco-labeling and information asymmetry: a comparison of five eco-labels in the Netherlands. J. Clean. Prod. 16, 263–276.

Widerberg, O., 2016. Mapping institutional complexity in the Anthropocene. In: Pattberg, P., Zelli, F. (Eds.), Environmental Politics and Governance in the Anthropocene, Section 6. Routledge, London and New York.

Zelli, F., Pattberg, P., 2016. Complexity, responsibility and urgency in the Anthropocene. Section 14. In: Pattberg, P., Zelli, F. (Eds.), Environmental Politics and Governance in the Anthropocene. Routledge, London and New York.

CHAPTER 9

Setting up an international agreement[1]

9.1 Introduction

Ideally, the approach for solving the problem of geologically scarce mineral resources is global. Geological scarcity is not a local or a regional problem; it is a problem of humanity as a whole, particularly for future generations. However, the climate change problem shows how difficult and time consuming it can be to agree on a workable and practical solution that is acceptable for all countries despite the fact that the technical pathways are known and feasible. Nevertheless, a global approach is the ideal way to safeguard geologically scarce mineral resources for future generations. In addition, countries and regions may decide to go faster and implement unilateral measures to save geologically scarce mineral resources in advance of a global agreement.

The idea of an international agreement to cope with depleting resources is not new. In 2002, *The Oil Depletion Protocol* was drafted (Heinberg, 2006).

In Chapter 8, we described 11 policy instruments to foster a responsible use of scarce primary resources. We concluded that an international agreement on extraction quotas for geologically scarce mineral resources is one of the most effective instruments to achieve the goal of a sustainable use of scarce resources. But, even without an agreement on extraction quotas, most of the other 10 policy instruments will also be more effective in a framework of coordination and cooperation by the international community.

The question arises why there has not been a similar urgency in the negotiation of an international agreement in response to the exhaustion of

---

[1] This chapter is based on the publication of Henckens, M.L.C.M., Driessen, P.P.J., Ryngaert, C.M.J., Worrell, E., 2016. International agreement on the conservation and sustainable use of geologically scarce mineral resources. Res. Pol. 49, 92–101.

Governance of the world's mineral resources
ISBN 978-0-12-823886-8
https://doi.org/10.1016/B978-0-12-823886-8.00029-4

minerals as there has been with environmental issues such as biodiversity and climate change. The explanation might be that mineral resource exhaustion does not directly codetermine the *"safe operating space for humanity"* (terminology from Rockström et al., 2009), but it is primarily an economic problem (with the exception of mineral resources that are essential for life, such as phosphate). Nevertheless, this economic problem may become serious for future generations, if no action is taken. Once the ores of a mineral are exhausted, extraction of that mineral from the earth's crust will become 10–1000 times more expensive (Steen and Borg, 2002).

One could argue that the price mechanism of the free market system could automatically lead to a sufficient reduction of the use of geologically scarce mineral resources due to the inevitable price increase that results from growing scarcity (Dasgupta and Heal, 1979). However, so far, the increasing costs of extraction due to declining ore grades, increasing depths of the mines, more remote mining areas, and smaller ore bodies were neutralized by improving technology (Skinner, 2001; Bardi, 2013; Bleischwitz, 2010). For a long period of time, the real prices of minerals have not increased. Moreover, as shown in Chapter 6, the price development of geologically scarce minerals does not differ from the price development of geologically non-scarce minerals. The conclusion that, for the time being, real prices are not changing is supported by Krautkraemer (1998), Cuddington (2010), and Fernandez (2012). Hence, the market does not yet reflect the large differences of geological scarcity of mineral resources. It remains unclear how closely before exhaustion of a mineral resource the market will react on geological scarcity by structural and permanent price increases of the depleting mineral. It remains also unclear whether, at the near-exhaustion stage, technological development will be able, again, to keep prices down at the same level as nowadays. It is also referred to the essay of Tilton (2003) in this respect. Summarizing, it is not certain whether the geologically scarcest mineral resources will be sufficiently saved for future generations, if humanity does not take measures to achieve a sustainable use of these resources. It is obvious that for a global issue like (future) resources scarcity, policy measures can best be prepared and executed in a global context.

In this section, we focus on the establishment of an international agreement on a production quota system for selected scarce mineral resources. First, we will discuss two different systems, which can be used to achieve and maintain agreed global production quotas for a geologically scarce mineral resource: a system of *cap-and-trade* and a *taxing system* (Section 9.2). In Section 9.3, we will discuss the goals of an international quota system in more detail. An international agreement must be fair for all countries, including resource countries and developing countries. To ensure fairness, we discuss the border conditions of an international agreement on resources production quotas in Section 9.4. The costs and the financial flows in connection with the Agreement are analyzed in Section 9.5. In Section 9.6 we discuss the necessity of the establishment of an International Competence Center on Mineral Resources Management. Finally, in Section 9.7, we present the core elements of a possible agreement. A full draft text for an agreement on the conservation and sustainable use of geologically scarce resources is included in the Annex to this Chapter.

9.2 Regulation of the extraction rate of geologically scarce mineral resources

Once a short list has been determined of the geologically scarcest mineral resources and priority minerals have been selected from that list, extraction regulation goals and policy instruments need to be agreed upon. The question is which policy instrument is most appropriate in this framework. It is generally accepted that market-oriented incentives are more efficient than a command and control approach that directly mandates what businesses or individuals should or should not do (Vogler, 2009; Helm et al., 2003; European Economics, 2008; Gerlagh and Van der Zwaan, 2006; Olmstead and Stavins, 2012; Molyneaux et al., 2010; Goulder and Parry, 2008). The major market-oriented incentives are taxing the production and/or use of geologically scarce mineral resources and "cap-and-trade" systems.

Applied on resource extraction, a cap-and-trade system fixes the maximum amount that is allowed to be extracted, but it allows for flexibility in the pricing of the extracted resource. A tax system influences the price of the extracted resource but leaves the extracted quantity uncertain. There is much literature comparing the two systems with each other. This literature is mostly centered on the merits of either system for greenhouse gas emission reduction. Criteria that are being used have the following three dimensions (i.e., Konidari and Mavrakis, 2007; Mees et al., 2014):

- Performance (e.g., goal achievement, effectiveness, etc.)
- Political acceptability (e.g., cost efficiency, equity/fairness, flexibility, stringency for non-compliance, legal certainty/predictability/credibility, transparency, controversy, etc.)
- Ease of implementation (e.g., feasibility, accountability, transparency, degree of complexity, etc.)

Some authors come to the conclusion that cap and trade is better (Murray et al., 2009; Keohane, 2009). According to other authors, taxing systems are better (Avi-Yonah and Uhlmann, 2009). Several authors plea for hybrid systems, combining taxes and cap-and-trade systems, including price floors and price ceilings, banking, and borrowing (Molyneaux et al., 2010; Mandell, 2008; Vogler, 2009). According to others (Goulder and Perry, 2008; Goulder and Schein, 2013), no instrument is best along all criteria. The conclusion is that a comparison of taxing systems and "cap-and-trade" systems does not unambiguously lead to a clear conclusion that either of the two systems is better.

This Chapter elaborates on a global cap-and-trade system to achieve a sustainable extraction of geologically scarce mineral resources. The argumentation for this choice is that cap and trade offers greater certainty that the required extraction regulation goals are achieved. The cap-and-trade system is successfully applied in the framework of the US Acid Rain Program for the emission reduction of sulfur dioxide, and it is also already being applied at an international scale in the EU Emission Trading System for the reduction of the emission of greenhouse gases. So far, taxing systems for environmental purposes are only employed at a national scale.

Striving after a global cap-and-trade approach does not necessarily hamper or withhold concerned user countries to formulate and implement their own resource saving policies in advance. This will have the advantage of offering flexibility to States as to which policies they pursue, such as policies based on taxing or other instruments such as directly imposing or promoting substitution of geologically scarce mineral resources in selected applications and recycling and arranging (voluntary) agreements with or between sectors of industry or society. To the extent that a global cap-and-trade system would be implemented, such national or regional fall back options may be loosened or abolished again, may work in parallel, or complement a global regime.

9.3 The objective of an international agreement

The objective of an international agreement is to achieve the conservation and sustainable use of geologically scarce mineral resources. Hence, the agreement must address the following issues: (1) the adoption of a definition of the sustainable extraction rate of mineral resources, (2) the selection of scarce mineral resources concerned for which extraction regulation should be considered with priority, (3) the sustainable extraction rate of the selected mineral resources, including the required time span within which the required extraction regulation must be achieved, and (4) the contribution of each of the producing countries. These issues will be elaborated in the following subsections.

9.3.1 Definition of the sustainable extraction rate of mineral resources

A definition for the sustainable extraction rate of mineral resources has been discussed and proposed in Chapter 5: "*The extraction rate of a mineral resource is sustainable, if a world population of 10 billion can be provided with that resource for a period of at least 200/500/1000 years in a way that during that period every country can enjoy the same service level of the resource as developed countries in 2020 for an affordable price.*"

The time period needs to be agreed by the signatories of the agreement.

9.3.2 Selection of priority mineral resources

It is obvious that an international agreement must include the scarcest mineral resources. According to Chapter 3, the eight scarcest resources are (in alphabetic order): antimony, bismuth, boron, copper, gold, indium, molybdenum and silver, which, according to our estimate, may be exhausted within a century. Additionally, other criteria can be taken into account in the selection, such as the extent that a mineral is essential for life, its economic importance, its substitutability, and its recycling potential.

Elements that are essential for life and that cannot be substituted for other elements may need priority compared to elements that are not essential for life, for instance, minerals used in fertilizer and micronutrients. This concerns boron, copper, and molybdenum. In this framework, the exhaustion of phosphate, though not considered in this book, will certainly need special attention as well.

The majority of antimony's applications are in flame retardants. Although antimony is a scarce element, it can be relatively easily substituted in its flame retardant applications. Molybdenum, however, is essential for the production of stainless steel, and so far, molybdenum seems to be hardly substitutable in this application. Therefore, even though molybdenum may be less scarce than antimony, it may get more priority than antimony.

The economic importance of an element for society in general or for specific countries depends on the strategic value of the applications. The overall weighing of these various factors is subjective and is influenced by the economic interests of the involved parties, so the priority setting will necessarily be a political process. The political and societal insights with respect to priority setting may change over time. Therefore, an international agreement on the conservation and sustainable use of geologically scarce mineral resources will need to incorporate a priority setting procedure, which is evaluated periodically. Due to the potentially changing nature of the priority setting and the notoriously difficult processes involved in adopting amendments to treaties, the details of such a procedure are best elaborated in a separate protocol to ensure that it can be more easily altered to adapt to new realities.

9.3.3 Extraction regulation goal and phasing down scheme

How fast must, or can, the required extraction regulation objective be reached? The answer depends on how fast society can change to production of substituting products, more material efficiency, and a higher recycling rate without too much destruction of capital. Looking to history, lessons may be learned from the phasing out of ozone-depleting substances, the phasing out of asbestos, the ban of the use of certain chemicals in certain applications (e.g., polychlorinated biphenyls and cadmium), the transition to cleaner and more economical cars, the emission reduction of greenhouse gasses, and the emission reduction of acidifying air pollutants (e.g., SO_2 and NO_x).

Without going into too much detail, one can say that a phasing down period will be in the order of 5—10 years, at a minimum. Determining factors are the time needed for the technical development of suitable substitutes and recycling technologies, plus the time needed to realize the necessary industrial facilities. Private companies must get sufficient time to amortize existing facilities in order to prevent too much financial loss. The feasibility of a phasing down scheme needs to be separately assessed for each priority mineral resource. Because phasing down schemes are specific and may differ per mineral, these schemes should be elaborated in a separate protocol per mineral. This will need to be reflected in any framework agreement.

9.3.4 Allocation of annual extraction quota to resource countries

For the minerals that are selected for extraction regulation, the capped annual quantities that may be extracted will need to be allocated between the resource countries according to their known reserves.

The globally agreed extraction regulation must go on, irrespective of the discovery of new reserves. This is necessary because the global extraction regulation scheme has already taken into account that most of the extractable resources will not yet have been discovered at the moment of the creation of the extraction regulation. That means that the quotas of mineral extraction that have been allocated to a country will need to decrease proportionally to the additional resource allocation to another country. Extra allocation to a certain resource country means less allocation to the other resource countries.

To prevent the allocation system from being too restrictive, it must allow resource countries to trade within the allocated quotas. By allowing trading of allocated extraction quotas, several objectives are achieved simultaneously: (1) flexibility of the system, (2) the final objectives are maintained, and (3) extraction will take place in countries and mines with the lowest extraction costs. If a resource country cannot deliver, for example, due to accidents, strikes, or geopolitical events, then the other resource countries may be allowed to buy (a part of) the extraction quotas allocated to that country.

Extraction allocation pro rata of proven reserves is a rational approach, but other criteria might be taken into consideration as well, such as extraction capacity, historical extraction share, domestic consumption, production costs, dependence on export, population, and external debt (see, e.g., the discussion within the Organization of the Petroleum Exporting Countries (OPEC) on a quota system for oil production (Sandrea, 2003)).

The reserves need to be evaluated regularly to assess whether these are still in accordance with the assumptions that were at the basis of the extraction regulation scheme. This is the task of a research body that must be installed as part of the international agreement on the conservation and sustainable use of geologically scarce mineral resources. The principle of annual extraction quota per resource country and the tradability of extraction quota among resource countries needs to be included in the framework agreement. The elaboration of the system requires further research and can be part of a separate protocol. In this framework, the experience with existing quotas systems, such as the OPEC system of quotas for oil production, the United Nations Framework Convention on Climate Change (UNFCCC) CO_2 emission quota system, the emission trading arrangements in the framework of UNFCCC, and the EU Emission Trading System, could provide a useful starting point.

9.4 Principles of an agreement on the conservation and sustainable use of geologically scarce mineral resources

The objectives of the international agreement refer to what, in Section 5.3.2., we have referred to as "*goal-orientated principles*" of international environmental agreements. "*Goal-orientated principles*" are the principles that are directly connected with the seriousness of the problem to be solved, such as the principle of sustainable use of resources. These principles need to be distinguished from the "*acceptability principles*" of international environmental agreements, which are related to the architecture and execution of the agreement, including issues such as the sovereign right and equity principles. This Section 9.4 deals with these "*acceptability principles,*" hereinafter referred to as "*principles.*"

Principles are preconditions of an agreement. Without adequate principles accepted by the parties of the agreement, the objectives of an agreement cannot be achieved. The aim of these principles is to satisfactorily comply with the justified interests of the partners of the agreement. Relevant principles for an agreement on the conservation and sustainable use of geologically scarce mineral resources are the sovereign right principle, the common concern of mankind principle, ethical principles (intragenerational equity, priority for the special situation, and needs of developing countries and fairness), responsibility assignment principles (such as the common but differentiated responsibilities [CDR] principle), and the polluter pays principle.

Most of these principles relate to burden sharing in connection with the agreement. The formulation and elaboration of these principles is essentially a political process and can therefore end up being a bottleneck of any agreement. In the below sections, we will elaborate on a number of these acceptability principles.

9.4.1 The sovereign right principle

The sovereign right principle is considered a general principle of international law (Sands et al., 2012). In 1962, the United Nations General Assembly adopted Resolution 1803 (XVII) on the "Permanent Sovereignty over Natural Resources." This happened within the framework of the decolonization process. The resolution "*provides that States and international organizations shall strictly and conscientiously respect the sovereignty of peoples and nations over their natural wealth and resources in accordance with the Charter of the*

United Nations and the principles contained in the resolution. These principles are set out in eight articles concerning, inter alia, *the exploration, development and disposition of natural resources, nationalization and expropriation, foreign investment and other related issues"* (United Nations, 2015). The sovereign right to exploit resources includes the right to be free from interference over their exploitation.

Principle 21 of the Stockholm Declaration (1972) applies the principle of sovereignty over resources providing that *"States have, in accordance with the Charter of the United Nations and the principles of international law, the sovereign right to exploit their own resources pursuant to their own environmental policies, and the responsibility to ensure that activities within their jurisdiction or control do not cause damage to the environment of other States or of areas beyond the limits of national jurisdiction."*

The second part of Principle 21 of the Stockholm Declaration, regarding the responsibility not to cause environmental damage to the environment of other States, does not appear to be directly relevant in the context of the conservation and sustainable use of geologically scarce mineral resources. Mining activities could cause damage to the environment, so reduction of mining activities is beneficial to the environment. Exhaustion of resources causes economic problems in the first place but—as such—no direct damage to the environment.

The wording of the Stockholm Declaration on sovereign rights is repeated in the Rio Declaration (1992) and is considered a basic obligation in international environmental law. The "sovereignty over resources" principle comes back in various forms in many international treaties (for instance, the Ramsar Convention on Wetlands, 1971; the International Tropical Timber Agreement, 1983; the Basel Convention on Wastes, 1989; the UN Framework Convention on Climate Change, 1992; and the Biodiversity Convention, 1992).

9.4.2 The common concern of mankind principle

The principle of *"sovereignty over natural resources"* is counterbalanced by the notion of *"common concern of mankind."* This principle is included in, for example, the preamble of the 1992 UNFCCC (*"Change in the Earth's climate and its adverse effects are a common concern of mankind"*) and in the preamble of the 1992 Biodiversity Convention (*"biological diversity is a common concern of mankind"*). According to the report of the 2014 Washington Conference of the International Law Association on legal principles relating to climate change, the language "*common concern of humankind*" in the preamble of the UNFCCC implies that permanent sovereignty should be exercised for the benefit of humankind, which consists of present and future generations. This point of view is shared by other scholars (Brunnée, 2007; Schrijver, 1997; Perrez, 2000).

Although, so far, exhaustion of geologically scarce mineral resources has not been specifically identified as a "*common concern of mankind*," the "*common concern of mankind*" principle is generally accepted as an approach to address global problems that are not directly related with public goods or "*common heritage*." The principle provides states with a "*legitimate interest in resources of global significance and a common responsibility to assist in their sustainable development*" (Cottier et al., 2014).

9.4.3 Common but differentiated responsibilities

Principle 7 of the 1992 Rio Declaration introduced the concept of CDR: "*In view of the different contributions to global environmental degradation, States have common but differentiated responsibilities. The developed countries acknowledge the responsibility that they bear in the international pursuit of sustainable development in view of the pressures their societies place on the global environment and of the technologies and financial resources they command.*" According the 1992 UN Framework Convention on Climate Change, parties should act to protect the climate system "*on the basis of equality and in accordance with their common but differentiated responsibilities and respective capabilities.*"

Stone (2004) distinguishes three versions of the CDR principle: (1) rational bargaining CDR, (2) equitable CDR, and (3) inefficient CDR. In this context "efficient" is meant in the sense of being Pareto positive: at least one party is better off and no party is worse off.

In the rational bargaining CDR, the negotiators pursue their own advantage. Outcomes are always Pareto positive. The equitable CDR introduces conditions for the outcome of the bargaining process, for example, that the poor must be better off at the end of the negotiations. The overall result is still Pareto positive, meaning that no party is worse off. In the inefficient CDR, it is accepted that some parties may be worse off, for example, the rich parties. In the latter example of an "inefficient CDR," there is a net wealth transfer from rich to poor.

9.4.4 Polluter pays principle in view of exhaustion of geologically scarce minerals

In the framework of exhaustion of geologically scarce mineral resources, the consumer can be considered as the "polluter." With an agreement on the conservation and sustainable use of geologically scarce mineral resources, eventually, the consumer will pay a considerably higher price for products that contain the respective scarce material or for products in which the original scarce material has been substituted. According to Principle 16 of the 1992 Rio Declaration on Environment and Development, *"national authorities should endeavor to promote the internalization of environmental costs and the use of economic instruments, taking into account the approach that the polluter should, in principle, bear the cost of pollution, with due regard to the public interest and without distorting international trade and investment."*

The conclusion is that the price mechanism of the free market system in combination with the measures proposed in Section 9.4.6 will automatically lead to compliance with the internationally accepted responsibility assignment principles:

- Poor countries will be spared compared to rich countries.
- Poor citizens, both in poor and rich countries, will be spared compared to rich citizens.

9.4.5 Compensation of resource countries and establishment of an annually fixed resource price

Without the cooperation of the resource countries, it will be very difficult or even impossible to substantially reduce the extraction of geologically scarce resources within a limited period of time. The system must include a mechanism that makes resource countries wholeheartedly stand behind an agreed extraction regulation goal. If not, there is a serious risk that an extraction regulation that is not supported by all resource countries leads to flooding of the market with scarce resources, decreasing their price and frustrating the objectives of the agreement.

When a resource country loses sovereignty over certain portions of its natural resources (and the related income) by being obliged to reduce the extraction of these resources for the purpose of serving a common concern of mankind, it should be compensated. Compensation of resource countries is justified because not only their sovereign rights but also their income are affected. The price increase that is probably caused by a reduced extraction (increase) of resources is not certain and may not be sufficient to compensate for the decreased production and export of resources. Without a guaranteed compensation for lost income, there is a substantial risk that resource countries may not want to participate in the international agreement, in which they play a crucial role.

The compensation principle is already being brought in practice in the UN REDD compensation programme on Reducing Emissions from Deforestation and Forest Degradation, which compensates (developing) nations for not logging their forests. The compensation for extraction regulation of mineral resources is discounted in an increased resource price. To compensate the resource countries, the resource price needs to be increased proportionally to the decrease of the extraction rate. Therefore, from the moment on that extraction regulation is implemented, a resource tonnage price needs to be fixed annually, directly reflecting the imposed extraction decrease.

The compensation of resource countries must be such that their income *with* an agreement on the conservation and sustainable use of geologically scarce resources is equal to their estimated income *without* such an agreement. A resource country should not get more compensation than the income that it would have received without international agreement. Corrections of the general approach may be necessary because the reserves in a country may be near to exhaustion. It is not necessary to compensate a country for lost income when this country would not have had this income anyway. The principle of the compensation mechanism needs to be part of the framework agreement. The detailed elaboration of the mechanism can be arranged in a separate protocol, since the outcome will be partly the result of political negotiations and may change over time.

9.4.6 Addressing the special situation and needs of developing countries

The solution of the geological scarcity problem may encompass substantial extra costs for all countries, including poor countries. Increasing scarcity of mineral resources is mainly caused by developed countries. Developed countries were able to generate welfare for their people and to build up a physical infrastructure co-based on the massive extraction and use of mineral resources. When limiting the further extraction of geologically scarce mineral resources, developing countries would be confronted with substantially higher costs for these resources, even though they have not yet been able to build up an infrastructure at the same level as developed countries and the majority of people in their societies have not yet been able to enjoy the services provided by these resources to the same extent of an average citizen of a developed country. It will be difficult to expect that developing countries would unconditionally agree with a system that leads to higher costs for their inhabitants without considering that they are not responsible for geological scarcity in the first place. Moreover, without the consent of these countries, an international agreement on the conservation and sustainable use of geologically scarce mineral resources cannot be globally ratified. This means that the role and position of developing countries needs special attention. The interest of the developing countries in this perspective is twofold: (a) that their access to geologically scarce mineral resources remains attainable and (b) that the costs for solving the scarcity problem are acceptable for the developing countries from a historical perspective.

(a) Equitable distribution of geologically scarce mineral resources

Geologically scarce mineral resources can be considered as part of the *"ecological space for humankind."* The right to equitably share the ecological space of mankind can be considered a fundamental human right (Hayward, 2006). From a certain point on, when a resource becomes very scarce, it looks justified to take steps to equitably share such a resource. There are several ways in which geologically scarce mineral resources could be equitably distributed to countries, for example:

-   Equal amount per capita
-   Equal amount per unit of GDP
-   Grandfathering: the distribution is based on the amount of resource that is used in a country in a reference year

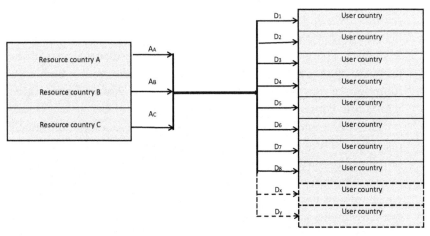

Figure 9.1 From extraction to distribution. *A*, allocated amount of resources that may be extracted by resource countries. *D*, distributed amount of resources to user countries.

- Contraction and convergence: convergence from the status quo to equal per capita amounts of the scarce resource over an agreed period of time
- Contraction and convergence with accounting of past use, for example, from 1990. Countries that have used more than the global average use per capita until an agreed moment in time have to reduce their consumption by this amount in a later period

Pan et al. (2014) provide an overview of 20 alternative allocation schemes for CO_2 emission rights which—in principle—are also applicable for the ways in which geologically scarce mineral resources can be distributed. The distribution system will be the subject of bargaining during the genesis of the international agreement. The agreed distribution system will be laid down in a separate protocol.

(b) Lower costs for developing countries

Each of the user countries—developed or developing—will pay the same fixed tonnage price for the amount of resources that is distributed to it. This resource tonnage price is annually fixed by an international body in the framework of the agreement and is universal for all countries. User countries will be allowed to trade the allocated quota of mineral resources. They may sell the resource for a price that the market (usually mineral processing companies) is prepared to pay. Because of the global extraction regulation, scarcity is artificially increased during the phasing down period. The market price of the resource, which the user countries will be able to receive, will probably become higher than the fixed tonnage price that the user countries have paid. The gross effect of this system will be that the costs of the proposed system are relatively higher for countries with a high consumption of the scarce material and relatively lower for countries with a low consumption of the scarce material.

The per capita consumption of mineral resources is positively related to GDP (Graedel and Cao, 2010). The per capita consumption of mineral resources in developing countries is lower than the per capita consumption of mineral resources in developed countries. With an equal amount per capita distribution system, this implies that—under the agreement, and initially, as long as they are in a development stage—developing countries would get more mineral resources distributed to them than they actually use or need. This creates a net profit for these countries. In this way, developing countries are compensated for the higher costs of the services of the resource in the future and for their contribution to saving of geologically scarce mineral resources currently and in the past.

The extraction quotas that are allocated to resource countries must be distinguished from the distribution quotas of extracted resources that are allocated to user countries (Fig. 9.1). In the proposed system, both types of quotas are tradable. The allocated quotas that are extracted by resource countries may be traded between resource countries; the resources that are distributed to user countries after extraction may be traded on the market.

Although geologically scarce mineral resources are equitably distributed to countries, it will remain the sovereign right of each country to determine how to manage the distribution of costs and benefits in its own country for its own citizens.

9.5 Costs

The costs of an international agreement on the conservation and sustainable use of geologically scarce mineral resources can be split up in three elements:

- The higher costs for substituents, increased recycling, and better material efficiency compared to the current situation and the higher market price for the original resource: These costs will vary per mineral resource and per application, and it is not easy to make a precise estimate of these costs in general. The optimal mix of substitution, material efficiency measures, and recycling will differ per resource and per application. Moreover these costs will depend on the required sustainable extraction rate. This type of costs will be paid by the ultimate consumers of the products, within which the scarce resources are included. This is fair because in this way the extra costs will be distributed according to the use of the resource. The economical consumer will incur fewer costs than the wasting consumer.
- The additional costs to compensate the resource countries for their loss of income: These costs will be incurred by the user countries and must be included in the annually fixed tonnage price that the user countries pay to the resource countries (Section 9.4.6). The annually fixed tonnage price is paid to the administrative body that is in charge of the execution of the agreement and transferred by the administrative

body to the resource countries. In return for the paid compensation costs, the user countries should gradually and proportionally become owner of the saved reserves. In this way, from a certain moment on, the remaining reserves are owned by the user countries and compensation does not need to be paid anymore.

- The costs for the international administrative bodies that will be in charge of the implementation and monitoring of the agreement: These are called the transaction costs. Usually these types of costs will not be permitted to exceed some percent of the total market value of the original amount of resources on an annual basis. These costs must be shared by the user countries in proportion to the amount of resource distributed to each of them. These costs are paid together with the fixed tonnage price to the international body that is in charge with the execution of the agreement.

The buyers (the processing industry) will pay the market price to the user countries. However, it will be necessary to establish a minimum for the market price for preventing the market price to become lower than the annually fixed tonnage price. The minimum price should be equal to the annual fixed tonnage price plus the transaction costs. The purpose of setting a minimum price is to ensure that a user country, whatever the market price will be, will not suffer any costs if it sells all scarce resources that it received through the distribution system.

The mechanism is presented is Fig. 9.2.

The financial impact of the proposed system on developing countries as compared to the impact on developed countries is elaborated via a concrete example in Table 9.1. The conclusion is that the system will, in principle, lead to a relative profit for countries with a low per capita use of the resource compared to countries with a high per capita use of the resource.

9.6 Necessary first step: establishing an International Competence Center on Mineral Resources Management

Global governance of the use of mineral resources requires thoughtful technical and scientific preparation at global scale. We believe that a prerequisite for a future global mineral resources governance system, including an Agreement on the Conservation and Sustainable Use of Mineral

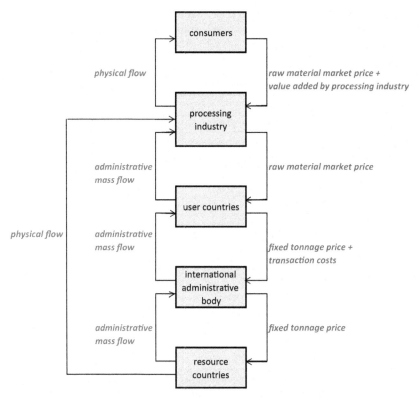

Figure 9.2 Physical mass flows, administrative mass flows, and financial flows in the context of the proposed international agreement on the conservation and sustainable use of geologically scarce mineral resources.

Resources, is to establish an International Competence Center on Mineral Resources Management, following the example of the Intergovernmental Panel on Climate Change, which has raised global attention for climate change and awareness of the urgency of addressing this problem and has galvanized and coordinated thousands of scientists to address the problem. Another positive example of international coordination is the Ozone Secretariat, which has been established in the framework of Montreal Protocol on Substances that Deplete the Ozone Layer.

The main tasks of an International Competence Center on Mineral Resources Management would be (Bringezu et al., 2016) as follows:
- Keeping estimates of extractable global resources up to date
- Monitoring global resources use

Table 9.1 Expenses and income of a poor country (A) and a rich country (B) connected with the use of a geologically scarce mineral resource. Assumptions are (1) distribution to user countries in year x: 1kg/capita; (2) fixed price to be paid to resource countries (including transaction and compensation costs): 1 US$ per kg raw material; (3) market price of raw material: 2 US$ per kg; and (4) product price: 4 US$ per kg equivalent of the raw material.[2]

	Expense	Income	Net costs
Country A with 100 million people and an annual resource use in products of 0.5 kg/capita	- Payment for distributed resource: $1*10^8$ US$ - Payment for products: $2*10^8$ US$ - Total expense: $3*10^8$ US$	Sale of distributed resource to processing industry: $2*10^8$ US$	$1*10^8$ US$
Country B with 100 million people and an annual resource use in products of 5 kg/capita	- Payment for distributed resource: $1*10^8$ US$ - Payment for products: $20*10^8$ US$ - Total expense: $21*10^8$ US$	Sale of distributed resource to processing industry: $2*10^8$ US$	$19*10^8$ US$

Assuming an original market price of 2 US$ per kg of the material, the original annual cost of (low use) country A was 50 million x 2 = 100 million US$. After the implementation of the agreement, the net costs are 100 million as well. Country B (high use) originally paid 500 million x 2 = 1 billion US$ annually for the raw material. After the agreement, this is 1.9 billion US$, which is almost twice as much.
[2]Kg equivalent of raw material represents the value of the services that originally (before extraction regulation) was delivered by 1 kg of raw material.

- Establishing a database of extractable global resources estimates and global resources use
- Developing a Global Mineral Resources Management Program

The nucleus of the Competence Center could be the International Resources Panel on the Sustainable Use of Natural Resources, set up in 2005 on the initiative of the European Commission and United Nations Environment Programme (European Commission, 2005). Without the application of other instruments described below, establishing such a Center alone will not have a big impact, but the Center is a prerequisite for prioritizing, elaborating, and creating the other policy instruments and for making them effective.

9.7 Core elements of an international agreement on the conservation and use of geologically scarce mineral resources

The present section examines the setup of an international agreement on the conservation and sustainable use of geologically scarce mineral resources. The focus is on the objectives, principles, setup, and mechanisms of the agreement.

9.7.1 Objectives

The objective of an international agreement on the conservation and sustainable use of geologically scarce mineral resources is *the sustainable extraction rate of geologically scarce mineral resources to ensure that geologically scarce mineral resources are equitably distributed between the current generation and future generations.*

9.7.2 Basic principles

The principles, on which an agreement on the conservation and sustainable use of geologically scarce mineral resources is based, are as follows:
- The sovereign right principle
- The concern to mankind principle
- Compensation of the resource countries for their willingness to reduce the extraction of geologically scarce mineral resources
- Equitable distribution of geologically scarce mineral resources to the world's countries

9.7.3 Setup

The agreement encompasses a cap-and-trade system and consists of the following:
- A priority setting methodology that results in a list of geologically scarce mineral resources, of which the extraction must be regulated with priority
- A procedure for the determination of a sustainable production level, an extraction regulation goal, and a phasing down scheme for each of the selected mineral resources
- A system for setting annual extraction quotas of the selected mineral resources and the allocation of these quotas to the resource countries
- A system of equitable distribution of geologically scarce mineral resources to user countries for a fixed price per ton

9.7.4 Financial mechanisms

The financial mechanism of the proposed agreement consists of the following: ´

- A system of compensation of resource countries by user countries for reducing the extraction of geologically scarce mineral resources
- A system to annually fix a resource price that includes the compensation for the resource countries and the transaction costs for the implementation of the international agreement
- A system that makes the user countries owner of the not extracted mineral resources to the extent that they have paid compensation to the resource countries
- The right for the resource countries to trade the extraction quotas between the resource countries
- The right for the user countries to sell the distributed resources on the free market
- An international body for the conservation and sustainable use of geologically scarce resources that is responsible for the transfer and appropriate administration of the necessary payments to the resource countries and for inspection, monitoring, evaluation, and research

9.7.5 Protocols

For each of the above system design elements and financial mechanisms, separate protocols will be needed to elaborate the agreed issues.

9.7.6 Institutional bodies

The following institutional bodies will need to be setup:
a. Conference of Parties for international cooperation and decision taking
b. Secretariat
c. Administrative body for scientific and technological advice: the International Competence Center on Mineral Resources Management
d. Body on implementation, monitoring, and evaluation

A full draft text of a framework Agreement on the Conservation and Sustainable Use of Geologically Scarce Mineral Resources is included in the Annex to this section.

9.8 Further research needed

In this Chapter, the main lines of an international agreement on the conservation and sustainable use of geologically scarce mineral resources were laid out. The details of the proposed agreement need further research. This concerns especially the following:

- The factors that (may) hamper the genesis of an international agreement on the conservation and sustainable use of geologically scarce mineral resources and how these hampering factors could be addressed
- An analysis of the interests of various partners to the agreement (resource countries and user countries, developed countries, and developing countries) and other stakeholders (mining companies, processing industry)
- How to deal with countries that do not ratify the agreement. The draft framework agreement in the Annex contains a tentative article on this subject, but this approach needs more research
- The methodologies for priority setting of geologically scarce mineral resources for the establishment of sustainable production levels and the determination of the extraction regulation goals and for the phasing down schemes
- The system for the allocation of annual extraction quotas to resource countries
- The system of distribution of regulated mineral resources to user countries
- The system of compensation of resource countries for their loss of export opportunities. In this framework, existing compensation schemes, such as the REDD compensation scheme, need to be evaluated
- The setup of a system of periodic evaluation of extractable reserves per resource country
- The ownership system of non-extracted resources
- The lessons that may be drawn from the creation, the implementation, and the execution of existing international environmental agreements and other relevant mechanisms, such as existing quota systems (oil-OPEC, CO_2-UNFCCC), emission trading schemes (UNFCCC and EU ETS), and compensation schemes (REDD)
- A study on how to harmonize an international agreement on the conservation and sustainable use of geologically scarce resources with existing WTO agreements.

References

Avi-Yonah, R.S., Uhlmann, D.M., 2009. Combating Global Climate Change: why a carbon tax is a better response to global warming than cap and trade. Stanford Environ. Law J. 28 (3), 1—50.

Bardi, U., 2013. Plundering the Planet. 33rd report for the Club of Rome.

Bleischwitz, R., 2010. International economics of resource productivity. Int. Econ. Pol. 7, 227−244.

Bringezu, S., Potocnik, J., Schandl, H., Lu, Y., Ramaswami, A., Swilling, M., Suh, S., et al., 2016. Multi-scale governance of sustainable natural resource use - Challenges and opportunities for monitoring and institutional development at the national and global level. Sustainability 8, 778. https://doi.org/10.3390/su8080778.

Brunnée, J., 2007. Common areas, common heritage and common concern. In: Bodansky, D., Brunnée, J., Hey, E. (Eds.), Oxford Handbook of International Environmental Law, 2007. Oxford University Press, Oxford.

Cottier, T., Aerni, P., Matteotti, S., de Sépibus, J., Shingal, A., 2014. The principle of common concern and climate change. In: NCCR Trade Regulation, Swiss National Centre of Competence in Research, Working Paper No 2014/18, June 2014. http://www.nccr-trade.org/fileadmin/user_upload/nccr-trade.ch/wp5/publications/Cottier_et_al_Common_Concern_and_Climate_Change_Archiv_final_0514.pdf.

Cuddington, J.T., 2010. Long term trends in the real real process of primary commodities: inflation bias and the Prebisch-Singer hypothesis. Resour. Pol. 35, 72−76.

Dasgupta, P., Heal, G., 1979. Economic Theory and Exhaustible Resources. University Press, Cambridge.

Europe Economics, January 2008. A Comparison of the Costs of Alternative Policies for Reducing UK Carbon Emissions. www.europe-economics.com.

European Commission, 2005. Thematic Strategy on the sustainable use of natural resources. Communication from the Commission to the Council, the European Parliament, the European Economic and Social Committee of the Regions. COM(2005) 670 final.

Fernandez, V., 2012. Trends in real commodity prices: how real is real? Resour. Pol. 37, 30−47.

Gerlagh, R., Van der Zwaan, B., 2006. Options and instruments for a deep cut in CO_2 emissions: carbon dioxide capture or renewables, taxes or subsidies. Energy J. 27 (3), 25−48.

Goulder, H.L., Parry, W.H., 2008. Instrument choice in environmental policy. Rev. Environ. Econ. Pol. 2 (2), 152−174.

Goulder, L.H., Schein, A., 2013. Carbon Taxes vs. Cap and Trade: A Critical Review. Working Paper 19338 of the National Bureau of Economic Research, Cambridge, MA. http://www.nber.org/papers/w19338.

Graedel, T.E., Cao, J., 2010. Metal spectra as indicators for development. Proc. Natl. Acad. Sci 107, 4920905−4920910.

Hayward, T., April 2006. Human rights versus emission rights: climate justice and the equitable distribution of ecological space. In: Global Justice and Climate Change Conference. Institute for Ethics and Public Affairs, San Diego State University.

Heinberg, R., 2006. The Oil Depletion Protocol. A Plan to avert oil wars, terrorism, and economic collapse, Clairview.

Helm, D., Hepburn, C., Mash, R., 2003. Credible carbon policy. Oxf. Rev. Econ. Pol. 19 (3), 2003.

International Law Association, 2014. Legal Principles Relating to Climate Change. Washington Conference.

Keohane, N.O., 2009. Cap and trade, rehabilitated: using tradable permits to control U.S. greenhouse gases. Rev. Environ. Econ. Pol. 1−21.

Konidari, P., Mavrakis, D.A., 2007. multi-criteria evaluation method for climate change mitigation policy instruments. Energy Pol. 35, 6235−6257.

Krautkraemer, J.A., 1998. Nonrenewable resources scarcity. J. Econ. Lit. 36 (4), 2065–2107.

Mandell, S., 2008. Optimal mix of emissions taxes and cap-and-trade. J. Environ. Econ. Manag. 56, 131–140.

Mees, H.L.P., Dijk, J., Van Soest, D., Driessen, P.P.J., Van Rijswick, M.H.F.M., Runhaar, H., 2014. A method for the deliberate and the deliberate selection of policy instrument mixes for climate change adaptation. Ecol. Soc. 19 (2), 58.

Molyneaux, L., Foster, J., Wagner, L., December 2010. Is There a More Effective Way to Reduce Carbon Missions? Energy and management Group, School of Economics, University of Queensland, Brisbane.

Murray, B.C., Newell, R.G., Pizer, W.A., 2009. Balancing cost and emissions certainty: an allowance reserve for cap-and-trade. Rev. Environ. Econ. Pol. 1–20.

Olmstead, S.M., Stavins, R.N., 2012. Three key elements of a post-2012 international climate policy architecture. Rev. Environ. Econ. Pol. 6 (1), 65–85.

Pan, X., Teng, F., Ha, Y., Wang, G., 2014. Equitable access to sustainable development: based on the comparative study of carbon emission rights allocation schemes. Appl. Energy 130, 632–640.

Perrez, F.X., 2000. Cooperative Sovereignty: From Independence to Interdependence in the Structure of International Environmental Law, 2000. Kluwer Law International, The Hague.

Rockström, J., Steffen, W., Noone, K., Persson, A., Stuart Chapin III, F., Lambin, E.F., Lenton, T.M., Scheffer, M., Folke, C., Schellnhuber, H.J., Nykvist, B., de Wit, C.A., Hughes, T., van der Leeuw, S., Rodhe, H., Sörlin, S., Snyder, P.K., Costanza, R., Svedin, U., Falkenmark, M., Karlberg, L., Corell, R.W., Fabry, V.J., Hansen, J., Walker, B., Liverman, D., Richardson, K., Crutzen, P., Foley, J.A., September 24, 2009. A safe operating space for humanity. Nature 461.

Sandrea, R., July 28, 2003. OPEC's challenge: rethinking its quota system. Oil Gas J. 29. http://www.ogj.com/articles/print/volume-101/issue-29/general-interest/opecs-challenge-re thinking-its-quota-system.html. (Accessed January 2016).

Sands, P., Peel, J., Fabra, A., MacKenzie, R., 2012. Principles of International Environmental Law, 2012. Cambridge University Press.

Schrijver, N., 1997. Sovereignty over Natural Resources: Balancing Rights and Duties. Cambridge University Press, Cambridge. Reprinted in 2008.

Skinner, B.J., 2001. Long run availability of minerals, keynote talk at a workshop "Exploring the resource base". In: Washington DC, April 22–23, 2001, Resources for the Future.

Steen, B., Borg, G., 2002. An estimation of the cost of sustainable production of metal concentrates from the Earth's crust. Ecol. Econ. 42, 401–413.

Stone, C.D., 2004. Common but differentiated responsibilities in international law. Am. J. Int. Law 98, 276–301.

Tilton, J.E., 2003. On borrowed time? Assessing the threat of mineral exhaustion. Miner. Energy - Raw Mater. Rep. 18 (1), 33–42.

United Nations, October 5, 2015. http://legal.un.org/avl/ha/ga_1803.html.

Vogler, N., 2009. Policy Option Issues for CO_2 Emissions. Nova Science Publishers, Inc., New York.

ANNEX to Chapter 9

Draft Framework Agreement on the Conservation and Sustainable Use of Geologically Scarce Mineral Resources

Preamble

The Parties to *this* Agreement

Acknowledging that the conservation and sustainable use of geologically scarce mineral resources is a common concern of mankind,

Concerned about the exhaustion of a number of geologically scarce mineral resources, such as, but not limited to, antimony, bismuth, boron, copper, gold, indium, molybdenum, rhenium and silver,

Aware of the urgent need to reduce the extraction of geologically scarce mineral resources,

Bearing in mind the interests of future generations,

Reaffirming the principle of sovereignty of States in international cooperation to address the exhaustion of mineral resources,

Noting the principle of common but differentiated responsibilities in addressing the exhaustion of mineral resources, and emphasizing the need to take into account the interests of developing States,

Recognizing the need for suitable substitutes for geologically scarce mineral resources, and of recycling technologies,

Noting the suitability of market mechanisms to address global governance failures, while recognizing that the sole reliance on market mechanisms may not automatically lead to the timely and sufficient reduction of the use of geologically scarce mineral resources.

Have agreed as follows:

Article 1—definitions

For the purpose of this Agreement:

1. The extraction rate of a mineral resource is defined as sustainable if a world population of 10 billion people can be provided with the resource for a period of at least 200/500/1000 years[3] in a way that during that period every country can enjoy the same service level of that resource as the developed world in 2020 for an affordable price.

---

[3] Time period to be established by the Conference of Parties.

2. The Resource State is defined as the State on whose territory extraction of mineral resources takes place.
3. The User State is defined as the State on whose territory mineral resources are used, consumed, or processed. Every State is considered to be a user State.

Article 2—objective

The ultimate objective of this Agreement and any related legal instruments that the Conference of the Parties may adopt is to achieve, in accordance with the relevant provisions of the Agreement, the conservation and sustainable use of geologically scarce mineral resources.

Article 3—principles

In their actions to achieve the objective of the Agreement and to implement its provisions, the Parties shall be guided, inter alia, by the following:

1. The Parties commit themselves to the conservation and sustainable use of geologically scarce mineral resources for the benefit of present and future generations of humankind, on the basis of equity and in accordance with their common but differentiated responsibilities and respective capabilities.
2. The specific needs and special circumstances of developing country Parties should be given full consideration.
3. Market mechanisms, in particular, the tradability of extraction quota by resource States and distribution quota by user States, can ensure efficient conservation and sustainable use of geologically scare mineral resources, as well as equity.

Article 4—priority setting and sustainable extraction level

At its first review session, the Conference of the Parties will decide on the definition of a sustainable extraction level and on which minerals should receive priority as far as extraction regulation is concerned.

For each of the selected minerals, the Conference of Parties will establish an extraction regulation goal. The decisions on priorities and goals will be laid down in a separate Protocol.

Article 5—phasing down scheme

The Parties shall phase down the extraction of geologically scarce mineral resources. At its first review session, the Conference of the Parties will adopt a phasing down scheme for each selected mineral, to be laid down in separate Protocols. A phasing down period lasts at least 5 years. In making its determination, the Conference will take into account such factors as the time needed for the technical development of suitable substitutes and recycling technologies, as well as the time needed to realize the necessary new industrial facilities and to amortize existing facilities.

Article 6—extraction regulation

1. The Parties shall regulate the rate of extraction so as to render extraction sustainable. To this effect, the Parties shall allocate annual extraction quota to the resource States. At its first review session, the Conference of the Parties shall determine an allocation system for extraction quota and the annual extraction quotas. The allocation system and the annual extraction quotas will be laid down in separate Protocols.
2. The extraction quotas are tradable among resource States.

Article 7—compensation of resource states

1. The Parties shall compensate the resource States for the loss of income as a result of the extraction limitations implemented pursuant to Article 6 of the Agreement.
2. The resource States shall receive full compensation for the costs of extraction regulation. Full compensation means that the total income of a resource State generated by resource extraction must remain approximately equal to the income that would have been generated without agreement.
3. Compensation will be ensured through a fixed resource tonnage price, annually set by the Conference of the Parties on the basis of advice given by the Subsidiary Body for Scientific and Technological Advice. Compensation payments will be administered by the Subsidiary Body for Implementation.
4. At its first review session, the Conference of the Parties will make arrangements to implement these provisions.

Article 8—distribution of extracted resources to user States

1. The amount of extracted resources shall be distributed to user States on the basis of an equitable formula to be adopted by the Conference of the Parties at its first review session and laid down in a separate Protocol.
2. User States shall pay a fixed tonnage price, decided in accordance with the procedure set out in Article 7.3 of this Agreement, for the amount of resources distributed to it. Payments are administrated by the Subsidiary Body for Implementation.
3. User States are allowed to trade the mineral resources distributed to them.

Article 9—control of trade with non-parties

1. Each State party shall ban the export of geologically scarce minerals falling within the scope of this Agreement to any State not party to the Agreement.
2. Each State party shall ban the import of geologically scarce minerals falling within the scope of this Agreement from any State not party to the Agreement.
3. For the purposes of this Article, the term "State not party to this Agreement" shall include, with respect to a particular mineral, a State that has not agreed to be bound by the Agreement.

Article 10—Conference of the parties

1. A Conference of the Parties is hereby established.
2. The Conference of the Parties, as the supreme body of this Agreement, shall keep under regular review the implementation of the Agreement and any related legal instruments that the Conference of the Parties may adopt, and shall make, within its mandate, the decisions necessary to promote the effective implementation of the Agreement. To this end, it shall:
   (a) Adopt the necessary Protocols to implement this Agreement, in particular with respect to priority setting (Article 4), the phasing down scheme (Article 5), the extraction regulations (Article 6), a system for allocation of annual extraction quota to resource States (Article 6), the compensation of resource States (Article 7), and a system of distribution of extracted resources to user States (Article 8);

(b) Periodically examine the obligations of the Parties and the institutional arrangements under the Agreement, in the light of the objective of the Agreement, the experience gained in its implementation; and the evolution of scientific and technological knowledge;

(c) Assess the implementation of the Agreement by the Parties and the overall effects of the measures taken pursuant to the Agreement;

(d) Review reports submitted by its subsidiary bodies and provide guidance to them;

(e) Agree upon and adopt, by consensus, rules of procedure and financial rules for itself and for any subsidiary bodies;

(f) Exercise such other functions as are required for the achievement of the objective of the Agreement as well as all other functions assigned to it under the Agreement.

Article 11—secretariat

1. A secretariat is hereby established.

2. The functions of the secretariat shall be:

(a) To make arrangements for sessions of the Conference of the Parties and its subsidiary bodies established under the Agreement and to provide them with services as required;

(b) To compile and transmit reports submitted to it;

(c) To facilitate assistance to the Parties, on request, in the compilation and communication of information required in accordance with the provisions of the Agreement;

(d) To prepare reports on its activities and present them to the Conference of the Parties.

3. The Conference of the Parties, at its first session, shall designate a permanent secretariat and make arrangements for its functioning.

Article 12—subsidiary body for scientific and technological advice

1. A subsidiary body for scientific and technological advice is hereby established to provide the Conference of the Parties and, as appropriate, its other subsidiary bodies with timely information and advice on scientific and technological matters relating to the Agreement. This body shall be open to participation by all Parties and shall be multidisciplinary. It shall

comprise government representatives competent in the relevant field of expertise. It shall report regularly to the Conference of the Parties on all aspects of its work.

2. Under the guidance of the Conference of the Parties, and drawing upon existing competent international bodies, this body shall:
   (a) Provide assessments of the state of scientific knowledge relating to the sustainable extraction of mineral resources;
   (b) Evaluate whether the extractable reserves remain in accordance with the assumptions that were the basis of the extraction regulation scheme;
   (c) Prepare scientific assessments on the effects of measures taken in the implementation of the Agreement;
   (d) Respond to scientific, technological, and methodological questions that the Conference of the Parties and its subsidiary bodies may put to the body.
3. The functions and terms of reference of this body may be further elaborated by the Conference of the Parties.

Article 13—subsidiary body for implementation

1. A subsidiary body for implementation is hereby established to assist the Conference of the Parties in the assessment and review of the effective implementation of the Agreement. This body shall be open to participation by all Parties and comprise government representatives who are experts on matters related to the sustainable extraction of geologically scarce mineral resources. It shall report regularly to the Conference of the Parties on all aspects of its work.
2. Under the guidance of the Conference of the Parties, this body shall:
   (a) Administer payments made by user States in accordance with Article 8.2 and compensation payments made to resource States in accordance with Article 7 section;
   (b) Monitor compliance by the resource States with the extraction limitations imposed on the basis of Article 6 section;
   (c) Assess the overall aggregated effect of the steps taken by the Parties to reduce the extraction of mineral resources in accordance with the extraction limitations agreed on by the Parties, on the basis of Article 6 section of the Agreement;
   (d) Assist the Conference of the Parties, as appropriate, in the preparation and implementation of its decisions.

Article 14—financial arrangements

1. The operational expenses of the Secretariat and the various subsidiary bodies established on the basis of this agreement shall be covered by a transaction fee imposed on the user States.
2. The transaction fee shall be calculated as a percentage of the fixed tonnage price per mineral.
3. The Conference of the Parties will decide on the amount of the transaction fee per mineral, taking into account the number of resource States, the total size of the market, and the number of scarce resources for which extraction regulation is required.

Article 15—international cooperation

1. Parties shall share information and knowledge regarding the reserves of extractable resources on their territory with other Parties, as well with the Subsidiary Body for Scientific and Technological Advice.

Article 16—settlement of disputes

1. In the event of a dispute between any two or more Parties concerning the interpretation or application of the Agreement, the Parties concerned shall seek a settlement of the dispute through negotiation or any other peaceful means of their own choice.
2. When ratifying, accepting, approving, or acceding to the Agreement, or at any time thereafter, a Party may declare in a written instrument submitted to the Depositary that, in respect of any dispute concerning the interpretation or application of the Agreement, it recognizes as compulsory ipso facto and without special agreement, in relation to any Party accepting the same obligation:
   (a) Submission of the dispute to the International Court of Justice, and/or
   (b) Arbitration in accordance with procedures to be adopted by the Conference of the Parties as soon as practicable, in an annex on arbitration.
3. The provisions of this Article shall apply to any related legal instrument that the Conference of the Parties may adopt, unless the instrument provides otherwise.

Article 17—amendments to the agreement

1. Any Party may propose amendments to the Agreement.
2. Amendments to the Agreement shall be adopted at an ordinary session of the Conference of the Parties.
3. The Parties shall make every effort to reach agreement on any proposed amendment to the Agreement by consensus. If all efforts at consensus have been exhausted, and no agreement reached, the amendment shall as a last resort be adopted by a three-fourths majority vote of the Parties present and voting at the meeting.

Article 18—protocols

1. The Conference of the Parties may, at any ordinary session, adopt protocols to the Agreement.
2. The requirements for the entry into force of any protocol shall be established by that instrument.
3. Only Parties to the Agreement may be Parties to a protocol.

Article 19—Depositary

The Secretary-General of the United Nations shall be the Depositary of the Agreement and of protocols adopted in accordance with Article 15 section.

Article 20—signature

This Agreement shall be open for signature by States Members of the United Nations or of any of its specialized agencies or that are Parties to the Statute of the International Court of Justice.

Article 21—ratification, acceptance, approval, or accession

The Agreement shall be subject to ratification, acceptance, approval, or accession by States. It shall be open for accession from the day after the date on which the Agreement is closed for signature. Instruments of ratification, acceptance, approval, or accession shall be deposited with the Depositary.

Article 22—entry into force

The Agreement shall enter into force on the 90th day after the date of deposit of the 50th instrument of ratification, acceptance, approval, or accession.

Article 23—reservations

No reservations may be made to the Agreement.

Article 24—withdrawal

At any time after 3 years from the date on which the Agreement has entered into force for a Party, that Party may withdraw from the Agreement by giving written notification to the Depositary.

Article 25—authentic texts

The original of this Agreement, of which the Arabic, Chinese, English, French, Russian, and Spanish texts are equally authentic, shall be deposited with the Secretary-General of the United Nations.

Epilogue

The main question asked in this book is whether it is possible to keep providing sufficient mineral resources for a global population of 10 billion people; whether this is possible for a period of at least 200 years but preferably 1000 years; whether this is possible while simultaneously realizing a global service level of the mineral resources that is equal to the service level in developed countries in 2020; and, finally, whether this is possible while maintaining affordable mineral resource prices. In other words, how can the *sustainable use* of mineral resources be realized?

A part of the answer is that, for many mineral resources, exhaustion is not yet really a prospect. On the other hand, a small number of mineral resources are so scarce in comparison to the annual extraction rate that humanity needs to take measures to conserve them for future generations. In this book, we have analyzed which measures can be taken for the 13 mineral resources with the relatively highest scarcity, while simultaneously realizing a global service level of these resources equal to the service level of these resources in developed countries in 2020. These resources are, in alphabetical order: antimony, bismuth, boron, chromium, copper, gold, indium, molybdenum, nickel, silver, tin, tungsten, and zinc.

Table 1 presents an overview of the achievability of the sustainability goals resulting from the analyses in Sections 7.2–7.14.

We conclude the following.

If we assume the lowest estimates for the ultimately available resources:
-   A sustainability ambition of 1000 years can only be achieved for boron, tungsten, and zinc, but not for the ten other materials considered.
-   A sustainability ambition of 500 years can be achieved not only for boron, tungsten, and zinc, but also for antimony, chromium, and indium, in the case of boron without taking any measures; however, it is still not achieved for the other seven considered materials.
-   A sustainability ambition of 200 years can be achieved for all materials except for molybdenum, and nickel, and in the case of boron and tungsten without taking any measures.

Table 1 The achievability of the sustainability goals for the 13 scarcest raw materials.

	Sustainability ambition (years of availability)					
	Estimates of ultimately available resources[a]					
	1000		500		200	
	Low	High	Low	High	Low	High
Antimony	−b	+b	+	+	+	+
Bismuth	−	−	−	+	+	++b
Boron	−	+	+	++	++	++
Chromium	−	++	−	++	+	++
Copper	−	++	−	++	+	++
Gold for technological applications	−	++	−	++	+	++
Indium	−	++	+	++	+	++
Molybdenum	−	−	−	−	−	+
Nickel	−	+	−	++	−	++
Silver for technological applications	−	−	−	+	+	++
Tin	−	+	−	+	+	++
Tungsten	+	+	+	+	++	++
Zinc	+	+	+	++	+	++

[a]See Sections 7.2–7.14.
[b]−, not achievable; +, achievable with measures; ++, achievable without measures

If we assume the highest estimates for the ultimately available resources:

- A sustainability ambition of 1000 years cannot be achieved for bismuth, molybdenum, silver, and tin; for the other nine materials the sustainability ambition can be achieved, and in the case of gold, indium even without taking any measures.
- A sustainability ambition of 500 years can be achieved for all materials except for molybdenum and tin, and in the case of boron, chromium, gold, indium, nickel, and zinc without taking any measures.
- A sustainability ambition of 200 years can be achieved for all materials, and in the case of bismuth, boron, chromium, copper, gold, indium, nickel, silver, tungsten, and zinc without taking any measures.

Bismuth, molybdenum, nickel, silver, and tin therefore seem to be the materials for which sustainable production is most difficult to achieve.

It is important to consider the results from the perspective that we made two quite optimistic assumptions:

1. To calculate the ultimately available amount of resources, we assumed that resources will be extractable to a depth of 3 km in the Earth's continental crust. The more optimistic this assumption is for a particular resource, the more difficult the sustainability ambitions will be to achieve.

2. We assumed that the service level of mineral resources will have a ceiling that is equal to the service level of raw materials in developed countries in 2020. However, this is questionable for at least some of the considered raw materials, for example for indium. As the service level in developed countries increases, people in the rest of the world will understandably claim the right to increase their consumption level accordingly. This means that production growth will last longer and/or go faster, and that sustainability goals will be more difficult to achieve.

The main conclusion is that humanity should not wait any longer before taking economizing measures with respect to the scarcest mineral resources identified in this book. The more time that is taken for the implementation of serious conservation measures, the more questionable it becomes that future generations will be able to enjoy the material prosperity enjoyed by people in developed countries in 2020.

Technically, it is possible to realize sustainable production rates to a certain extent. However, without accompanying policy measures at a global scale, the technology potential will not be used as long as there is no positive economic outcome for companies and citizens.

Glossary

Abundance Average concentration of a mineral resource in the earth's crust

Antisolorant Chemical for protecting materials against the impact of solar radiation; color stabilizer

Apparent consumption Mine production + secondary refined production + imports (concentrates and refined metal) − exports (concentrates and refined metal) + adjustments for government and industry stock changes

Beneficiation The operation of removing the gangue from the ore to produce a concentrate and tailings

Bimodal distribution Distribution with two peaks

By-products When different products are extracted from the same mine, the products with a relatively small financial contribution to the output of the mine are considered by-products or companion metals. The main product is called host-product or carrier metal.

Carrier metals See host products

Choke price The price level at which the demand for a commodity for a given application will fall to zero because a more adequate (better and/or cheaper) substitute is available

Coproducts Materials mined together and generating a similar financial contribution to the mining operation

Commodity A raw material or primary agricultural product that can be bought and sold

Common but differentiated responsibilities The developed countries acknowledge the special responsibility that they bear in view of the pressures their societies place on the global environment and of the technologies and financial resources they command

Common concern of mankind Sovereignty should be exercised for the benefit of mankind, which consists of present and future generations

Companion metals See by-products

Concentrate Product of ore processing

Consumption of resources End use of resources; includes primary plus secondary resources

Consumption The amount of a resource in final consumer products

Critical raw materials Raw materials combining a high economic importance with a high supply risk. Criticality has two dimensions: risk of supply disruption and vulnerability of the economic system for such a disruption

Cumulative availability curve See cumulative supply curve

Cumulative supply curve The cumulative supply curve is a concept that reflects how the cumulative supply of a mineral could vary over all time with the extraction costs

Decolorant Chemical having the property of removing color

Depletion See exhaustion

Downcycling Downcycling is the phenomenon that an element is recycled together with another element and loses its specific functions, while permanently included in the flow of the other element

Enamel Smooth, durable, vitreous coating on metal, glass, or ceramics

End-of-life recycling (efficiency) rate Proportion of a material in end-of-life products, which is recycled

Enrichment factor The ratio between the ore grade and the upper crustal abundance of a mineral resource

Exergy The minimum energy costs involved in producing a mineral resource with a specific chemical composition and grade from common rock containing the average abundance of the resource in the continental earth's crust

Exhaustion Lack of profitability or feasibility of further extraction of a mineral resource due to (a combination of) financial, environment, energy, climate change, waste, water use, and social factors

Exponential growth of production Production growth, which is proportional to the production at any moment; 2% annual growth of production p in year 0 means that the production in year 1 is $1.02 \times p$, in year 2: $1.02 \times 1.02 \times p$, in year 3: $1.02 \times 1.02 \times 1.02 \times p$, etc.

Extractable global resources Estimate of the amount of resources in the upper 3 km of the earth's crust on the basis of the relation of this amount with the upper crustal abundance

Fining agent Chemical added to glass melt to eliminate gas seeds in the final product

Fixed stock paradigm The earth's crust contains a finite, hence depletable amount of resources

Galvanizing Galvanizing provides a zinc coating on another metal, mostly steel, to protect the metal against corrosion

Gangue Material that surrounds, or is closely mixed with, a wanted mineral in an ore deposit

Grade The concentration of a resource in an ore

Growth scenario 1 The annual production increase of mineral resources between 2015 and 2050 is equal to the annual production increase between 1980 and 2015. The annual production increase between 2050 and 2100 is half of the assumed production increase between 2015 and 2050. After 2100, no further production increase is assumed

Growth scenario 2 The annual production increase of mineral resources between 2015 and 2100 is equal to the annual production increase between 1980 and 2015. The annual production increase between 2100 and 2200 is half of the assumed production increase between 2015 and 2100. After 2200, no further production increase is assumed

Heat stabilizers Chemical protecting plastics against heat

Home scrap Scrap that is reutilized within the plant where the material is produced

Host product/metal When mining materials/metals together, the material/metal with the highest financial contribution is considered the host product, also called carrier product/metal. The other products are called by-products or companion products/metals

Hubbert curve The bell-shaped curve presenting the annual oil production of a well over time

Identified resoucrces Resources whose location, grade, quality, and quantity are known or estimated from specific geologic evidence. Identified resources include economic, marginally economic, and subeconomic components

In-use dissipation Loss of the resource to the environment through its consumption, for example, by use in fertilizers or in detergents

In-use stock The amount of the resource, which is in use. The in-use stock is determined by the lifetimes of the products in which the resource is contained

Intergenerational equity principle The legal principle that future generations may have a legitimate expectation of equitable access to planetary resources

Intragenerational equity principle The equitable distribution of resources, costs, and benefits between people and peoples of the same generation

Karat Indicates the proportion of gold in an alloy based on a total of 24 parts

LED Light-emitting diode

Major elements Elements with an average abundance in the earth's crust of >0.1 weight % (1000 ppm)

Manganese crusts Centimeter to decimeter thick pavements of manganese and iron oxides on the flanks of sea mounts at water depths of 1000—2500 m

Manganese nodules Centimeter to decimeter size lumps of manganese and iron oxides occurring on ocean floors at depths of about 5500 m

Marginal costs The additional costs to produce one extra unit of a product

Material efficiency Material efficiency reflects the quantity of services that can be provided by a given amount of a material

Mineral resources governance The set of norms, values, and rules through which mineral resources are managed

Mineral resources policy The set of concrete goals for achieving a more sustainable use of mineral resources in the interest of future generations

Mineral A mineral is a pure inorganic substance that occurs naturally in the earth's crust. All of the earth's crust, except the rather small proportion of the crust that contains organic material, is made up of minerals

Mineralogical barrier Economic limit grade for extractability

Minor elements Elements with an average abundance in the earth's crust of <0.1 weight % (1000 ppm)

Modal distribution Distribution with one peak

Moderately scarce elements Elements of which the ultimately available resources may be exhausted within 500—800 years according to growth scenario 1

New scrap See prompt scrap

Nonrenewable resource A nonrenewable resource is a resource, which is not replaced naturally. Examples are resources like copper, gold, oil, gas, and coal

Nonscarce elements Elements of which the ultimately available resources will not be exhausted within 900 years according to growth scenario 1

Occurrence See abundance

Old scrap Resources recycled from end-of-life products

Opportunity cost paradigm When the price for a mineral commodity is rising, society has to consider what to give up in order to obtain an additional ton of that scarce commodity. The price mechanism will foster substitution and technological innovations

Ore Enriched deposit of a mineral

Overall recycling efficiency rate The proportion of a metal in end-of-life products, new scrap, and other metal-bearing residues, which is recycled

Overburden Waste rock or materials overlying an ore or mineral body that are displaced during mining without being processed

P-value The probability that a correlation is determined by chance. If the P-value of a correlation is smaller than 0.05, then the correlation is considered significant

Peak mineral The bell-shaped curve supposed to present the (global, regional, national, and single mine's) annual production of a mineral resource over time

Peak oil The bell-shaped curve supposed to present the (global, regional, national, and single mine's) annual production of oil over time

Peak theory According to the peak theory, growth and decline of the extraction of a mineral resource will develop in a more or less bell-shaped way. There may be several peaks or one single long lasting peak

Policy instruments Nontechnical means to foster the achievement of policy goals

Polluter pays principle The polluter should bear the costs of pollution

Primary resources Virgin resources directly originating from a mine and not having been consumed and recycled before

Prompt scrap Resources recycled from the fabrication and manufacturing stages

Raw material consumption Includes primary and secondary resources

Raw material use Includes only primary resources

Recovery efficiency The ratio of the actual production of a resource and the total amount of that resource contained in the mined ore

Recycled content Proportion of secondary resources in end use

Recycling input rate See recycled content

Refractory Material retaining strength and form at high temperatures, such as fire bricks

Renewable resource A renewable resource is a resource that can be used repeatedly and replaced naturally. Examples include oxygen, fresh water, solar energy, and biomass

Reserve base That part of an identified resource that meets specified minimum physical and chemical criteria related to current mining and production practices, including those for grade, quality thickness, and depth

Reserve That part of the reserve base that could be economically extracted or produced at the time of determination

Resource intensity The amount of water, energy, and chemicals consumed, the amount of greenhouse gas emitted, and the amount of waste rock produced per unit of a mineral produced

Resource optimist Resource optimists have a strong believe that humanity will be able to cope with the effects of depletion of resources. They support the so-called opportunity cost paradigm

Resource pessimist Resource pessimists state that the earth is finite and so the amount of resources is finite as well. So resource pessimists believe that it is a matter of time that supply cannot meet demand anymore. They support the so-called fixed stock paradigm

Resource productivity See material efficiency

Resource A concentration of naturally occurring solid, liquid, or gaseous material in or on the earth's crust in such a form and amount that economic extraction of a commodity from the concentration is currently or potentially feasible

Resources governance The set of norms, values, and rules through which mineral resources are managed

Resources policy instruments Concrete measures for implementation of resources policy

Resources policy The set of concrete goals for achieving a more sustainable use of mineral resources in the interests of future generations

Scarce elements Elements of which the ultimately available resources may be exhausted within a century according to growth scenario 2, and within 400 years according to growth scenario 1

Scarcity of resources Scarcity of resources is reflected by the period of time until exhaustion of the resource. This is the ratio of ultimately available resources and the annual production of the virgin resource taking into account the annual production growth

Seafloor massive sulfide Polymetallic sulfides produced by seafloor hot springs

Secondary resources Recycled resources; these may consist of old scrap, prompt (or new) scrap, and home scrap

Service level The quantity of services delivered per unit of a raw material. An improvement of the service level of a raw material with 25% means that with the same quantity of raw material, 25% more services can be delivered or that for delivering the same services, 25% less of the raw material is required

Slag Glasslike by-product left over after separation of desired mineral from its ore (i.e., by smelting)

Solder Alloy used to connect metallic parts

Sovereign right principle The sovereignty of peoples and nations over their natural wealth and resources

Strong sustainability concept Human-made capital and natural capital are different categories that are not interchangeable

Sustainable development Sustainable development is the kind of development that meets the needs of the present without compromising the ability of future generations to meet their own needs

Sustainable extraction The extraction of a mineral resource is sustainable, if a world population of 10 billion can be provided with that resource for a period of at least 200/500/1000 years in a way that during that period every country can enjoy the same service level of that resource as in developed countries in 2020 for an affordable price

Sustainable production See sustainable extraction

Tailings Rock stripped of valuable minerals

Tectonic diffusion model Model developed by Kesler and Wilkinson (2008) for estimating the distribution of ores in the earth's crust. The model is based on the continuous movement of the earth's crust and calibrated by comparing the results with the real presence of ores in the earth's crust

Total global deposits The estimate of available resources based on an extrapolation of the results of a thorough assessment in 1998 by the US Geological Survey of undiscovered deposits of gold, silver, lead, and zinc in the United States

Ultimately available resources The highest of two estimates: the extractable global resources or the total global deposits

Use of resources Use of primary (virgin) resources

UV stabilizers Chemicals protecting plastics against ultraviolet radiation

Weak sustainability concept Elements of sustainable development are interchangeable as long as economic development and welfare as a whole do not diminish

List of abbreviations

ABS	Acrylonitrile butadiene styrene
Ag	Silver
Al	Aluminum
As	Arsenic
Au	Gold
B	Boron
Ba	Barium
Be	Beryllium
Bi	Bismuth
Cd	Cadmium
CIGS	Copper—indium—gallium—selenide
Co	Cobalt
Cr	Chromium
EC	European Commission
EEE	Electrical and electronic equipment
EFRA	European Flame Retardant Association
EGR	Extractable global resources
EHS	Environment, health, and safety
EoL	End-of-life
EROEI	Energy return on energy investment
EU	European Union
EU ETS	European Union Emission Trading System
Fe	Iron
Ga	Gallium
GATT	General Agreement on Tariffs and Trade
GDP	Gross domestic product
Ge	Germanium
GER	Gross Energy Requirement
GFRP	Glass fiber—reinforced plastic
Gt	Giga tons = billion tons = 1000 Mt
HDI	Human Development Index
Hg	Mercury
IFC	International Finance Corporation
IMA	International Molybdenum Association
In	Indium
IPCC	Intergovernmental Panel on Climate Change
IR	Identified resources
ISO	International Standardization Organization

ITA	International Tin Association
ITO	Indium tin oxide
ITRI	International Tin Research Institute
kt	kiloton = 1000 tons
LCA	Life cycle assessment
LCD	Liquid crystal display
Li	Lithium
ME	Material efficiency
Mg	Magnesium
Mn	Manganese
Mo	Molybdenum
Mt	Mega tons = million tons = 1000 kt
Nb	Niobium
Ni	Nickel
OECD	Organisation for Economic Co-operation and Development
OPEC	Organization of Petroleum Exporting Countries
Pb	Lead
PCB	Polychlorinated biphenyl
PCB	Printed circuit board
PCZ	Prime Fe—Mn Crust Zone
PE	Polyethylene
PET	Polyethylene terephthalate
PGM	Platinum group metals
PP	Polypropylene
ppb	Parts per billion
ppm	Parts per million
PSS	Product—service systems
PV	Photovoltaic
PVC	Polyvinyl chloride
RB	Reserve base
Re	Rhenium
REACH	EU Directive on the Registration, Evaluation, Authorisation and Restriction of Chemicals
REDD	Reducing emissions from deforestation and forest degradation
REE	Rare-earth elements
RER	Recycling efficiency rate
RIR	Recycling Input Rate
Sb	Antimony
Se	Selenium
SMS	Seafloor massive sulfide
Sn	Tin

Sr	Strontium
t	tons = 1000 kg
Ta	Tantalum
TGD	Total global deposits
Ti	Titanium
UCA	Upper crustal abundance
UN	United Nations
UNCTAD	United Nations Conference on Trade and Development
UNDP	United Nations Development Programme
UNEP	United Nations Environment Programme
UNFCC	United Nations Framework Convention on Climate Change
UNREDD	United Nations Program on Reducing Emissions from Deforestation and Forest Degradation
URR	Ultimately recoverable resources
USD	United States dollar
USGS	United States Geological Survey
UV	Ultraviolet
V	Vanadium
W	Tungsten
WEEE	Waste of electrical and electronic equipment
WGI	World Governance Index
WTO	World Trade Organization
Zn	Zinc

Index

Printed in the United States
by Baker & Taylor Publisher Services